T0047507

HELP YOUR KIDS WITH GEOGRAPHY

A UNIQUE STEP-BY-STEP VISUAL GUIDE

DK

DK LONDON
Senior Editor Scarlett O'Hara
Project Editors Edward Aves, Tom Booth, Ben Davies, Abigail Mitchell
Senior Art Editor Elaine Hewson
Designers and Illustrators Amy Child, Shahid Mahmood
Cartography Simon Mumford
Jacket Editor Emma Dawson
Jacket Designer Surabhi Wadhwa-Ghandhi
Jacket Design Development Manager Sophia MTT
Producer, Pre-production Rob Dunn
Senior Print Producer Jude Crozier
Managing Editor Christine Stroyan
Managing Art Editor Anna Hall
Publisher Andrew Macintyre
Art Director Karen Self
Design Director Phil Ormerod
Publishing Director Jonathan Metcalf

DK DELHI
Senior Editors Janashree Singha, Arani Sinha
Editors Devangana Ojha, Tanya Singhal, Nandini D. Tripathy
Assistant Editor Rishi Bryan
Senior Art Editor Ira Sharma
Project Art Editor Vikas Sachdeva
Art Editors and Illustrators Anukriti Arora, Sourabh Challariya, Shipra Jain, Jomin Johny
Assistant Art Editors and Illustrators Sonali Mahthan, Adhithi Priya, Shreya Singal
Managing Editor Soma B. Chowdhury
Senior Managing Art Editor Arunesh Talapatra
Senior Picture Researcher Surya Sankash Sarangi
Picture Researchers Nishwan Rasool, Rituraj Singh
Picture Research Manager Taiyaba Khatoon
Illustrators, Digital Operations Delhi Manjari Rathi Hooda, Nain Rawat, Rohit Rojal, Alok Singh
Senior DTP Designer Vishal Bhatia
DTP Designers Nityanand Kumar, Rakesh Kumar, Bimlesh Tiwary
Production Manager Pankaj Sharma
Pre-production Manager Balwant Singh
Jacket Designers Priyanka Bansal, Suhita Dharamjit
Jackets Editorial Coordinator Priyanka Sharma
Managing Jackets Editor Saloni Singh

THIRD EDITION
Editor Sarah Carpenter
US Editor Jennette Elnaggar
Designer Annabel Schick
Managing Editor Carine Tracanelli
Managing Art Editor Anna Hall
Production Editor Jacqueline Street-Elkayam
Senior Production Controller Poppy David
Jacket Designer Stephanie Cheng Hui Tan
Senior Jackets Coordinator Priyanka Sharma Saddi
Jackets Design Development Manager Sophia MTT

This American Edition, 2023
First American Edition, 2019
Published in the United States by DK Publishing
1745 Broadway, 20th Floor, New York, NY 10019

23 24 25 26 27 10 9 8 7 6 5 4 3 2 1
001–335612–June/2023

Published in Great Britain by Dorling Kindersley Limited

A catalog record for this book is available from the Library of Congress.
ISBN 978-0-7440-8075-9

Printed and bound in Malaysia

For the curious
www.dk.com

FSC
www.fsc.org
MIX
Paper | Supporting responsible forestry
FSC™ C018179

This book was made with Forest Stewardship Council ™ certified paper–one small step in DK's commitment to a sustainable future. For more information go to www.dk.com/our-green-pledge

CONSULTANTS

DR. DAVID LAMBERT

Dr. David Lambert is Emeritus Professor of Geography Education at UCL Institute of Education, London. He graduated from the University of Newcastle, completing a PGCE at the University of Cambridge and a PhD at the University of London. A secondary school teacher for 12 years, he has written award-winning textbooks and published widely on the curriculum, pedagogy, and assessment of geography in education. He was appointed chief executive of the Geographical Association in 2002 and became Professor of Geography Education in 2007. His recent books include *Learning to Teach Geography* (3rd Edition, 2015) and *Debates in Geography Education* (2nd Edition, 2017). He led the EU-funded GeoCapabilities project from 2013 to 2017 (www.geocapabilities.org).

DR. SUSAN GALLAGHER HEFFRON

Dr. Susan Gallagher Heffron is an independent education consultant with a PhD in Curriculum and Instruction from the University of Nebraska at Lincoln. She has extensive experience in geography education, and has worked on national and international projects for geography teachers' professional development. A coeditor of *Geography for Life: The National Geography Standards*, 2nd Edition, she also serves on the editorial board for *The Geography Teacher*. She taught for 14 years, followed by 6 years as faculty at higher education institutions.

CONTRIBUTORS

JOHN WOODWARD

John Woodward has written more than 50 books and hundreds of articles on the natural world. His work with DK includes titles such as *Geography Encyclopedia*, *Ocean Encyclopedia*, *Oceans Atlas*, *Eyewitness Climate Change*, *Eyewitness Water*, and *SuperEarth*. He works as a regular volunteer on a wildlife reserve near his home in southern England and helps manage local conservation projects designed to increase biodiversity and resilience to climate change.

JOHN FARNDON

John Farndon has an MA in Geography from the University of Cambridge and is the author of hundreds of books on science and nature. He has been shortlisted five times for the Young People's Science Book Prize for books such as *How the Earth Works*. He is also the author of the acclaimed *Atlas of Oceans* and *The Wildlife Atlas*.

FELICITY MAXWELL

Felicity Maxwell has a BSc in Geology and Botany as well as an MSc in Geology from Victoria University of Wellington, New Zealand. She also has an MSc in Environmental Management and Technology from Oxford Brookes University, UK. Felicity has worked for many years for government organizations and as a consultant on land and biodiversity protection in New Zealand.

SARAH WHEELER

Sarah Wheeler graduated from the London School of Economics in 1981 with a degree in Human Geography. She is an Outstanding Geography Teacher currently teaching in a grammar school in the south of England. Sarah has been a senior examiner for an examinations board for over 30 years and has been a writer and consultant for several publications, including the Hodder AQA GCSE textbook and study guide.

ARTHUR MORGAN

Arthur Morgan grew up in London and has a BA in Geography from the University of Manchester. He is interested in urban geography and wrote his dissertation on affordable housing provision. He is an active volunteer in international projects such as the building of the Transcaucasian Trail, a hiking path connecting Georgia and Armenia.

Introduction

There is no bigger subject than geography—because it is about the whole world!

Geography is one of the major subjects in the school curriculum around the world. This is because we generally agree how important it is for everyone to have some knowledge and understanding of how the world works, and of the global system.

A passion for geography demonstrates a real curiosity about the workings of our wonderful planet, and this is a great advantage in today's fast-changing world.

This book sets out to explain the essentials of the subject and helps parents help their kids with their geography homework. It covers the key areas taught in schools and will refresh the memory of parents who haven't studied the subject since they were in school.

Geography is not just an accumulation of facts and figures. The subject tackles a range of ideas, some of which are complicated (tectonic plates, weathering and erosion, and ecosystems), and some that are constantly developing (globalization, sustainability, and climate change). All these concepts are about our natural world, how human beings relate to each other, and how people and their environment interact.

Help Your Kids with Geography is like no other geography book. It is packed full of the information you need to make sense of the world.

It is a book to excite your curiosity and address several global issues head on. It encourages you to think geographically about the world, to form a view and, I hope, to argue. This does not mean to squabble or simply disagree, but to listen to different viewpoints and accept that in many geographical matters there are no single stories but many different perspectives.

I used to be a school teacher and I was a parent of young children. I hope this book will provide plenty of opportunities for adults and young people to read, share ideas, and talk about the Earth as our home. It will even help you take on the world! That is geography's power.

David Lambert

DAVID LAMBERT
EMERITUS PROFESSOR OF GEOGRAPHY EDUCATION
UCL INSTITUTE OF EDUCATION

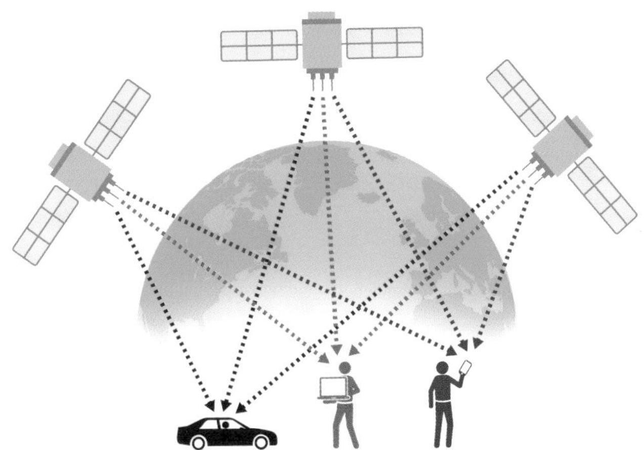

Contents

3 PRACTICAL GEOGRAPHY

What is geography?

GEOGRAPHY IS THE STUDY OF THE WORLD WE LIVE IN.

Geographers study the Earth's landscapes, its atmosphere, its natural environments, and its people. They look in particular at where these things are and why, and how they change over time.

The origins of geography

The word "geography" comes from Ancient Greek—"geo" means "earth" and "graph" means "writing." Ancient Greek scholars were interested in where their homeland was in relation to other places and what these different places were like. They made maps to give them a picture of what the world around them looked like.

Our planet is roughly spherical in shape and a globe represents it that way.

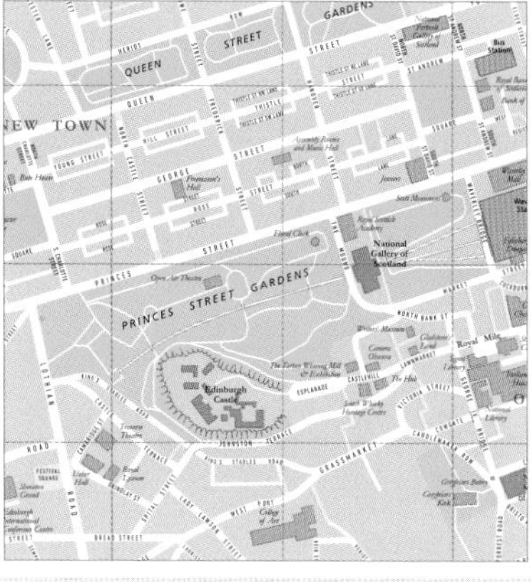
QUEEN
STREET
PRINCES STREET
GARDENS
Royal Mile

◁△ **Understanding the world**
Maps and globes have always been central to geography. Geographers use them to study the distribution of places and how they relate to each other.

Physical geography

The subject of physical geography studies the non-human parts of the Earth—its landscapes and rocks, its atmosphere, and its rivers, lakes, and oceans—as well as the plants and animals that inhabit these places. It is similar to Earth science, but physical geography is more concerned about where things are. There are many different branches of physical geography.

△ **Biogeography**
Biogeographers look at where plants and animals live. They are especially interested in biomes—large regions where particular communities of plants and animals live.

△ **Geology and geomorphology**
Geologists study rocks and minerals, and the Earth's crust and interior. Geomorphologists study landforms and processes that shape the landscape.

△ **Meteorology and climatology**
Meteorologists look at the atmosphere and try to forecast the weather. Climatologists study climates—the average weather in each region of the world.

Human geography

The topic of human geography is about where and how people live. It studies how people interact with each other and with different kinds of environments. Human geographers are interested in the environments people create for themselves, both in rural areas and in cities.

Urban geography
Cities are a subject of study for urban geographers. They try to understand why and how things change in cities, and how cities make links across the world.

Economic and social geography
Economic geographers study where economic activities (such as industry and farming) take place. Social geographers look at the distribution of different groups of people.

Population geography
Population geographers are interested in where people are born, where they die, and how they move about. They also study how populations change.

Practical skills

Geographers have a large area of study, so they need to develop a wide range of skills. Understanding how to use maps of various kinds is crucial. Geographers also need to know how to observe and measure things. Statistical skills (processing data as numbers) are very important for geographers.

△ **Location and direction**
Geographers need to know where things are. They rely on maps, satellite position systems (GPS), and compasses to pinpoint location and direction.

Q.1
Q.2
Q.3
Q.4
Q.5

△ **Surveying**
For human geographers, a survey is a range of questions designed to find data about people. For physical geographers, it's a measurement of the landscape.

Fieldwork

When geographers go outdoors to study the landscape or the human environment, it is called fieldwork. For many, this is the most enjoyable part of geography. It can take geographers to wild places in nature, such as high mountains or beautiful forests.

Thinking geographically

GEOGRAPHERS USE A GEOGRAPHICAL METHOD TO INVESTIGATE AND THINK ABOUT THE WORLD AROUND THEM.

Geographers try to establish where in the world things are, how they work, and how they fit into the bigger picture. Geographical thinking combines specific facts and figures with ideas or concepts. It is especially interested in finding links, patterns, and relationships.

Core knowledge

Geographical thinking is built on core knowledge—that is, basic facts and figures about the world, such as the names and sizes of the continents, where the main rivers and mountain ranges are, what the main layers of the atmosphere are, which the biggest cities are, and so on. Core knowledge answers straightforward questions such as these:

What is the population of Belgium?

Where is Peru?

How deep is this lake?

What is a glacier?

What is the route of the Gulf Stream?

▽ **Finding connections**
Geography is a subject that provides a broad picture of the world around us and shows how different things are related to each other.

Geography relates the **local** to the **global.**

Geography links the **physical** to the **human.**

Geography connects **people** to their **environment.**

Geographers link **facts** and **concepts.**

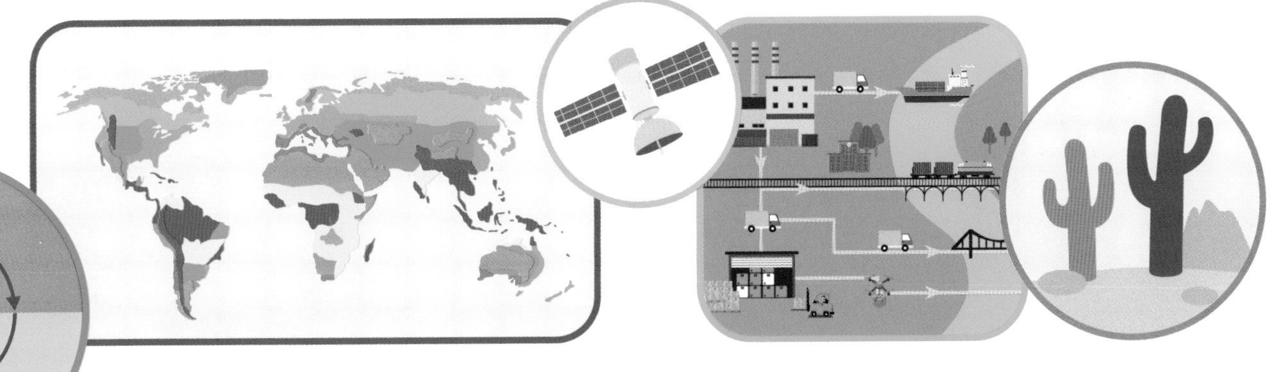

Conceptual knowledge

Besides knowing lots of facts and figures about the world, geographers need ideas to help make sense of these. Concepts such as urbanization, globalization, climate, and the hydrological cycle all fit into the big three geographical ideas: place, space, and environment. Most geographical knowledge can be organized under these headings, so they make good starting points for examining subjects in more detail.

PLACE

A place is an area or location that has been identified and named, usually by the people living there. It can be a single street or a whole continent.

A place can be described and researched in depth. A geographer might study its climate and geology, its soil, its population, and more. No two places are exactly the same. The special characteristics of one place may help give the geographer an understanding of the world as a whole.

Questions you might ask about place:

- What is the economic situation of this place? How does this relate to its natural resources?
- What are the benefits of choosing to farm in this specific place? What are the soil and climate like in this location?

SPACE

Space is the surface of the world in three dimensions. The word "spatial" means "about space." Space also includes links and patterns between different phenomena and places.

Geographers look for patterns of spatial variations, such as population density, and try to understand why they occur. They are interested in the extent of phenomena, such as the damage caused by an earthquake or pollution from a factory. The effect of space is also important, such as the distance between places of economic activity.

Questions you might ask about space:

- How do global production processes work? How do goods get from producers to consumers on a global scale?
- What are the consequences of an uneven distribution of wealth?

ENVIRONMENT

Our environment is our surroundings, both living and non-living. It can be natural; managed by humans (as in the case of farmland); or built, such as a city.

Geographers can study the ways humans interact with the world around them, their impact on it, and how humans can care for the environment. Environment can be better understood by looking at ecosystems—the natural systems where living things interact with each other.

Questions you might ask about environment:

- Why are environments such as deserts and coral reefs fragile? How can we preserve them?
- What renewable energy sources should we use?

Geography in action

GEOGRAPHY IS NOT JUST A SUBJECT FOR SCHOOL, IT HAS A WIDE RANGE OF PRACTICAL APPLICATIONS.

Geography provides a clear framework for understanding the world, and a guide for how to care for it too. Both physical and human geography offer a broad range of areas to study and fascinating jobs or activities to pursue.

Putting it into practice

Specialist skills developed for geography and related subjects have an important role to play in the way we interact with our environment – both the natural environment and the human environment. Geography provides an overview, but specialist skills provide detailed, hands-on knowledge of many fields. Here are just some of the varied jobs or activities a geographer might end up doing.

Physical geography

Geography provides a broad understanding of the world's natural environments, both the living (animals and plants) and the non-living (rocks and gases). With this knowledge it is possible to pursue a range of careers and interests.

Geologist
The study of rocks is made by geologists, who seek sources of valuable minerals.

Vulcanologist
Vulcanologists study volcanoes and try to predict when they might erupt.

Climatologist
A key role in our future is played by climatologists, who can warn of changes to the climate.

Soil scientist
A soil scientist analyzes soil to discover how best it can grow plants.

Meteorologist
A meteorologist focuses on the weather and tries to make accurate forecasts.

Park ranger
Parks are vital refuges for wildlife, and rangers help protect them.

Biogeographer
To help us know how best to protect the natural world, biogeographers study habitats.

Hydrologist
Controlling water resources and protecting areas from flooding is work carried out by hydrologists.

Human geography

Cities and other parts of the human environment are getting bigger and more complicated. To make the best choices in our planning, we need to understand how cities work. The skills of human geographers are becoming very important.

Geography for fun

The world outdoors becomes much more fun if you have some geographic skills to help you understand it better.

Town planner
Planning how cities look and how the land is used is the job of a town planner.

Surveyor
The world depends on accurately made maps, so surveyors are very important.

Politician
A geographer's understanding of how people and their environments work can help politicians make better decisions.

Orienteering
Orienteering is an activity in which you find your way in rough country using only a compass.

Planning a route
Knowing how to use a map properly helps you plan either the shortest or most scenic route.

Tour guide
A good knowledge of landscape and people will help tour guides inform visitors.

Business
Geographical knowledge helps businesses decide where to locate.

Exploring
Why not become an explorer going to wild places to increase our geographical knowledge of the world?

Sailing
Understanding the weather, tides, and ocean currents will help you if you take up sailing.

Becoming greener
An appreciation of our planet and its systems will help you choose to live a "greener" life and care for our planet.

Transportation planners
Locating roads and railroad lines needs a good knowledge of traffic flow and the landscape.

Aid worker
An understanding of how the world's wealth is distributed is used to target aid for poor countries.

Making your own maps
If you really want to know your way around, why not make your own maps?

Travelling the world
By travelling into other countries, you see new landscapes and learn about other cultures.

Gardening
Knowing things about your garden, such as its soil and drainage, can make you an expert gardener.

Teacher
Geography teachers play a key role in teaching students about the world they live in.

1

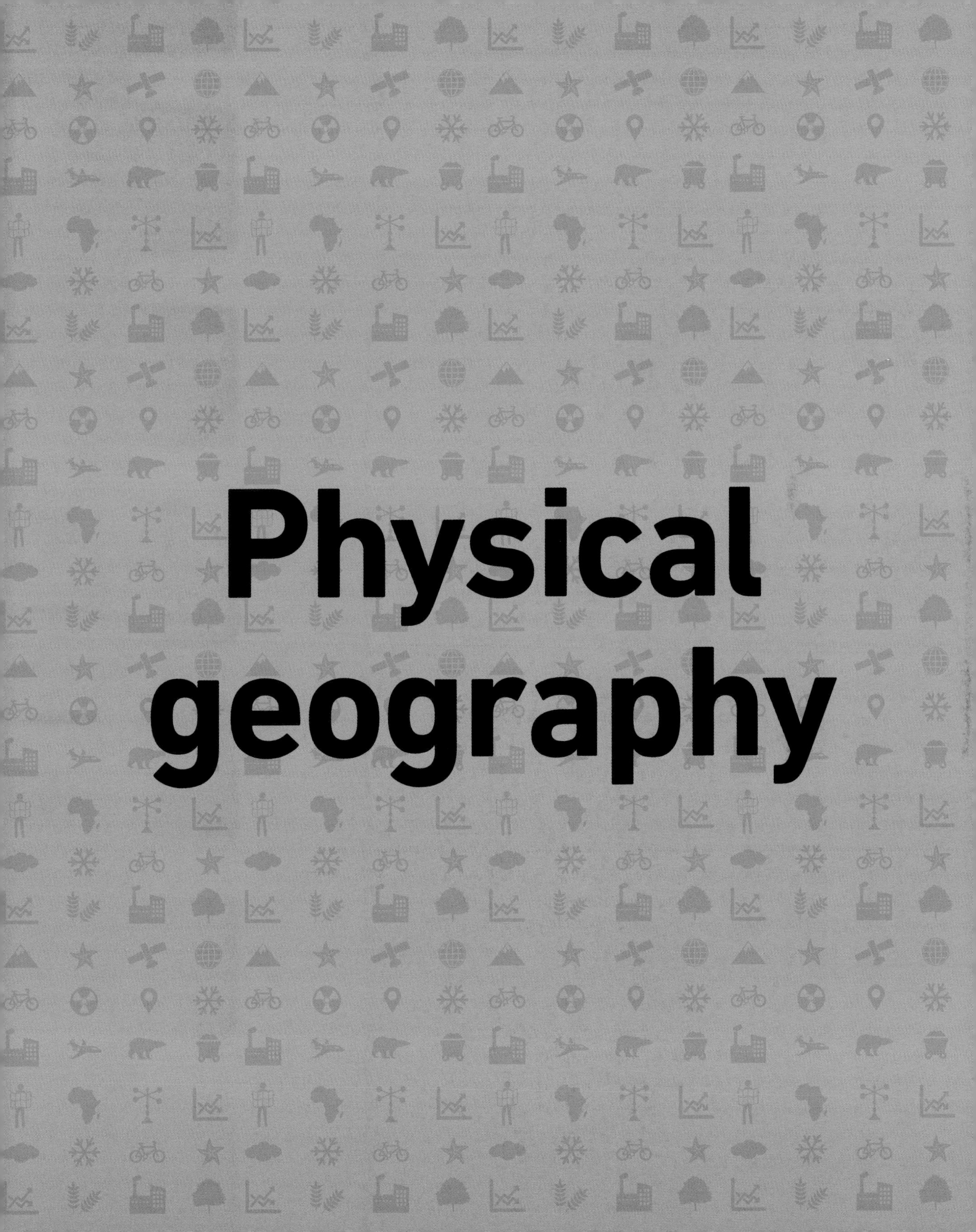

Physical geography

What is physical geography?

PHYSICAL GEOGRAPHY IS THE STUDY OF THE EARTH'S NATURAL SYSTEMS AND THE RELATIONSHIP BETWEEN THE ROCKS, WATER, ATMOSPHERE, AND LIVING THINGS.

Geographers are interested in how the planet Earth works. They study the formation of rocks and soils, its climate, how water and ice shape its surface, and the communities of animals and plants that live on it.

The Earth

Geographers study the history of the Earth, looking at the geological processes that have shaped our changing planet. They study how forces from deep within the Earth have powered the formation of huge mountains and deep trenches, and have moved continents over time.

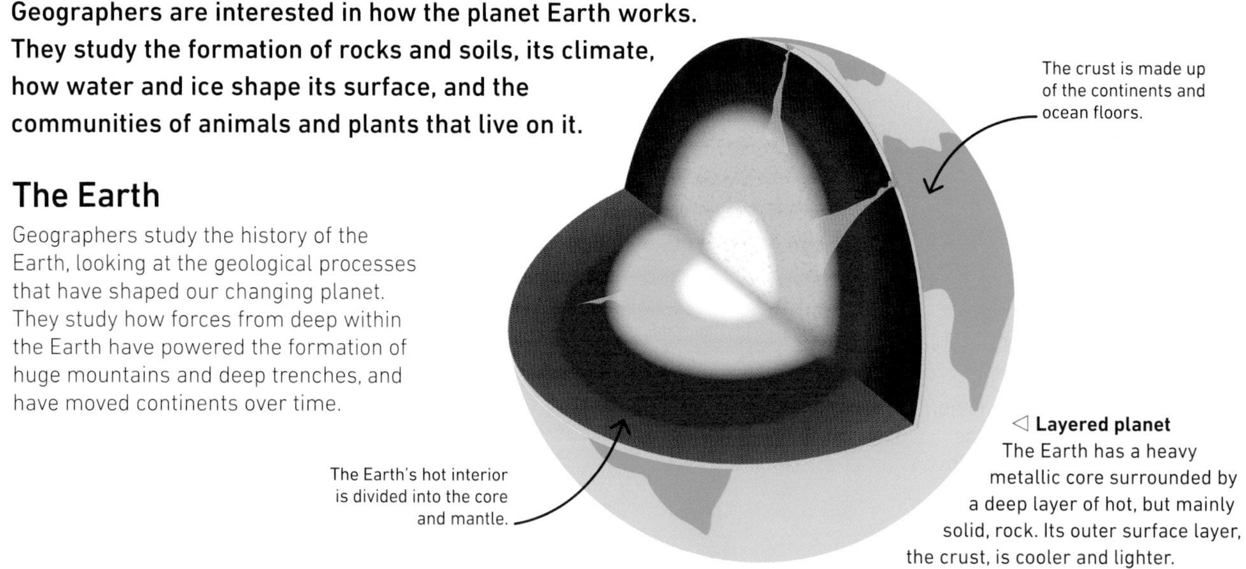
The crust is made up of the continents and ocean floors.
The Earth's hot interior is divided into the core and mantle.

◁ **Layered planet** The Earth has a heavy metallic core surrounded by a deep layer of hot, but mainly solid, rock. Its outer surface layer, the crust, is cooler and lighter.

Rocks and soil

The Earth is made up of different types of rock and soil. Geographers investigate the processes that form them. Some rocks are produced deep underground, while others are formed out of the fragments of other, older rocks pressed hard together. With time, rocks are weathered away, and some of the fragments are carried off by the forces of erosion to mix with the remains of living things and form soil.

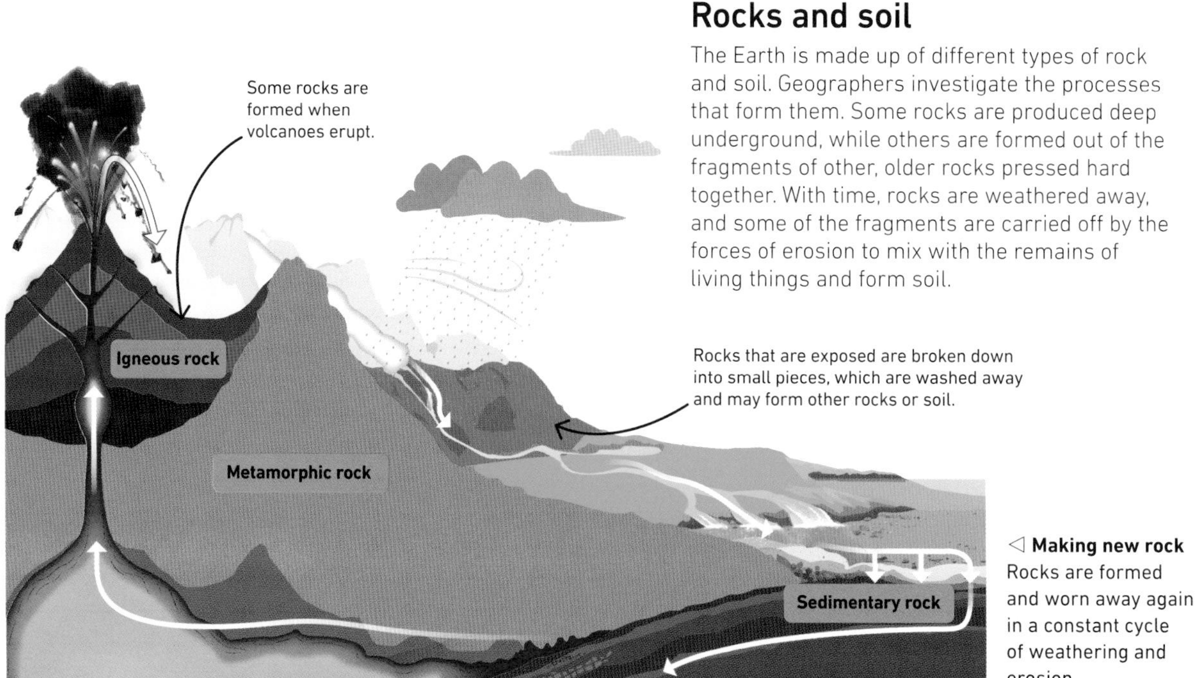
Some rocks are formed when volcanoes erupt.
Igneous rock
Metamorphic rock
Rocks that are exposed are broken down into small pieces, which are washed away and may form other rocks or soil.
Sedimentary rock

◁ **Making new rock** Rocks are formed and worn away again in a constant cycle of weathering and erosion.

The work of water

Water shapes our planet in different ways. Geographers study this process, looking at how rivers carve valleys and create broad floodplains, and at how the sea erodes and creates landforms. In cold climates, snow builds up to form glaciers that transform landscapes in even more dramatic ways.

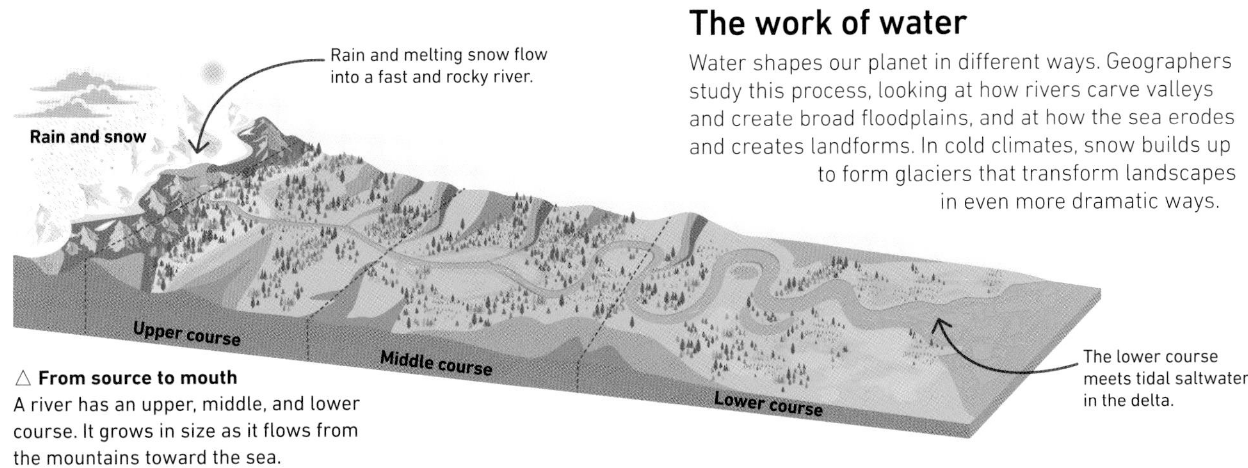
Rain and melting snow flow into a fast and rocky river.
Rain and snow
Upper course
Middle course
Lower course
The lower course meets tidal saltwater in the delta.

△ **From source to mouth**
A river has an upper, middle, and lower course. It grows in size as it flows from the mountains toward the sea.

Weather and climate

"Weather" describes the state of the atmosphere at a particular moment, while "climate" describes the average weather in a region over time. Geographers study both, looking at winds, rain, clouds, and other weather phenomena, and at how they affect the world around us.

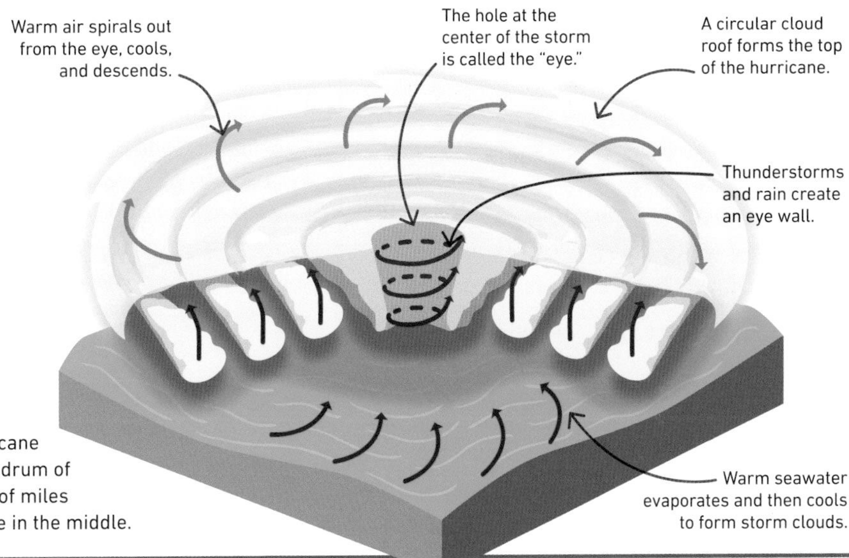
Warm air spirals out from the eye, cools, and descends.
The hole at the center of the storm is called the "eye."
A circular cloud roof forms the top of the hurricane.
Thunderstorms and rain create an eye wall.
Warm seawater evaporates and then cools to form storm clouds.

Inside a storm ▷
A full-blown hurricane is a giant rotating drum of clouds, hundreds of miles across, with a hole in the middle.

Biogeography

Geographers divide the world into "biomes," landscapes associated with particular types of animals, plants, and a specific climate. Biomes include deserts, different types of forest and grassland, as well as the seas and oceans. The study of these biomes is called biogeography.

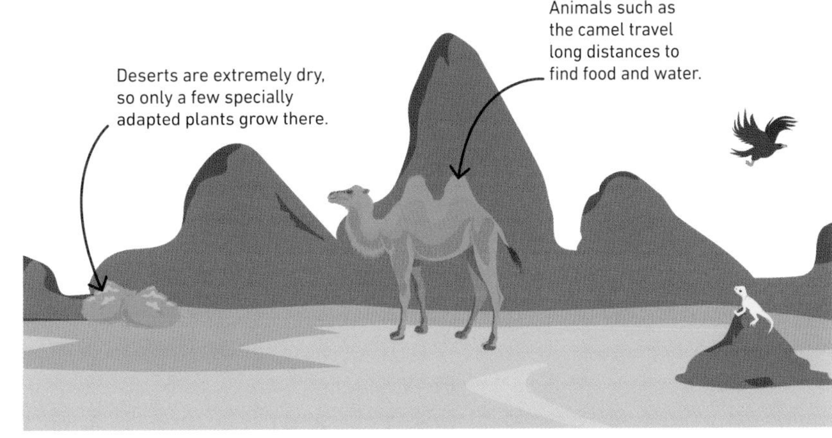
Deserts are extremely dry, so only a few specially adapted plants grow there.
Animals such as the camel travel long distances to find food and water.

Dry and dusty ▷
This illustration shows some of the plants and animals that are adapted to life in cold deserts.

Earth's history and geological time

THE GEOLOGICAL HISTORY OF THE EARTH IS DIVIDED INTO EONS, ERAS, AND PERIODS.

SEE ALSO

Earth's structure 22–23 ›
Volcanoes and hot springs 34–37 ›
Rocks and minerals 40–41 ›
Sedimentary rocks and fossils 47–49 ›
Ice ages 62–63 ›

The major geological milestones in the evolution of life forms on Earth happened over millions of years.

Timeline for life on the Earth

Each eon is divided into eras, and each era consists of periods. The table below gives information about the main divisions of the Phanerozoic eon, which started when complex life became abundant on Earth 541 million years ago. Fossils preserved in rocks reveal how life evolved.

According to scientists, the **age of** the **Earth** is about **4.5 billion** years.

Eon	Era	Period (millions of years ago)	What was happening
Phanerozoic	PALEOZOIC 541–252 million years ago	CAMBRIAN (541–485)	During the Cambrian period, animals with hard shells first became common in the oceans. This is called the "Cambrian explosion" of life.
		ORDOVICIAN (485–443)	Complex animals such as trilobites flourished in the oceans, but at this time there was no animal life on land—only a few very simple plants.
		SILURIAN (443–419)	Silurian oceans teemed with life, including early fish. Short plants, fungi, and small animals, such as millipedes, became abundant on land.
		DEVONIAN (419–358)	This has been called the "Age of fish" because at this time fish began to evolve into more types. Tall trees appeared on land, forming dense forests.
		CARBONIFEROUS (358–298)	Insects, such as giant dragonflies, flew through the forests and were hunted by some of the earliest land vertebrates—animals with backbones.
		PERMIAN (298–252)	The land formed a huge supercontinent with vast deserts. The period ended with a global catastrophe that destroyed 96 percent of all life forms.
	MESOZOIC 252–66 million years ago	TRIASSIC (252–201)	As life slowly recovered, the first small dinosaurs and airborne pterosaurs appeared. Some of these hunted the earliest furry mammals.
		JURASSIC (201–145)	Giant plant-eating dinosaurs roamed the forests, using their long necks to feed from treetops. They were prey for big meat-eating dinosaurs.
		CRETACEOUS (145–66)	Dinosaurs evolved into many different species, including birds, but all the giant dinosaurs were wiped out by a global disaster 66 million years ago.
	CENOZOIC 66–0 million years ago	PALEOGENE (66–23)	Surviving mammals evolved into bigger forms resembling modern rhinos. These replaced the dinosaurs, dominating life on land.
		NEOGENE (23–2)	Birds, mammals, reptiles, and others gradually evolved into modern animals. These included the first upright-walking ancestors of humans.
		QUATERNARY (2–Present)	The climate became cooler, causing a series of ice ages. The first true humans evolved in Africa, and gradually spread across the world.

The story of life

Life began on Earth about 3.8 billion years ago. For most of that time, it consisted of microscopic single-celled forms. But about 600 million years ago, the first complex, multicelled life forms appeared, which gradually evolved into the animals, fungi, and plants that exist today.

PRECAMBRIAN

Ancient volcanoes
Water vapor that erupted from volcanoes created the oceans where life evolved.

Comets
Icy comets that crashed to Earth may have contained the first elements of life.

CAMBRIAN

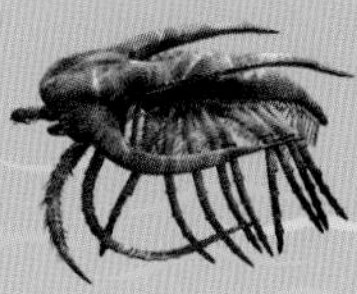

Marrella
This early animal lived in the sea and had a hard-shelled body with jointed legs and spines.

ORDOVICIAN

Sacabambaspis
This armored fish had very close-set eyes and no jawbones.

SILURIAN

Cooksonia
Dating from the middle of the Silurian period, this was one of the oldest plants with stems.

DEVONIAN

Drepanaspis
The head of this jawless, armored fish was covered with a broad, flattened shield.

CARBONIFEROUS

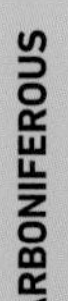

Lepidodendron
This early tree had a scaly bark and could grow to more than 100 ft (30 m) in height.

PERMIAN

Dimetrodon
The fossils of this sail-backed hunter show that it was related to the ancestors of mammals.

TRIASSIC

Eudimorphodon
Pterosaurs were flying reptiles that lived at the same time as dinosaurs.

JURASSIC

Cryolophosaurus
The fossil remains of this powerful hunter were found on Antarctica.

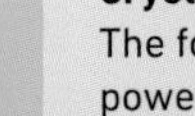

CRETACEOUS

Velociraptor
This feathered meat-eating dinosaur had a very sharp claw on each foot.

PALEOGENE

Uintatherium
This rhinoceros-sized animal was a big plant-eating "megaherbivore."

NEOGENE

Australopithecus afarensis
This early hominid could have been the first to walk upright 4 million years ago.

QUATERNARY

Homo neanderthalensis
This strongly built human species was adapted to life in icy climates.

Earth's structure

SEE ALSO
❮ 20–21 Earth's history and geological time
Moving plates and boundaries 24–25 ❯
Shifting continents 26–27 ❯
The atmosphere 74–75 ❯

OUR PLANET IS A GIANT SPHERE OF ROCK AND METAL SURROUNDED BY LAYERS OF WATER AND AIR.

The Earth is made up of three layers: the core, the mantle, and the crust. More than 70 percent of the crust is covered by oceans, and the rest forms the islands and continents where we live.

Earth's formation

Earth formed about 4.54 billion years ago from a cloud of rock, dust, and gas surrounding the newly formed sun. Gravity pulled this space debris together to form a sphere, which eventually melted and developed a layered structure of metal and rock. This structure then cooled down, finally becoming cool enough to support liquid oceans and an atmosphere.

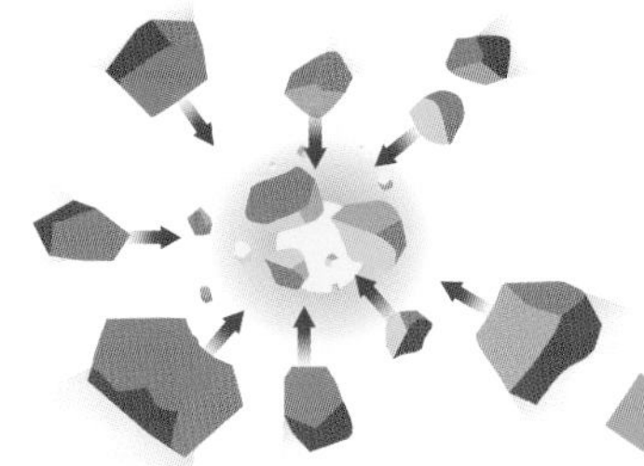

◁ **Accretion**
Rocks drifting in space started to be pulled together by their own gravity to form one large object, and eventually a planet. This growth by gradual accumulation of material is called accretion.

Meltdown ▷
As more rocks smashed into the planet, the energy from all these impacts converted to heat. The planet began to melt, and most of the heavy metal from its rocks sank to form the core.

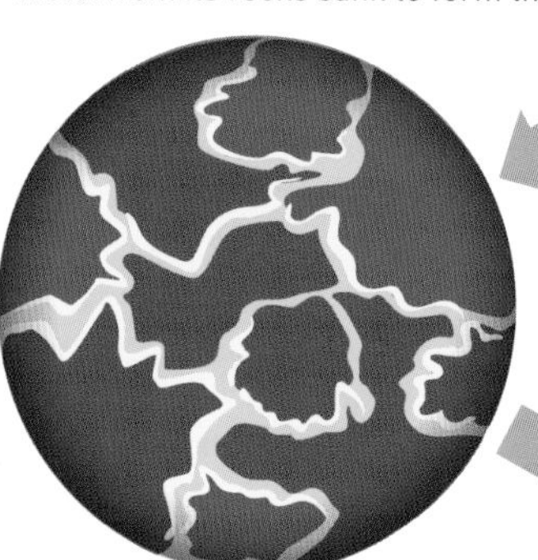

△ **Cooling**
When accretion slowed down, the planet cooled. Most of its rock solidified, forming a series of layers around the still-hot metal core.

△ **Oceans and atmosphere**
The gases that erupted from volcanoes formed the early atmosphere. Water vapor turned into clouds and then rain, which eventually filled the oceans.

REAL WORLD

Meteoric evidence

We cannot drill down to the Earth's core to see what it is made of, but scientists have deduced that it is mostly iron and nickel. This theory is supported by the fact that many meteorites, thought to be from the cores of planets destroyed billions of years ago, contain these metals. For example, this meteorite that fell on Russia in 1947 is made of iron.

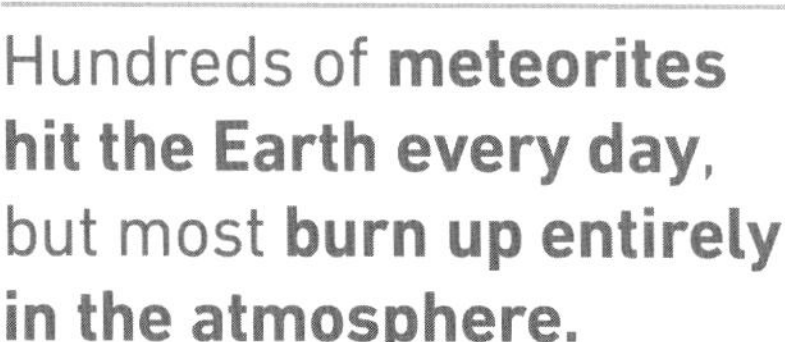

Hundreds of **meteorites hit the Earth every day**, but most **burn up entirely in the atmosphere.**

Layered planet

The Earth has a layered internal structure. Its gravity has pulled most of its heaviest, metallic elements down near the core, and the lightest elements exist as atmospheric gases. Rock lies in between, with the heaviest forming Earth's mantle. Some of the lighter ones make up the oceanic crust, and the lightest rocks of all form the continental crust, which lies above the ocean floors as dry land.

Inside Earth ▷
Planet Earth has a relatively thin crust that encloses a deep layer of hot rock called the mantle. The mantle surrounds the planet's heavy, metallic core.

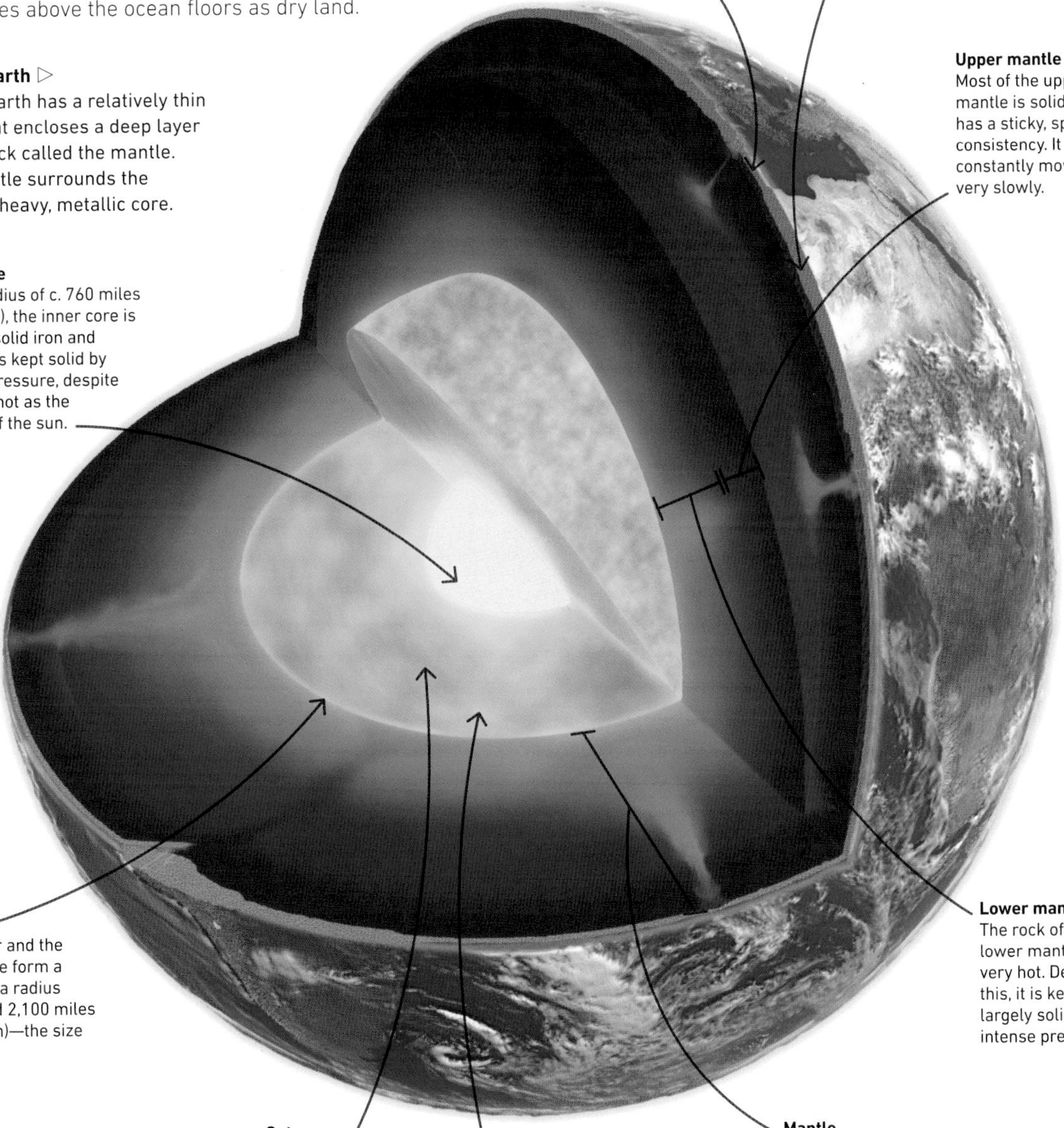

Oceanic crust
Much of the cool, brittle shell of the upper mantle is capped with lighter rocks that form the ocean floors. This oceanic crust is 3–6 miles (5–10 km) thick.

Continental crust
Lighter rocks are created by volcanic processes and form the core of continents. Together with layers of sedimentary rocks, they make up the continental crust, which is up to 27 miles (43 km) thick.

Upper mantle
Most of the upper mantle is solid but has a sticky, spongy consistency. It is constantly moving very slowly.

Inner core
With a radius of c. 760 miles (1,220 km), the inner core is made of solid iron and nickel. It is kept solid by intense pressure, despite being as hot as the surface of the sun.

Core
The inner and the outer core form a ball with a radius of around 2,100 miles (3,400 km)—the size of Mars.

Lower mantle
The rock of the lower mantle is very hot. Despite this, it is kept largely solid by intense pressure.

Outer core
Made of iron, nickel, and sulfur, the outer core is under slightly less pressure than the inner core. It remains a layer of liquid, molten metal.

Magnetic field
Swirling currents in the liquid outer core form Earth's magnetic field.

Mantle
This layer is 1,800 miles (2,890 km) thick and made up of heavy rock that contains a lot of iron and other metals. It is divided into the lower and upper mantle.

Moving plates and boundaries

SEE ALSO
❮ 22–23 Earth's structure
Shifting continents 26–27 ❯
Earthquakes and tsunamis 28–30 ❯
Volcanoes and hot springs 34–37 ❯

THE CRUST FORMS THE TOP OF EARTH'S BRITTLE OUTER LAYER, OR LITHOSPHERE.

The lithosphere is broken into rocky plates that fit together like a jigsaw, floating on the mantle. The plates move continually—slowly but powerfully, often triggering earthquakes or volcanic eruptions.

The Moving Shell

The moving plates of the lithosphere are giant slabs of rock shaped to the curve of the Earth. They are often thousands of miles across and average 62 miles (100 km) thick.

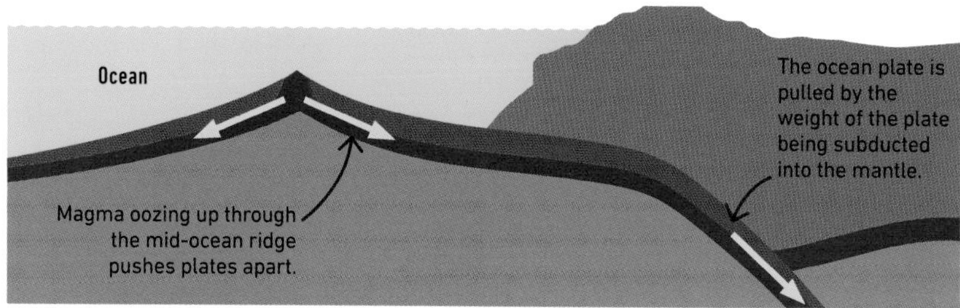
Ocean
Magma oozing up through the mid-ocean ridge pushes plates apart.
The ocean plate is pulled by the weight of the plate being subducted into the mantle.

△ **Why do plates move?**
Plates are pushed apart beneath the middle of the ocean by molten rock oozing up through long cracks to form a ridge (ridge push). They are also pulled from either edge by their own weight as they sink into the mantle (slab pull).

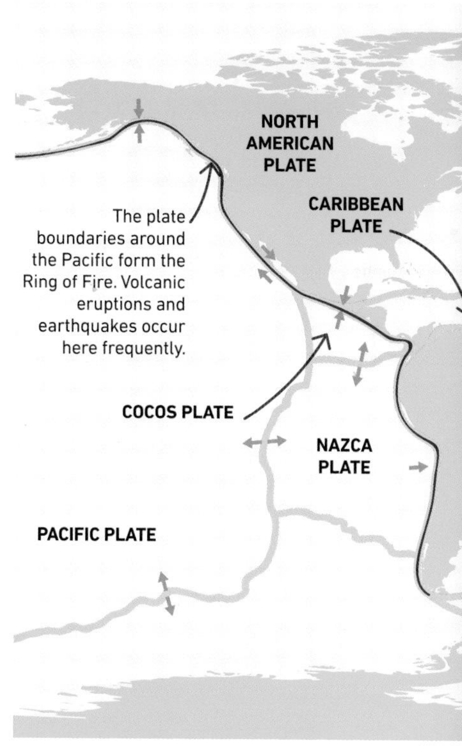
NORTH AMERICAN PLATE
CARIBBEAN PLATE
The plate boundaries around the Pacific form the Ring of Fire. Volcanic eruptions and earthquakes occur here frequently.
COCOS PLATE
NAZCA PLATE
PACIFIC PLATE

Plate boundaries

The tectonic plates move in different directions. Some move apart to form constructive plate boundaries, so called because they allow molten rock to rise up and create new crust. Others collide to form destructive boundaries, where the edge of one plate moves beneath another and is destroyed. There are some boundaries where the plates slide past each other.

Tectonic plates typically move as **fast** as a fingernail grows.

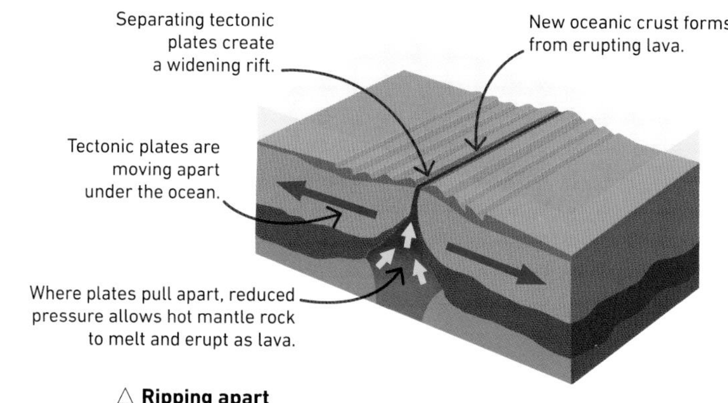
Separating tectonic plates create a widening rift.
New oceanic crust forms from erupting lava.
Tectonic plates are moving apart under the ocean.
Where plates pull apart, reduced pressure allows hot mantle rock to melt and erupt as lava.

△ **Ripping apart**
Where plates pull apart at constructive boundaries on the ocean floor, rifts open up. This reduces pressure on the hot rock below, allowing it to melt and erupt as lava that forms new crust.

Tectonic plates

The rocky plates of the lithosphere are called tectonic plates. There are seven giant plates and dozens of much smaller ones. These plates are moving apart in some places, especially beneath oceans, creating long rifts like the Mid-Atlantic Ridge. In other places, such as around the Pacific, they are pushing together, so one plate sinks (or subducts) beneath another, forming the highly volcanic subduction zones of the "Pacific Ring of Fire."

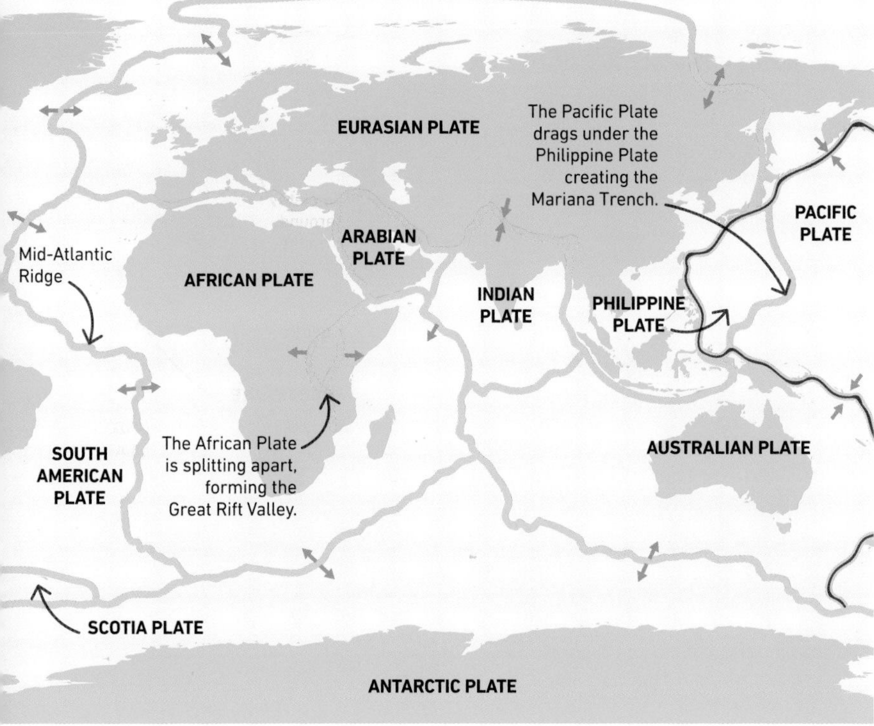
EURASIAN PLATE
The Pacific Plate drags under the Philippine Plate creating the Mariana Trench.
PACIFIC PLATE
ARABIAN PLATE
Mid-Atlantic Ridge
AFRICAN PLATE
INDIAN PLATE
PHILIPPINE PLATE
The African Plate is splitting apart, forming the Great Rift Valley.
AUSTRALIAN PLATE
SOUTH AMERICAN PLATE
SCOTIA PLATE
ANTARCTIC PLATE

REAL WORLD

Iceland's rift valley

Two tectonic plates are pulling apart at a boundary that divides the floor of the Atlantic Ocean, forming the Mid-Atlantic Ridge. Part of the ridge has been pushed above sea level to form Iceland. The rift between the plates passes through the island, creating a broad rift valley split by long cracks where the rocks are pulling apart.

KEY
- Plate boundary
- Ring of Fire
- Direction of plate movement

◁ **Fractured world**
Some tectonic plates are huge, such as the Eurasian Plate. Others are smaller, such as the Caribbean Plate. Most volcanoes erupt from the plate boundaries, which are also earthquake zones.

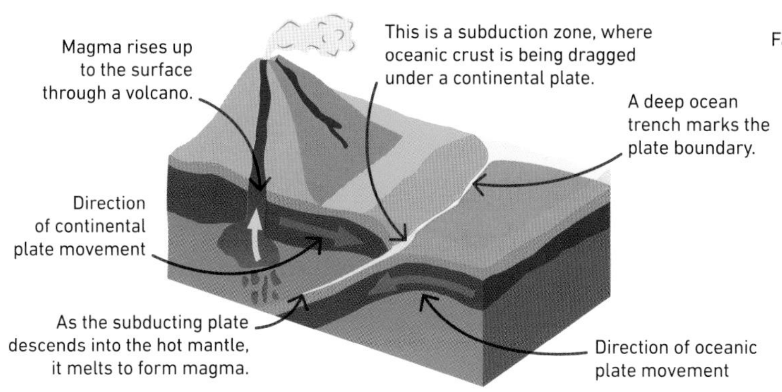
Magma rises up to the surface through a volcano.
This is a subduction zone, where oceanic crust is being dragged under a continental plate.
A deep ocean trench marks the plate boundary.
Direction of continental plate movement
As the subducting plate descends into the hot mantle, it melts to form magma.
Direction of oceanic plate movement

△ **Grinding together**
At destructive plate boundaries, oceanic crust is dragged down (subducted) into the mantle, grinding beneath continental plates. Deep ocean trenches mark these subduction zones.

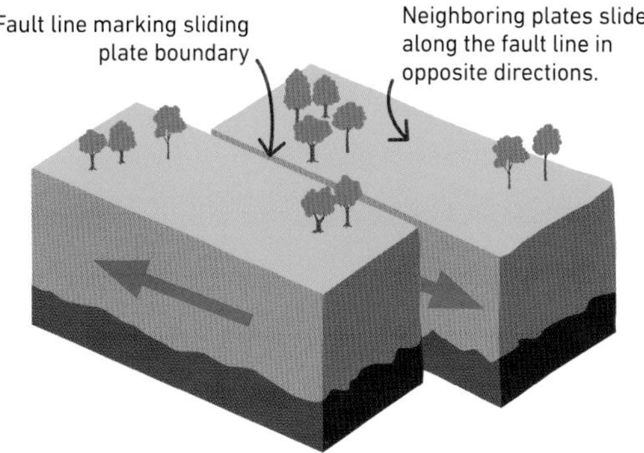
Fault line marking sliding plate boundary
Neighboring plates slide along the fault line in opposite directions.

△ **Sliding boundaries**
Tectonic plates can slide past each other without pulling apart or colliding. Most of the boundaries are on the ocean floors, but some are between continental plates.

Shifting continents

AS TECTONIC PLATES SLOWLY SHIFT, THEY CARRY CONTINENTS AROUND THE EARTH WITH THEM.

SEE ALSO
❮ 22–23 Earth's structure
❮ 24–25 Moving plates and boundaries
Continents and oceans 202–203 ❯

The continents are constantly on the move, jostling this way and that. The movement is slow but can be measured by satellite. North America and Europe are moving away from each other even now at 1 in (2.5 cm) per year.

As plates are pushed apart at the mid-ocean ridge, continents move farther apart.

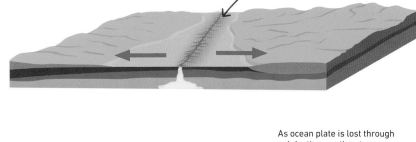

Floating continents

The continental crust is made of lighter rock than oceanic crust, so floats higher on the Earth's mantle and forms dry land. As the plates move around, heavier ocean crust between is pulled down into the mantle in places. But the continents are carried on top—slowly moving, splitting apart, and joining up again.

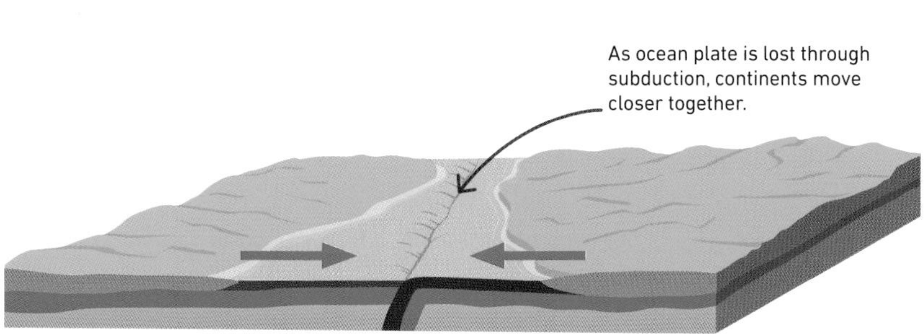
As ocean plate is lost through subduction, continents move closer together.

Push and pull ▷
Continents move apart where the oceanic crust between them expands, or move together where the oceanic crust is subducted.

Continental drift theory

In 1912, German scientist Alfred Wegener noticed that continents all fit together like pieces of a jigsaw. He suggested that millions of years ago the continents were joined in one "supercontinent." Wegener's idea was dismissed at the time but proven right in the 1960s, when scientists discovered that all of Earth's surface is broken into shifting tectonic plates.

India
Africa
Australia
Antarctica
South America

Match fit ▷
Wegener found strong matches between continents of similar rocks and fossils, all dating from around 250–300 million years ago.

KEY

Cynognathus fossils
Fossils of the land reptile *Cynognathus* have been found in South America and Africa, showing that the reptile could move easily between the two.

Lystrosaurus fossils
Fossils show the reptile *Lystrosaurus* lived 300 million years ago in southern Africa, India, and Antarctica.

Glossopteris fossils
Fossils of the fern *Glossopteris* have been found in all the southern continents, suggesting that they were once joined together.

Supercontinents

Scientists know there were at least seven supercontinents in the past, which have come together then split apart. The most recent was Pangaea, which formed around 300 million years ago. Around the time of the first dinosaurs, Pangaea slowly started to split into the continents we know today.

Pangaea

◁ **225 million years ago** Pangaea formed about 335 million years ago, as earlier continents were pushed together into a supercontinent. It existed for 160 million years.

Laurasia
Gondwana

◁ **150 million years ago** By the late Jurassic period, Pangaea had split up to form two smaller supercontinents, Laurasia in the north and Gondwana in the south.

North America
Europe
Asia
Africa
South America
Australia
Antarctica

◁ **Today** Modern continents formed in the Cretaceous period, when the last giant dinosaurs were alive, and slowly drifted to their present position.

Continental collision

Throughout the history of our planet, the continents have been splitting up and joining together. About 140 million years ago, part of east Africa broke away and was carried north across the Indian Ocean by tectonic plate movement. This island continent eventually collided with Asia about 40 million years ago and became India.

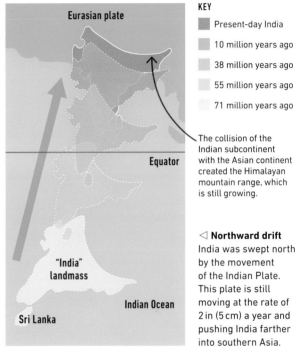

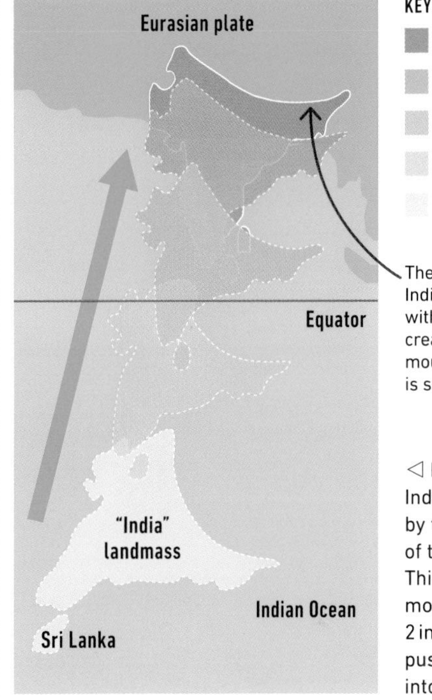
Eurasian plate
Equator
"India" landmass
Sri Lanka
Indian Ocean
KEY
Present-day India
10 million years ago
38 million years ago
55 million years ago
71 million years ago
The collision of the Indian subcontinent with the Asian continent created the Himalayan mountain range, which is still growing.

◁ **Northward drift** India was swept north by the movement of the Indian Plate. This plate is still moving at the rate of 2 in (5 cm) a year and pushing India farther into southern Asia.

Continental breakup

The tectonic plates are constantly on the move, carrying the continents along with them. The African Plate is splitting into two plates, at the Great Rift Valley in east Africa. In the future, the land to the east will become a separate island continent in the Indian Ocean.

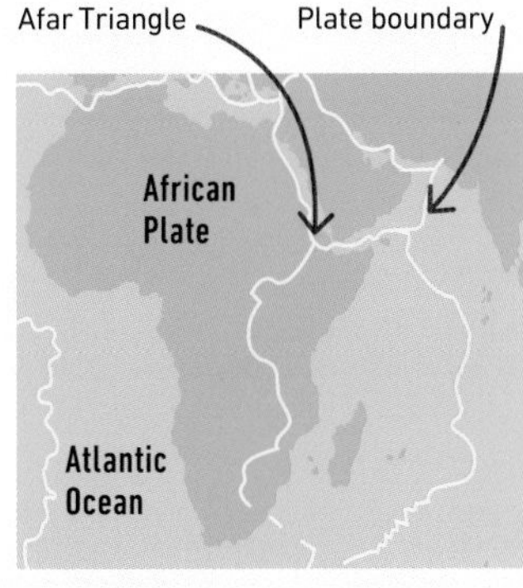
Afar Triangle
Plate boundary
African Plate
Atlantic Ocean

△ **Present-day Africa** The Great Rift Valley passes through Africa from the Afar Triangle in the north to Mozambique in the south.

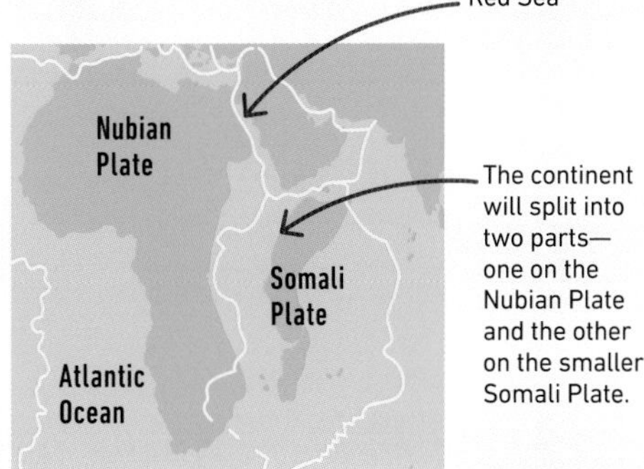
Red Sea
Nubian Plate
Somali Plate
Atlantic Ocean
The continent will split into two parts—one on the Nubian Plate and the other on the smaller Somali Plate.

△ **Africa in the future** As the rift opens up, its valley will become a long, narrow sea that broadens into an ocean, splitting Africa in two.

Earthquakes and tsunamis

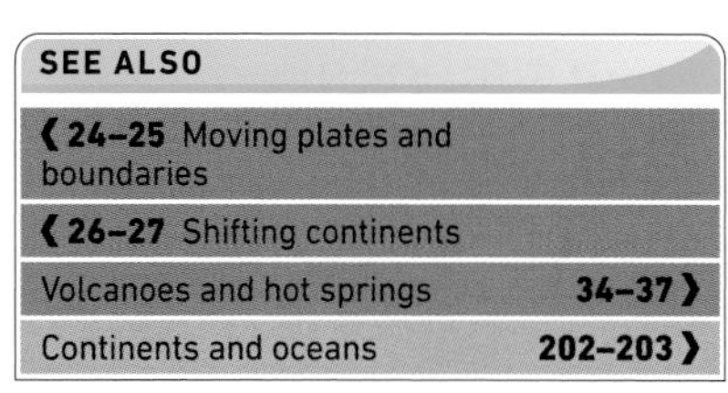

SEE ALSO
❮ 24–25 Moving plates and boundaries
❮ 26–27 Shifting continents
Volcanoes and hot springs 34–37 ❯
Continents and oceans 202–203 ❯

AS THE MOVING PLATES OF THE EARTH'S CRUST SHIFT, THEY CAN CAUSE EARTHQUAKES AND TSUNAMIS.

When the plates of the Earth's crust grind against each other, their edges become distorted, building up strain. Eventually they give way, and the shock can cause earthquakes and tsunamis.

Why earthquakes happen

Most earthquakes occur in regions where two plates of the Earth's crust are grinding together. The plates are always moving, very slowly. In some places, this creates steady slip along the faults that form plate boundaries, causing frequent, small tremors. But if the faults get locked, this builds up tension, causing them to give way and generate earthquakes.

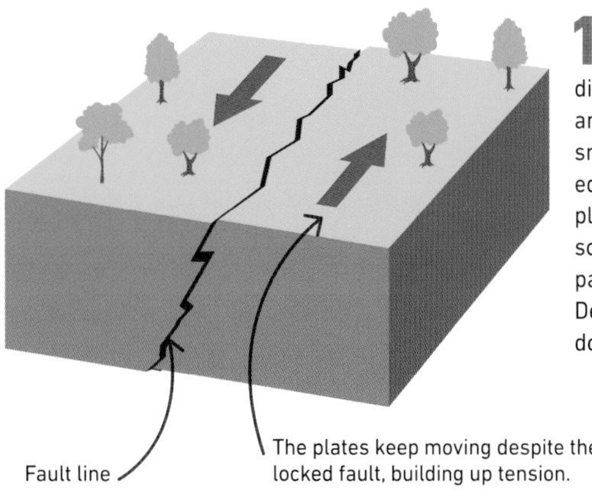

Fault line
The plates keep moving despite the locked fault, building up tension.

1 Locked plates
The fault lines that divide tectonic plates are rarely straight and smooth. The ragged edges of the moving plates lock together, so they cannot slip past each other. Despite this, the plates do not stop moving.

2 Distorted rocks
As the plates keep moving, the rocks on each side of the fault become distorted. They bend like springs, storing up energy. As long as the fault stays locked, the rocks will keep bending, building up the tension until, eventually, something snaps.

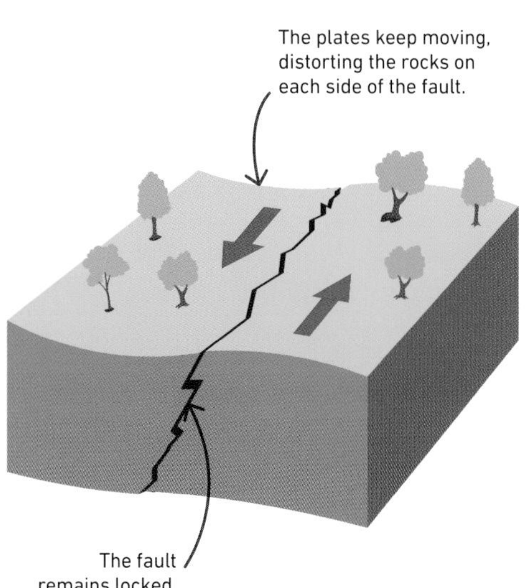

The plates keep moving, distorting the rocks on each side of the fault.
The fault remains locked.

Measuring earthquakes

Earthquake scientists use a network of seismometers around the world to register quake waves and get combined readings. In the past, they would rate them on the Richter scale. They now use the Moment Magnitude Scale for the biggest earthquakes. This combines Richter seismometer readings with observations, such as how much rocks move along a fault, to reveal the power of an earthquake.

▽ **Magnitude**
To compare earthquakes, scientists use a scale that depends partly on accurate measurement of the size or "magnitude" of vibrations using a seismograph. The most widely used scale is the Moment Magnitude Scale (Mw).

MAGNITUDE

0 1 2 3 4 5 6 7 8 9

Each step on the Mw scale is 10 times bigger than the one before.

About **500,000 earthquakes occur** worldwide **each year. Most** of them **cannot be felt.**

3 Snap and shock
When the fault finally gives way, the rocks spring back. All the movement that should have occurred over many years happens within a few minutes, causing an earthquake. Shock waves radiate from the earthquake focus (the fracture point), shaking the Earth like ripples on a pond.

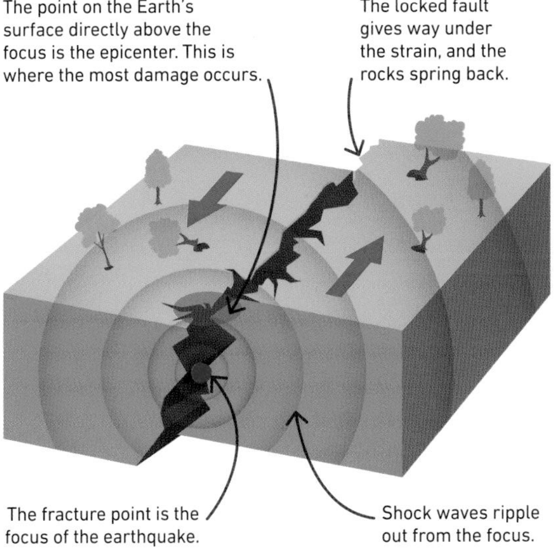
The point on the Earth's surface directly above the focus is the epicenter. This is where the most damage occurs.
The locked fault gives way under the strain, and the rocks spring back.
The fracture point is the focus of the earthquake.
Shock waves ripple out from the focus.

The Ring of Fire

Destructive earthquakes occur on the boundaries where tectonic plates are either colliding or sliding past each other. Most of these lie around the edges of the Pacific Ocean, forming a danger zone known as the Pacific Ring of Fire. It owes its name to the many volcanoes that erupt along the boundaries where plates are colliding.

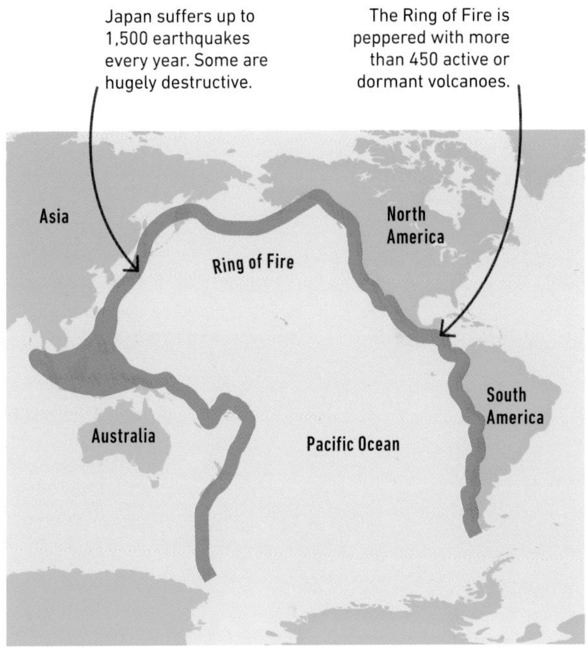
Japan suffers up to 1,500 earthquakes every year. Some are hugely destructive.
The Ring of Fire is peppered with more than 450 active or dormant volcanoes.
Asia
North America
Ring of Fire
South America
Australia
Pacific Ocean

△ **Earthquake zone**
About 80 percent of the world's most serious earthquakes and tsunamis occur along the Pacific Ring of Fire, which extends across Indonesia and the fringes of the Pacific Ocean.

Intensity ▷
You can assess the intensity on a scale that assesses the damage done by an earthquake locally. The effect depends on how near the earthquake center you are.

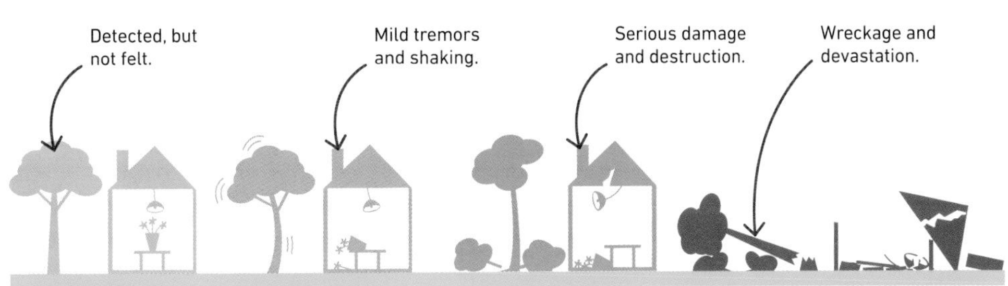
Detected, but not felt.
Mild tremors and shaking.
Serious damage and destruction.
Wreckage and devastation.

Level 1
An earthquake can be detected only by seismometers.

Levels 2–4
Tremors might be noticeable. Damage is likely to be caused by falling objects.

Levels 5–6
Tremors cause weak structures to shake, lift, and fracture.

Levels 7–9
There is tremendous loss of life and serious destruction of property over large areas.

Tsunamis

Tsunamis, sometimes called "tidal waves," are a series of enormous waves caused by a disturbance underwater. They can be set off by landslides, volcanic eruptions, or meteorites, but most start with undersea earthquakes. This is why the Pacific, with its earthquake zones, gets the most tsunamis.

A 1,700 ft (520 m) **tsunami** hit Lituya Bay in Alaska in **1958**.

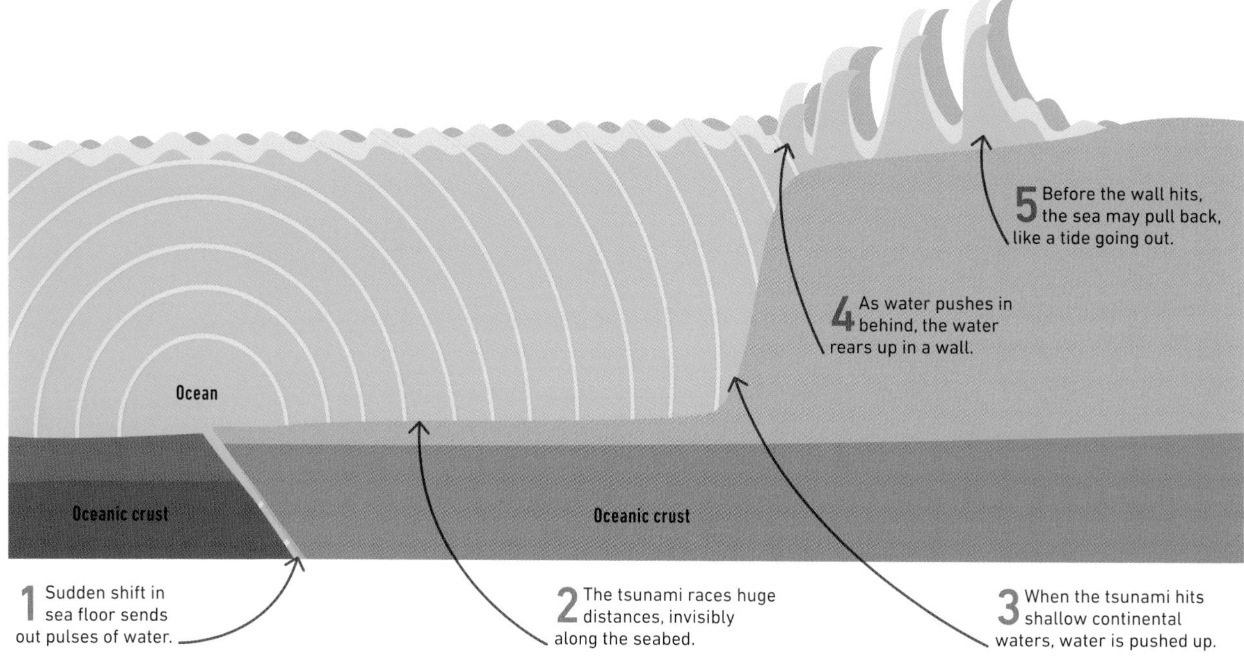

5 Before the wall hits, the sea may pull back, like a tide going out.
4 As water pushes in behind, the water rears up in a wall.
Ocean
Oceanic crust
Oceanic crust
1 Sudden shift in sea floor sends out pulses of water.
2 The tsunami races huge distances, invisibly along the seabed.
3 When the tsunami hits shallow continental waters, water is pushed up.

Fast as a jet plane

The deeper the water, the faster the tsunami. The Pacific Ocean is 13,000 ft (4,000 m) deep on average, so tsunamis speed along at over 500 mph (800 km/h). That means a tsunami generated in the Aleutian Islands may reach Hawaii in less than four and a half hours.

Depth (ft)	Speed (mph)	Wavelength (miles)
6,560	313	94
656	99	30
164	49	14
32	22	7

REAL WORLD

Tsunami warnings

Tsunamis can cause devastation when they hit land, so advance warning can save many lives by giving people the chance to flee to higher ground. There is a global network of warning centers that detect threatening earthquakes with seismographs and pick up on small tell-tale changes in the water level. The heart of the DART tsunami warning system is a network of buoys moored across the ocean. Each is linked to a Bottom Pressure Recorder (BPR) on the ocean floor that can pick up subtle changes in water pressure.

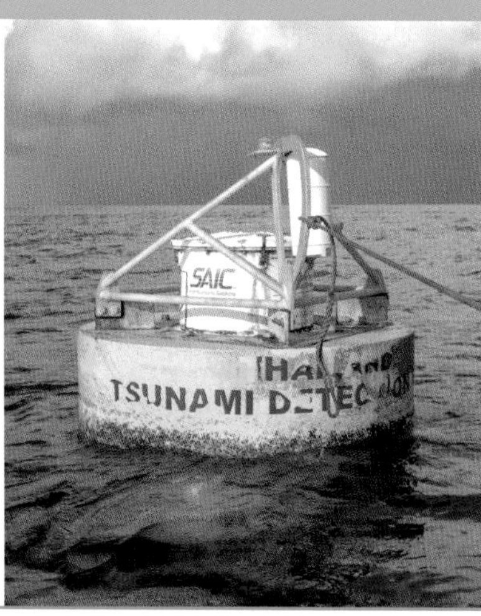

SAIC
TSUNAMI D

Mountain building

THE SAME FORCES THAT CAUSE EARTHQUAKES ALSO BUCKLE AND CRACK CONTINENTS, BUILDING MOUNTAINS.

SEE ALSO
‹ 22–23 Earth's structure
‹ 24–25 Moving plates and boundaries
‹ 26–27 Shifting continents
‹ 28–30 Earthquakes and tsunamis

Most mountain ranges are created by tectonic plates pushing together. They can also form where the continental crust is being pulled apart or where hard rocks are exposed by erosion.

Crumpled crust

Where oceanic crust is pushed beneath continental crust, creating a subduction zone, the edge of the continent is crumpled into fold mountains. These are squeezed upward by the pressure of the converging plates. Fold mountains are also dotted with volcanoes, fueled by magma that forms deep in the subduction zone.

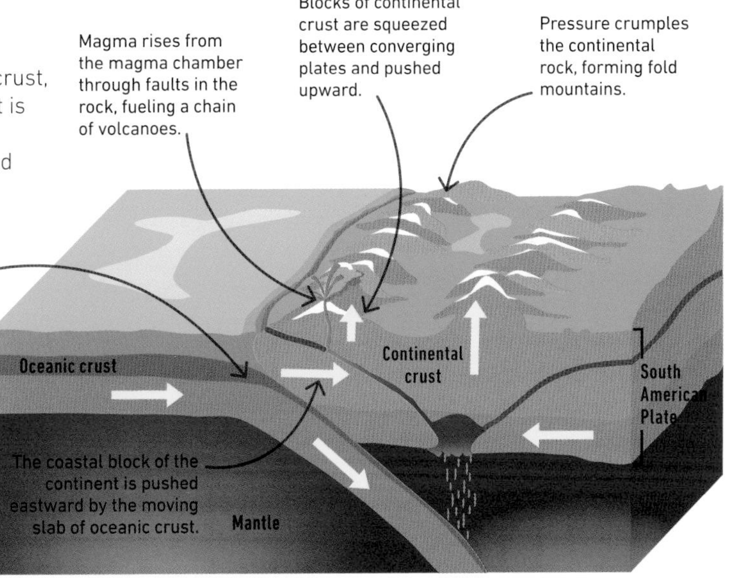
Magma rises from the magma chamber through faults in the rock, fueling a chain of volcanoes.
Blocks of continental crust are squeezed between converging plates and pushed upward.
Pressure crumples the continental rock, forming fold mountains.
Heavier oceanic crust bends and sinks beneath the lighter continental crust, creating a subduction zone.
Nazca Plate
Oceanic crust
Continental crust
South American Plate
The coastal block of the continent is pushed eastward by the moving slab of oceanic crust.
Mantle

KEY
Direction of movement

Formation of the Andes ▷
The Andes have been created above a subduction zone where the floor of the Pacific Ocean is grinding beneath the edge of South America. The pressure forces the continental rock upward.

Collision zones

Where two tectonic plates carrying continental crust collide, this causes massive crumpling of the crust, pushing up high mountains. Oceanic crust that once lay between the continents is dragged into the mantle deep below the collision zone. Meanwhile, lighter ocean-floor sediments are folded and pushed upward to form part of the mountain range.

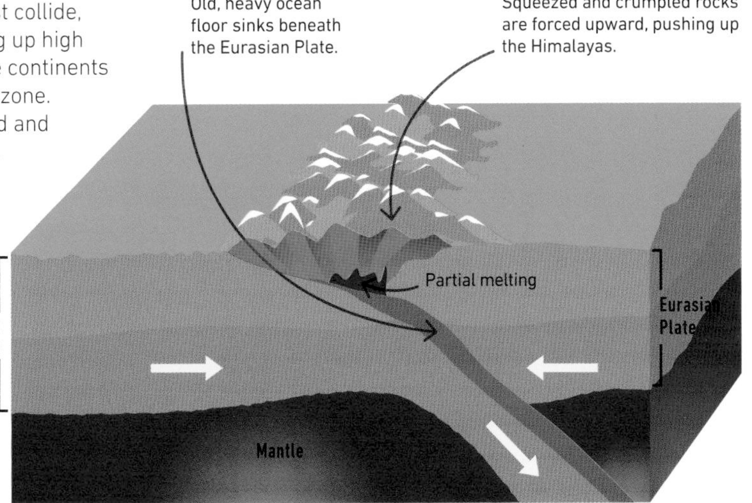
Old, heavy ocean floor sinks beneath the Eurasian Plate.
Squeezed and crumpled rocks are forced upward, pushing up the Himalayas.
Partial melting
Indian Plate
Eurasian Plate
Mantle

KEY
Direction of movement

Raising the Himalayas ▷
The highest mountain range on Earth has been formed by the collision of India with Asia. The ocean floor on the Indian Plate is slipping beneath the Eurasian Plate, while the pressure is pushing up the Himalayas and the Tibetan plateau.

»

Folding and faulting

Where mountains are built by tectonic plates pushing together, the rocks are squeezed and crumpled. Horizontal layers of rock—known as strata—can be distorted into dramatic folds, and even overturned. The rock can fracture, creating faults that make the strata slip out of alignment, and push the deeper layers of older rock up above the younger rock on top.

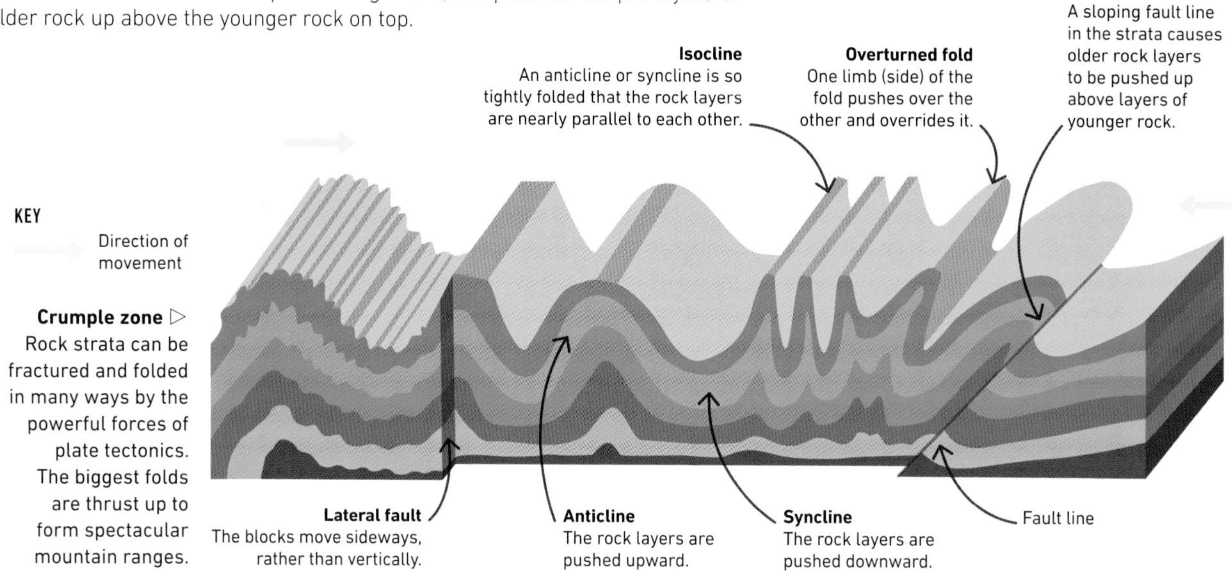
KEY
Direction of movement
Isocline
An anticline or syncline is so tightly folded that the rock layers are nearly parallel to each other.
Overturned fold
One limb (side) of the fold pushes over the other and overrides it.
Thrust fold
A sloping fault line in the strata causes older rock layers to be pushed up above layers of younger rock.
Lateral fault
The blocks move sideways, rather than vertically.
Anticline
The rock layers are pushed upward.
Syncline
The rock layers are pushed downward.
Fault line

Crumple zone ▷ Rock strata can be fractured and folded in many ways by the powerful forces of plate tectonics. The biggest folds are thrust up to form spectacular mountain ranges.

Block mountains

Another way that mountains can form is in places where the continental crust is being pulled apart or pushed together. Here, the tension makes the crust fracture, creating steeply sloping faults. Big blocks of crust slip down the fault planes to form rift valleys, often lined by steep cliffs. The sections of crust that have not subsided form block mountains. This type of landform is also known as basin and range topography.

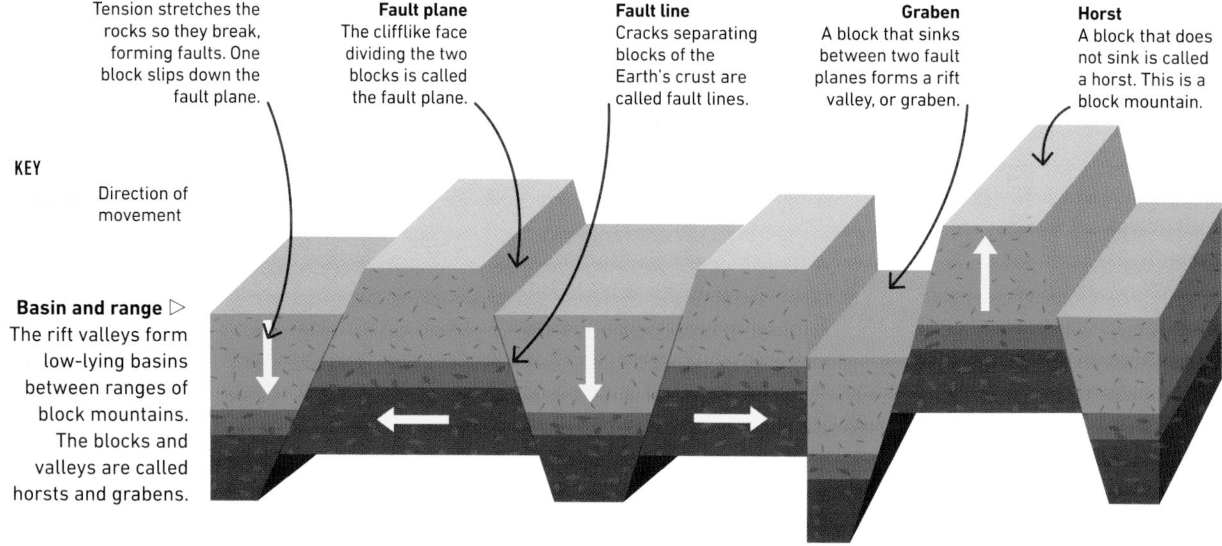
Tension stretches the rocks so they break, forming faults. One block slips down the fault plane.
Fault plane
The clifflike face dividing the two blocks is called the fault plane.
Fault line
Cracks separating blocks of the Earth's crust are called fault lines.
Graben
A block that sinks between two fault planes forms a rift valley, or graben.
Horst
A block that does not sink is called a horst. This is a block mountain.
KEY
Direction of movement

Basin and range ▷ The rift valleys form low-lying basins between ranges of block mountains. The blocks and valleys are called horsts and grabens.

Granite cores

During mountain building, molten rock (magma) often seeps up toward the surface from deep in the crust. Molten rock may either erupt from volcanoes or solidify below the ground to form huge masses of very hard granite called batholiths. Over time, the softer rock above may erode away, exposing the granite that resists erosion and forms mountains.

Batholiths extend over at least **40 sq miles (100 sq km)**, and many are **much larger**.

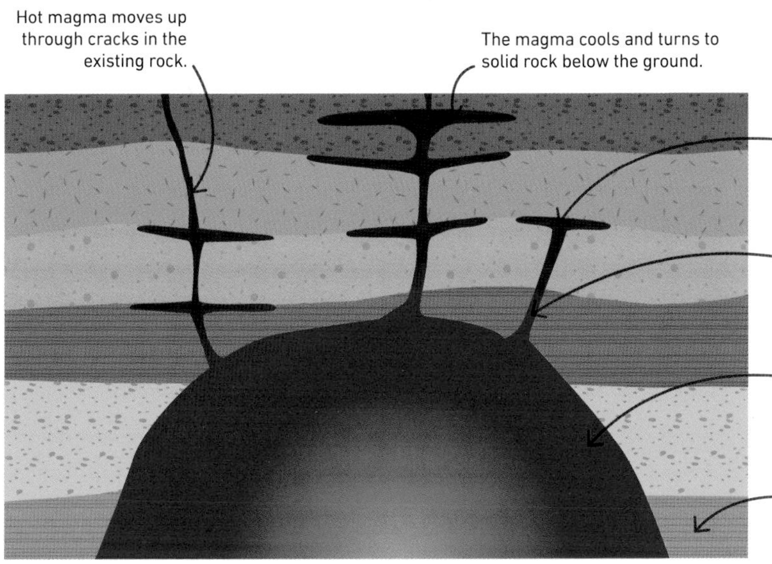
Hot magma moves up through cracks in the existing rock.
The magma cools and turns to solid rock below the ground.
Sill
Magma that flows between rock layers forms a sill.
Dike
Magma that cools in a crack forms a dike.
Batholith
A mass of hard granite forms where magma has replaced the original rock.
Bed of sediment

1 Buried batholith
The molten rock cools very slowly underground, gradually forming a batholith. Smaller masses cool more quickly to create dikes, sills, and similar buried rock formations.

Surrounding rock is worn away, so the ground level sinks.
Erosion of the surrounding rock exposes the batholith.

2 Exposed batholith
Over millions of years, the rock around the batholith is worn away. The hard rock of the batholith, however, gets eroded more slowly and survives above the surrounding ground level.

REAL WORLD

Sugarloaf Mountain

The spectacular Sugarloaf Mountain that rises almost 1,300 ft (400 m) above Rio de Janeiro, Brazil, is a mass of granite-type rock exposed by erosion. It is one of several similar peaks near the city. These steep, dome-shaped mountains are known as bornhardts, named after the geologist Wilhelm Bornhardt who first described them.

Volcanoes and hot springs

VOLCANOES FORM AS MAGMA AND OTHER MATERIALS FROM EARTH'S HOT INTERIOR ERUPT THROUGH THE CRUST.

All but a few active volcanoes lie close to plate margins, where there is a ready supply of magma (molten rock). Magma erupting on the surface is called lava. Some volcanoes ooze magma continually. Others explode suddenly, blasting out ash and fiery gases.

Inside a volcano

Some volcanoes are simply cracks in the crust, but the most distinctive are the cone-shaped "stratovolcanoes." Beneath such volcanoes, magma rises continually into a space called the magma chamber. The chamber fills up until eventually the pressure drives the magma to erupt through the volcano's narrow chimney or "vent," powered by expanding bubbles of steam and gas.

Stratovolcanoes ▷
Most volcanoes on land are stratovolcanoes, meaning they are made up of alternating layers of lava and volcanic ash.

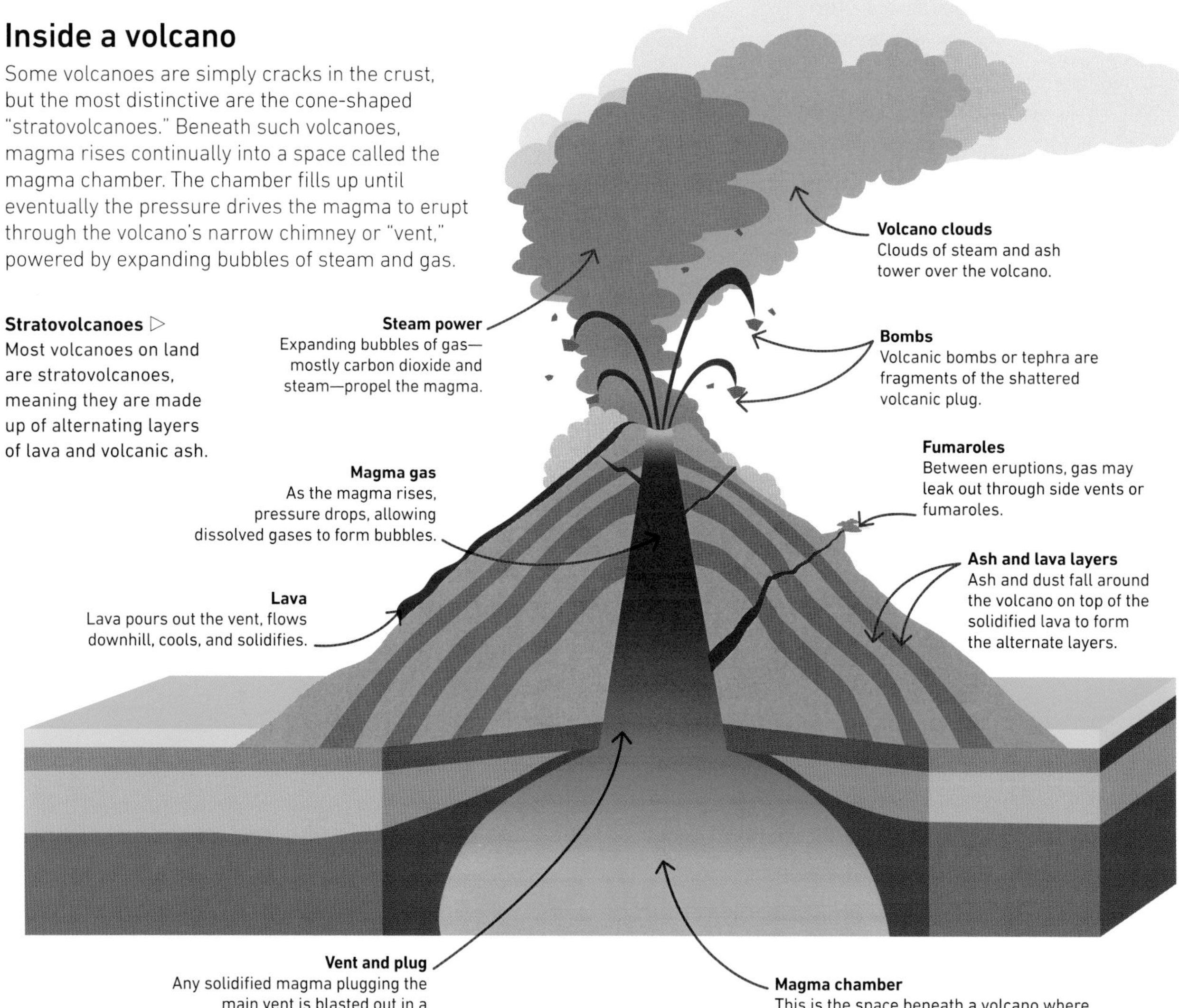
Steam power
Expanding bubbles of gas—mostly carbon dioxide and steam—propel the magma.
Volcano clouds
Clouds of steam and ash tower over the volcano.
Bombs
Volcanic bombs or tephra are fragments of the shattered volcanic plug.
Fumaroles
Between eruptions, gas may leak out through side vents or fumaroles.
Magma gas
As the magma rises, pressure drops, allowing dissolved gases to form bubbles.
Ash and lava layers
Ash and dust fall around the volcano on top of the solidified lava to form the alternate layers.
Lava
Lava pours out the vent, flows downhill, cools, and solidifies.
Vent and plug
Any solidified magma plugging the main vent is blasted out in a shower of ash and cinders.
Magma chamber
This is the space beneath a volcano where magma builds up before an eruption.

Kinds of volcanoes

Volcanoes can be classified either by the nature of their eruptions or by their shape and size. The kind of eruption, and the shape and size, depend on the nature of the magma, and the situation.

The extinct **Hawaiian** shield volcano, **Pūhāhonu**, is the **biggest** on Earth.

▽ **Effusive eruptions**
Most effusive eruptions occur where plates pull apart at mid-ocean ridges. Here, hot, fluid basic (non-acid) lava floods far and wide. They tend to ooze lava continually, rather than erupt occasionally.

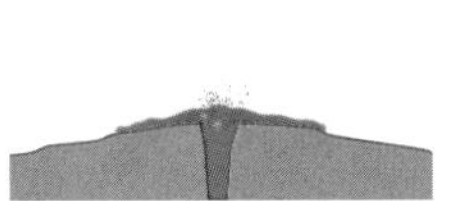

Icelandic
Where lava oozes from cracks

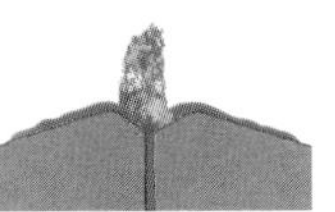

Hawaiian
Where lava spills from vents

▽ **Explosive eruptions**
Cool and sticky acidic magma unleashes explosive eruptions. The eruptions are sporadic and often violent. They tend to occur near subduction zones, or where continents collide.

Vulcanian
Are more sporadic and violent

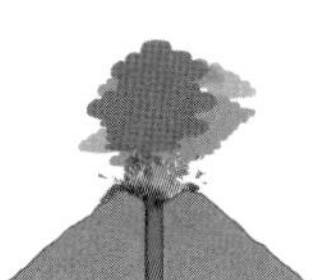

Strombolian
Are frequent but only mildly explosive

Pelean
Glowing avalanches of cinders and hot gas

Pillnian
Powerful blasts of gas and ash

Volcano forms

Volcanoes vary greatly in shape. The biggest by far are huge shield volcanoes, such as Hawaii's Mauna Kea. Stratovolcanoes often build high mountain peaks. The most common are small cinder cones.

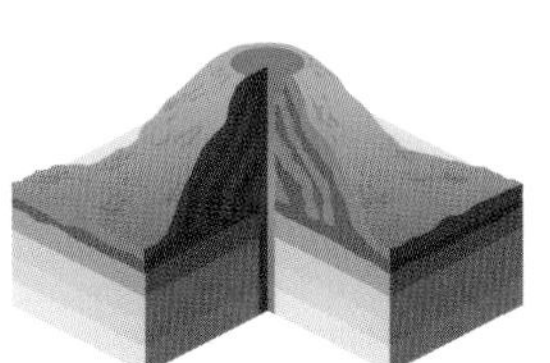

Cinder cones
Small mounds built entirely of cinders

Shield volcanoes
Flow out far and wide

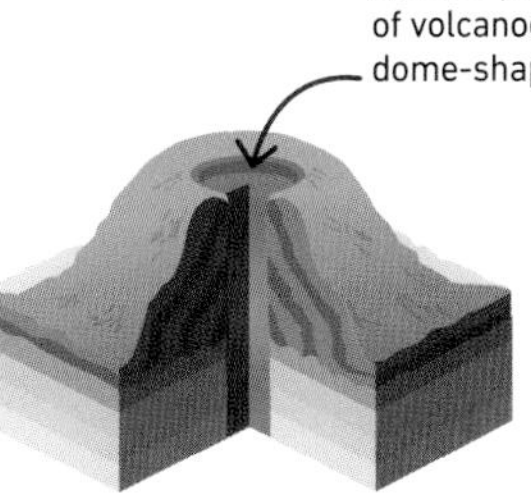

About 6 percent of volcanoes are dome-shaped.

Dome volcanoes
Built from lava too sticky to flow far

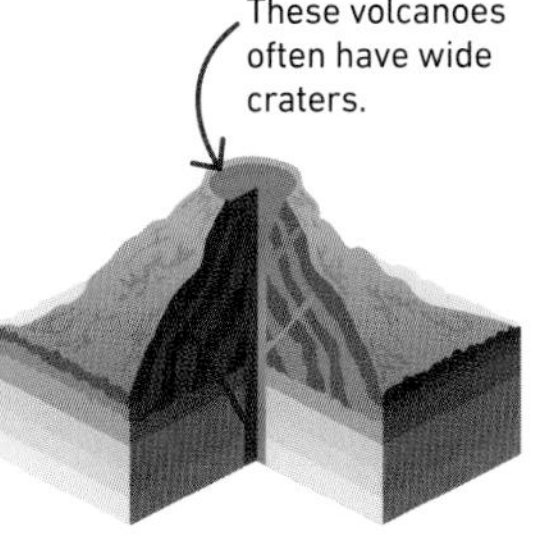

These volcanoes often have wide craters.

Stratovolcanoes
Cones with alternate lava and ash layers

Hot spots

Unlike most other volcanoes, hot spot volcanoes erupt far from plate margins where a plume of heat rises from deep in the mantle. As the plume comes up under the thin ocean plate, it melts rock to create a volcano.

▽ **Hawaiian hot spot**
The plume always stays in the same place, so as the plate moves over the top, it creates a chain of volcanic islands, like the Hawaiian chain.

Lithosphere
PACIFIC PLATE
Asthenosphere

Volcanoes deprived of their magma supply become islands.

Supervolcanoes

A few giant volcanoes have the power to blast lava and ash over huge distances, so they do not form typical volcanic cones. After such a massive eruption, the empty magma chamber below the ground collapses, forming a broad depression called a caldera. A new magma chamber may then form, leading to another gigantic eruption.

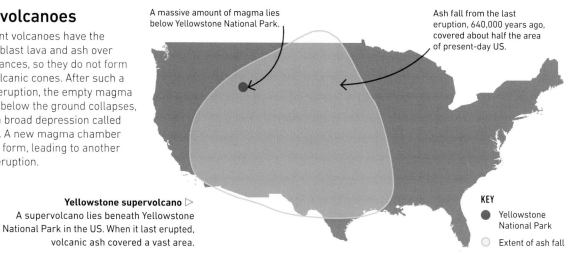

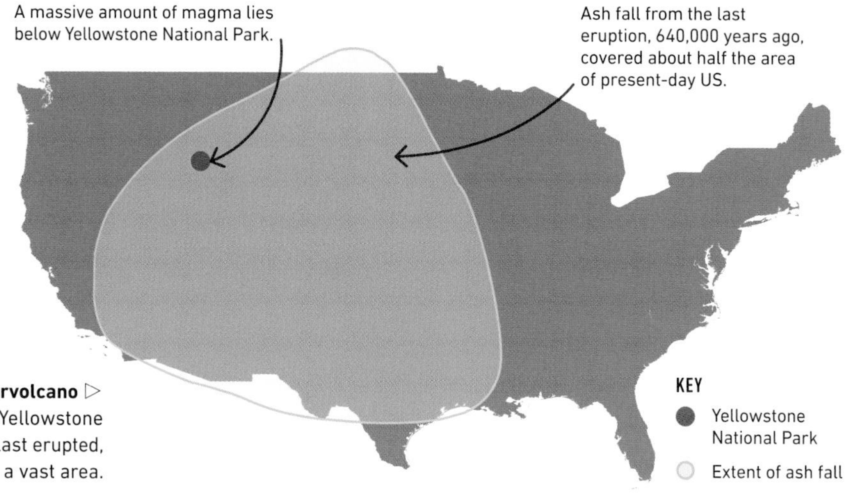
A massive amount of magma lies below Yellowstone National Park.
Ash fall from the last eruption, 640,000 years ago, covered about half the area of present-day US.
KEY
Yellowstone National Park
Extent of ash fall

Yellowstone supervolcano ▷ A supervolcano lies beneath Yellowstone National Park in the US. When it last erupted, volcanic ash covered a vast area.

Volcano: dead or alive?

Very few volcanoes erupt continuously. Even the most active ones have quiet periods, when no lava or gas erupts from the crater. Scientists talk about volcanoes being active, dormant (sleeping), and extinct (dead). But it is not always easy to be sure whether a volcano that is not erupting has just paused or has stopped altogether.

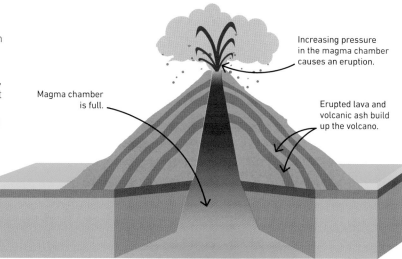

Increasing pressure in the magma chamber causes an eruption.

Magma chamber is full.

Erupted lava and volcanic ash build up the volcano.

1 Active volcanoes
A volcano erupting lava, ash, or smoke is clearly active. Volcano scientists may also spot signs that another eruption is due. Generally, it is said to be active if it has erupted in the last 10,000 years or so.

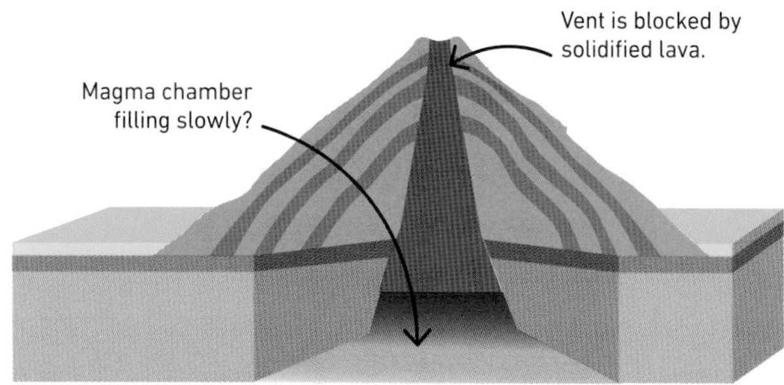
Vent is blocked by solidified lava.
Magma chamber filling slowly?

2 Dormant volcanoes
Any volcano that has not erupted for a long time, yet is not entirely quiet, is described as dormant. But magma may well be slowly building up deep down, ready to power another eruption.

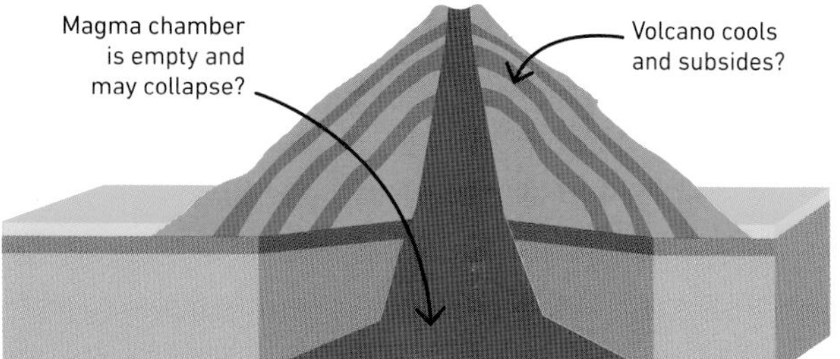
Magma chamber is empty and may collapse?
Volcano cools and subsides?

3 Extinct volcanoes
A volcano that has not erupted in tens of thousands of years is likely to be extinct. Maybe the magma has gone cold and solid. But magma may simmer under even an "extinct" volcano.

Living with volcanoes

Despite the dangers associated with volcanic regions, many people live close to volcanoes because volcanic rock breaks down to form very fertile soil for agriculture. There are also other benefits. However, while some volcanoes may seem safe to people who are unaware of the risks involved, they may still erupt any day.

At least **300 million** people live **within range** of an **active volcano**.

Advantages

Fertilizing ash
The ash clouds formed by volcanoes settle on the land and eventually form rich soils, ideal for growing crops.

Geothermal energy
In some volcanic regions, water heated by hot rock is used to make steam that powers electricity generators.

Tourism
An erupting volcano is a spectacular sight. Locals can earn money catering for tourists who enjoy sightseeing.

Disadvantages

Loss of life
In 1902, an eruption on Martinique island wiped out a whole town of 30,000 people. Living near volcanoes is risky.

Damage to property and the economy
Even if few people are killed by a volcanic eruption, the lava and ash can cause widespread destruction.

Damage to habitats and landscapes
Catastrophic eruptions can flatten forests and trigger tsunamis, threatening wildlife. Recovery can take a long time.

Hot springs and geysers

In highly volcanic regions such as supervolcano calderas (a large crater left when a volcano collapses in on itself), groundwater seeping down through cracks comes into contact with very hot rock. This usually makes the water boil back up to the surface as a hot spring. In some places, the hot water is held under pressure until it explodes out of the ground as a geyser, a fountain of boiling water and steam.

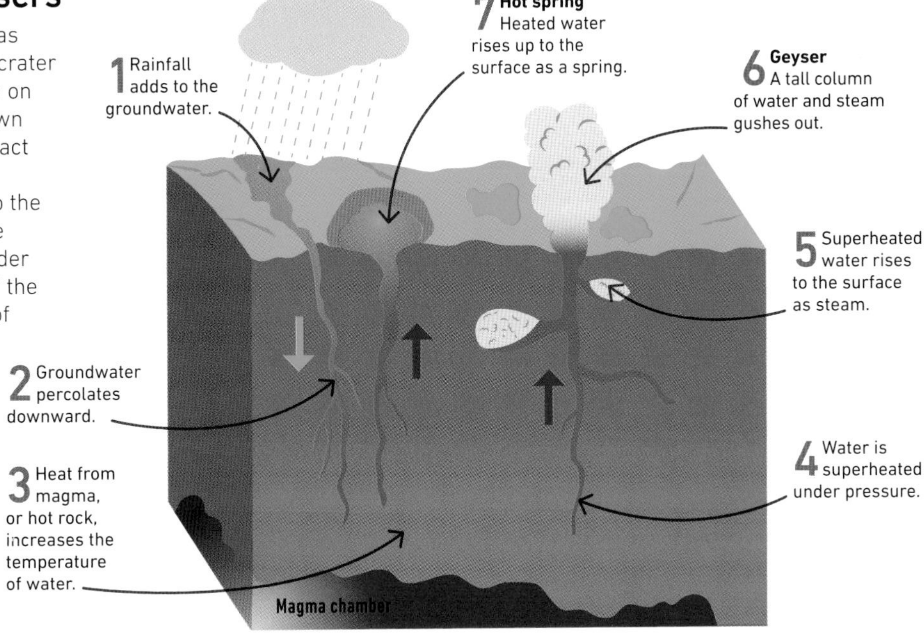
1 Rainfall adds to the groundwater.
2 Groundwater percolates downward.
3 Heat from magma, or hot rock, increases the temperature of water.
4 Water is superheated under pressure.
5 Superheated water rises to the surface as steam.
6 Geyser A tall column of water and steam gushes out.
7 Hot spring Heated water rises up to the surface as a spring.
Magma chamber

A geothermal system ▷
Hot springs and geysers occur in clusters called geothermal systems. The heat rising from magma chambers lying deep below the ground causes these thermal features.

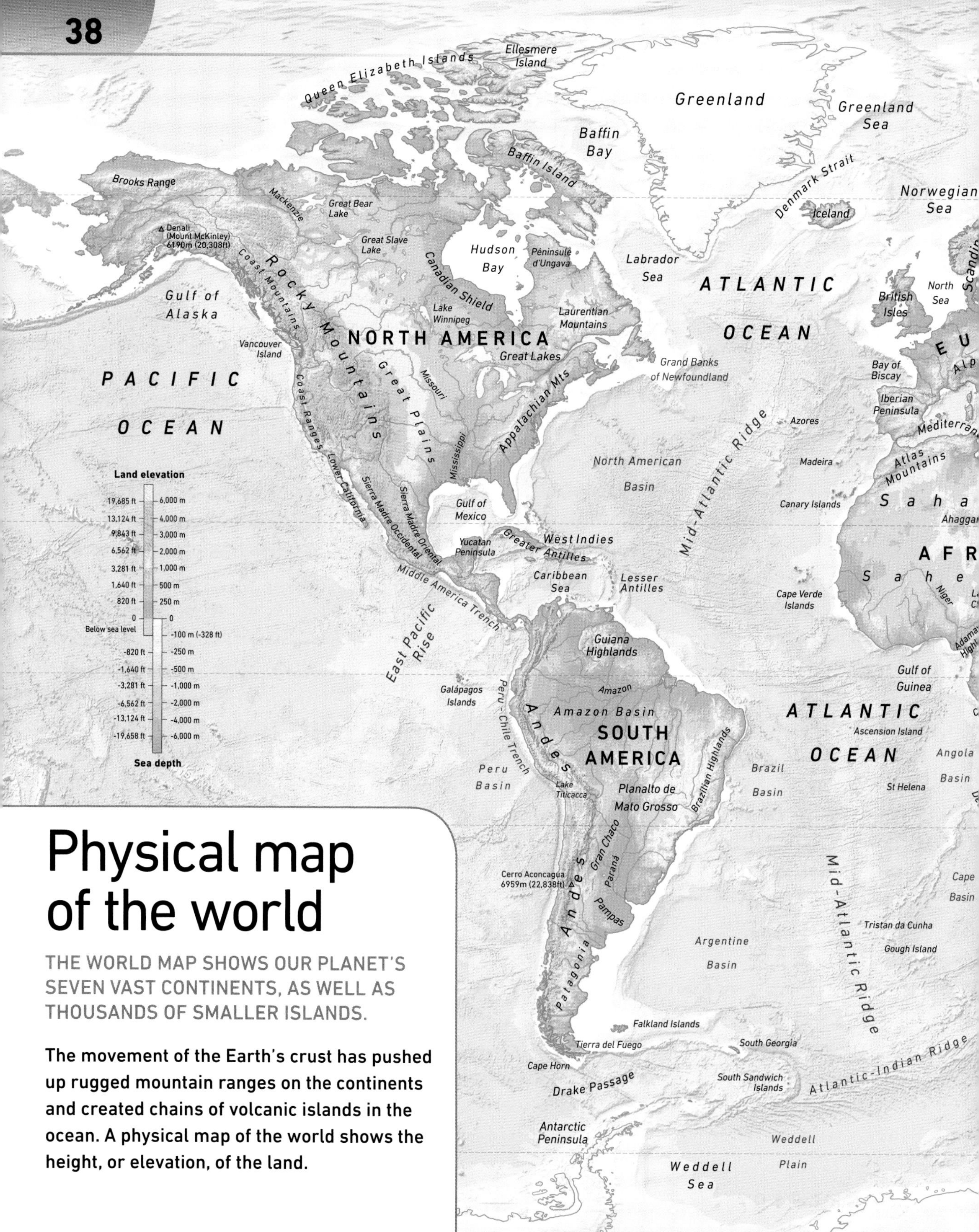
Queen Elizabeth Islands
Ellesmere Island
Greenland
Greenland Sea
Baffin Bay
Baffin Island
Denmark Strait
Norwegian Sea
Iceland
Brooks Range
Mackenzie
Great Bear Lake
Denali (Mount McKinley) 6190m (20,308ft)
Great Slave Lake
Hudson Bay
Péninsule d'Ungava
Labrador Sea
Canadian Shield
ATLANTIC OCEAN
British Isles
North Sea
Scandin
Coast Mountains
Rocky Mountains
Gulf of Alaska
Lake Winnipeg
Laurentian Mountains
NORTH AMERICA
Vancouver Island
Great Lakes
Grand Banks of Newfoundland
Bay of Biscay
EU
Alp
PACIFIC OCEAN
Great Plains
Missouri
Coast Ranges
Appalachian Mts
Mid-Atlantic Ridge
Iberian Peninsula
Azores
Mediterran
Mississippi
North American Basin
Madeira
Atlas Mountains
Land elevation
19,685 ft
6,000 m
13,124 ft
4,000 m
9,843 ft
3,000 m
6,562 ft
2,000 m
3,281 ft
1,000 m
1,640 ft
500 m
820 ft
250 m
0
Below sea level
-100 m (-328 ft)
-820 ft
-250 m
-1,640 ft
-500 m
-3,281 ft
-1,000 m
-6,562 ft
-2,000 m
-13,124 ft
-4,000 m
-19,658 ft
-6,000 m
Sea depth
Lower California
Sierra Madre Occidental
Sierra Madre Oriental
Gulf of Mexico
Canary Islands
Saha
Ahaggar
Yucatan Peninsula
Greater Antilles
West Indies
AFR
Middle America Trench
Caribbean Sea
Lesser Antilles
Sahe
Niger
Cape Verde Islands
East Pacific Rise
Guiana Highlands
Adamaw Highl
Gulf of Guinea
Galápagos Islands
Peru-Chile Trench
Amazon
Amazon Basin
Andes
SOUTH AMERICA
Brazilian Highlands
ATLANTIC OCEAN
Ascension Island
Angola Basin
Peru Basin
Lake Titicaca
Planalto de Mato Grosso
Brazil Basin
St Helena
Gran Chaco
Paraná
Cerro Aconcagua 6959m (22,838ft)
Pampas
Mid-Atlantic Ridge
Cape Basin
Tristan da Cunha
Argentine Basin
Gough Island
Patagonia
Falkland Islands
Tierra del Fuego
South Georgia
Cape Horn
Drake Passage
South Sandwich Islands
Atlantic-Indian Ridge
Antarctic Peninsula
Weddell Plain
Weddell Sea

Physical map of the world

THE WORLD MAP SHOWS OUR PLANET'S SEVEN VAST CONTINENTS, AS WELL AS THOUSANDS OF SMALLER ISLANDS.

The movement of the Earth's crust has pushed up rugged mountain ranges on the continents and created chains of volcanic islands in the ocean. A physical map of the world shows the height, or elevation, of the land.

ARCTIC OCEAN
Severnaya Zemlya
Franz Josef Land
Spitsbergen
New Siberian Islands
Novaya Zemlya
Laptev Sea
East Siberian Sea
Barents Sea
Kara Sea
Chukchi Sea
Yenisei
Central Siberian Plateau
Lena
Arctic Circle
West Siberian Plain
Khrebet Cherskogo
Bering Strait
Ural Mountains
Ob'
Siberia
Baltic Sea
North European Plain
Volga
Sea of Okhotsk
Bering Sea
Aleutian Basin
Kamchatka
Aleutian Islands
Lake Baikal
Aleutian Trench
ASIA
Sakhalin
Lake Balkhash
Altai Mountains
Amur
Carpathian Mts
Danube
Manchurian Plain
Gobi
Kuril-Kamchatka Trench
Emperor Seamounts
Mount El'brus 5642m (18,510ft)
Aral Sea
Chinook Trough
Black Sea
Caucasus
Tien Shan
Balkans Mts
Caspian Sea
Sea of Japan (East Sea)
Hokkaido
Northwest Pacific Basin
Anatolia
Pamirs
Kunlun Mountains
Yellow River
Honshu
Japan
Iranian Plateau
Hindu Kush
K2 8611m (28,251ft)
Plateau of Tibet
Yellow Sea
-430m (-1411ft)
Syrian Desert
Zagros Mountains
Himalayas
Yangtze
Kyushu
Bonin Trench
PACIFIC OCEAN
East China Sea
Persian Gulf
Indus
Mount Everest 8848m (29,029ft)
Ryukyu Islands
Libyan Desert
Thar Desert
Ganges
Taiwan
Tropic of Cancer
Hawaiian Islands
Mid-Pacific Mountains
Tibesti
Nile
Red Sea
Arabian Peninsula
Deccan
Philippine Sea
Central Pacific Basin
Hawai'i
Polynesia
Arabian Sea
Western Ghats
Eastern Ghats
Bay of Bengal
Mekong
South China Sea
Philippine Islands
Philippine Trench
Mariana Islands
Mariana Trench
Micronesia
Ethiopian Highlands
Gulf of Aden
Arabian Basin
Andaman Islands
-10,994m (-36,070ft)
Marshall Islands
Horn of Africa
Sri Lanka
Line Islands
Great Rift Valley
Maldive Islands
Nicobar Islands
Malay Peninsula
Melanesia
Caroline Islands
Congo Basin
Somali Basin
Borneo
Equator
Lake Victoria
Sumatra
Celebes
New Guinea
Phoenix Islands
Kilimanjaro 5895m (19,340ft)
Mid-Indian Basin
Ninetyeast Ridge
East Indies
Lake Tanganyika
Seychelles
Java Sea
Mount Wilhelm 4509m (14,793ft)
Solomon Islands
Java Trench
Java
Cook Islands
Arafura Sea
Lake Nyasa
Samoa
Timor Sea
Great Barrier Reef
INDIAN OCEAN
Wharton Basin
Coral Sea
Zambezi
Mozambique Channel
Madagascar
Fiji
Tonga
Vanuatu
Mauritius
Réunion
Great Sandy Desert
Tropic of Capricorn
Kalahari Desert
AUSTRALIA
New Caledonia
Great Victoria Desert
Simpson Desert
Great Dividing Range
Kermadec Trench
Drakensberg
Nullarbor Plain
Darling
Southwest Indian Ridge
Louisville Ridge
Tasman Sea
North Island
Cape of Good Hope
New Zealand
Crozet Basin
South Australian Basin
Bass Strait
Chatham Rise
Tasmania
Basin
Southeast Indian Ridge
Tasman Basin
South Island
Campbell Plateau
Kerguelen
South Indian Basin
Pacific-Antarctic Ridge
Enderby Plain
SOUTHERN OCEAN
Antarctic Circle
ANTARCTICA
SCALE
0
1,000
2,000
km
miles
0
1,000
2,000

Rocks and minerals

THE EARTH'S CRUST IS MADE OF MANY KINDS OF ROCKS, AND ROCKS ARE MADE OF DIFFERENT MINERALS.

SEE ALSO
❮ 22–23 Earth's structure
Igneous rocks 42–43 ❯
Metamorphic rocks 50–51 ❯
The rock cycle 52–53 ❯

Rocks are mostly made of tiny grains or crystals of minerals, packed tightly together. There are thousands of kinds of minerals. But most rocks are made from just a handful of minerals—feldspar, quartz, mica, olivine, pyroxene, and amphibole.

Some **minerals** take **thousands of years** to **develop**, while **others grow** in only a **few hours**.

Minerals

Minerals are natural solids that form crystals. Most of the mineral crystals in rock are very small, but here and there bigger crystals form, and some of these make beautiful gemstones. Each mineral has its own unique chemical recipe.

Gold

△ **Native elements**
Gold, diamond (carbon), and a few other chemical elements occur naturally in pure form. All other minerals are made from compounds—two or more elements combined.

Quartz

△ **Silicates**
The most common kinds of minerals are silicates, such as quartz and feldspar. They are made from compounds of silicon and oxygen with other elements, typically metals.

Ruby

△ **Oxides (and others)**
Other common mineral groups include oxides (compounds of oxygen with a metal) and carbonates (carbon and oxygen with a metal).

Fire opal

△ **Mineraloid**
Mineraloids look like minerals but are not built from crystals. They include substances made by living things such as amber (tree sap) and coal (fossilized plants).

Mineral hot spots

Rare and special minerals form under particular conditions, where chemicals get concentrated by natural processes. In particular, liquids cool slowly, in cavities, allowing crystals to gradually grow. In Malakialina, Malagasy Republic, beryl crystals grow as big as tree trunks.

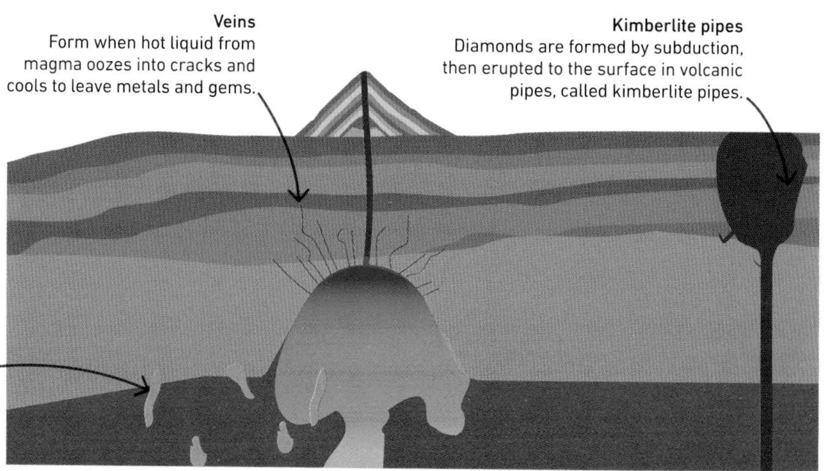

Veins
Form when hot liquid from magma oozes into cracks and cools to leave metals and gems.

Kimberlite pipes
Diamonds are formed by subduction, then erupted to the surface in volcanic pipes, called kimberlite pipes.

Pegmatites
When magma cools, the last, super-concentrated "cream" forms bands called pegmatites packed with treasures, such as aquamarines.

Ores and metals

Only a few metals, such as gold and copper, are found in pure form. Most are mixed into other minerals, called ores. To get the metal, the ores must first be mined or quarried, then processed to extract the metal—typically by "smelting" (heating until the metal runs out).

Iron oxide deposits ▷
The modern world relies on iron and steel, which are made by refining iron ores. Most iron ores have a distinctive reddy color, because they go rusty.

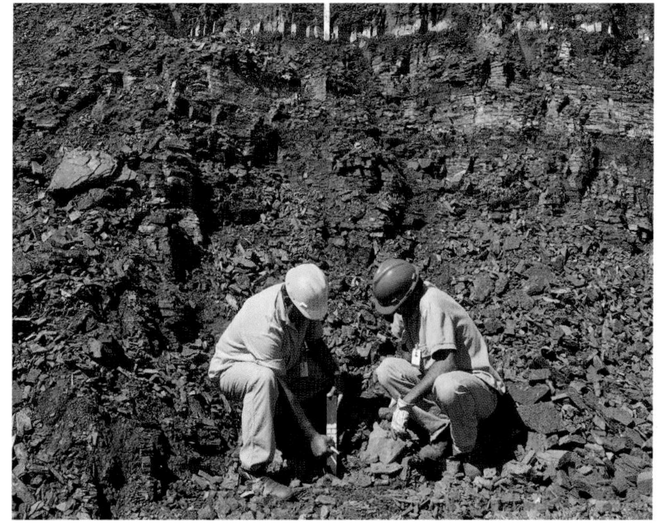

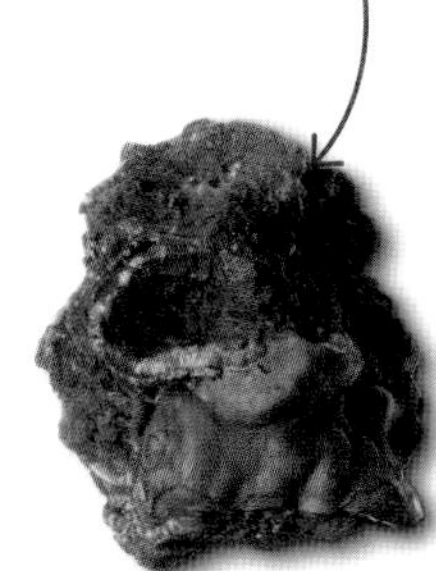

Hematite is the most important iron ore due to its high iron content and its abundance. It is mined extensively.

Identifying minerals

A mineralogist can often identify a mineral from the geological conditions where it is found. But further clues come from the mineral's combination of properties and by conducting a number of tests.

Needlelike crocoite

△ Habit
Many minerals grow in their own distinctive shape or "habit," including bladelike crystals, branches, and needles.

Metallic galena

△ Luster and color
A mineral's luster is its look or shininess. A few minerals can be identified by color alone, but most come in a variety of colors.

Flaky mica

△ Cleavage
Some minerals break into distinctive shapes, along planes of weakness, known as cleavage.

Talc

△ Hardness
A mineral's hardness is its resistance to being scratched. It is compared on a scale of one to ten. Talc has a hardness of one.

Cinnabar

△ Streak
When some minerals are scraped across the back of a tile, they always leave a streak of the same color.

Gold SG 19.3

Pyrite SG 5

△ Density
A key clue to a mineral's identity is Specific Gravity (SG)—its density compared to water. Gold is almost four times as dense as pyrite.

Igneous rocks

TOUGH, IGNEOUS ROCKS FORM WHEN MOLTEN MAGMA COOLS AND TURNS SOLID.

SEE ALSO
❮ 22–23 Earth's structure
❮ 40–41 Rocks and minerals
Weathering and erosion 44–46 ❯
The rock cycle 52–53 ❯

When magma erupts on the Earth's surface as lava or debris, it solidifies to form volcanic or "extrusive" igneous rocks. When it cools and turns solid underground, it makes "intrusive" igneous rock.

Igneous formations

Volcanic rock solidifies as a coating over the surface. But deep underground, magma can solidify in huge intrusive masses called plutons and giant domes called batholiths. These deep intrusive rocks are described as plutonic rock. In between, molten magma squeezes into spaces in the existing rock in all kinds of shapes, including sheets known as dikes and sills, and toadstool shapes called laccoliths.

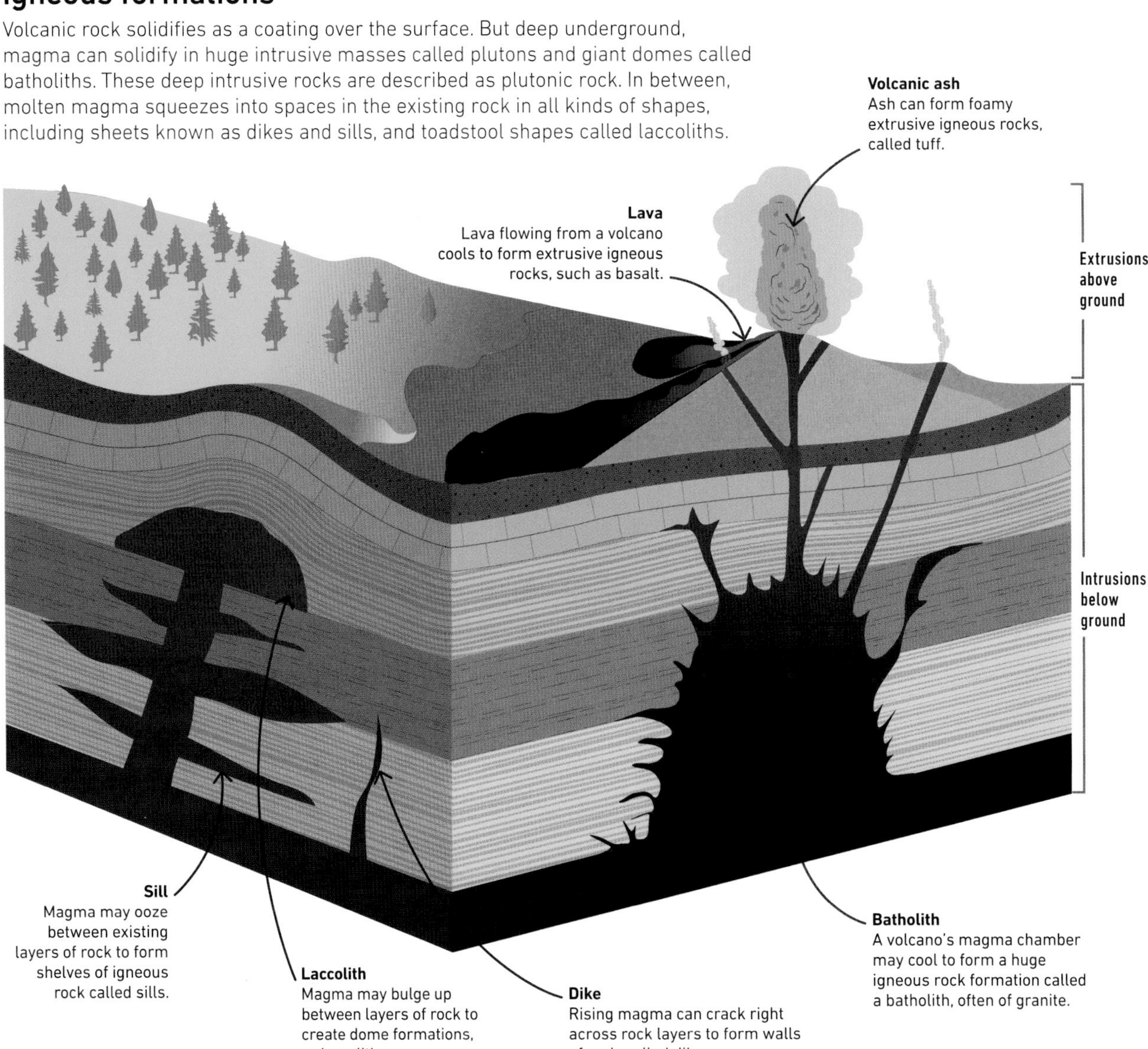
Volcanic ash
Ash can form foamy extrusive igneous rocks, called tuff.
Lava
Lava flowing from a volcano cools to form extrusive igneous rocks, such as basalt.
Extrusions above ground
Intrusions below ground
Sill
Magma may ooze between existing layers of rock to form shelves of igneous rock called sills.
Laccolith
Magma may bulge up between layers of rock to create dome formations, or laccoliths.
Dike
Rising magma can crack right across rock layers to form walls of rock, called dikes.
Batholith
A volcano's magma chamber may cool to form a huge igneous rock formation called a batholith, often of granite.

Types of igneous rocks

All igneous rocks form from magma. But hot magma that oozes up through the seafloor makes darker-colored rocks than the magma that gets contaminated by other rocks in continents. Also, deep, slowly cooling plutonic rocks have large grains or crystals, while volcanic rocks have fine grains, because in the open air lava cools too quickly for grains to grow.

Extrusive rock
Lava that cools within hours forms glassy obsidian. If it takes weeks to cool instead, it forms crystalline rocks, such as rhyolite.

Rocks that cool very fast have no crystals.

Obsidian

Rhyolite

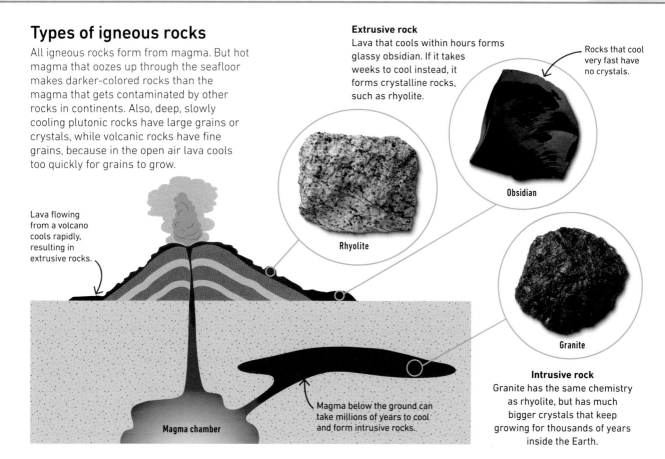

Lava flowing from a volcano cools rapidly, resulting in extrusive rocks.

Granite

Magma below the ground can take millions of years to cool and form intrusive rocks.

Magma chamber

Intrusive rock
Granite has the same chemistry as rhyolite, but has much bigger crystals that keep growing for thousands of years inside the Earth.

Interlocking crystals

Igneous rocks form as very hot, molten rock cools and starts to solidify. The various minerals in the melt form crystals, and as these crystals grow, they become interlocked in a rigid structure. This makes the resulting igneous rock very hard and resistant to the forces of weathering and erosion.

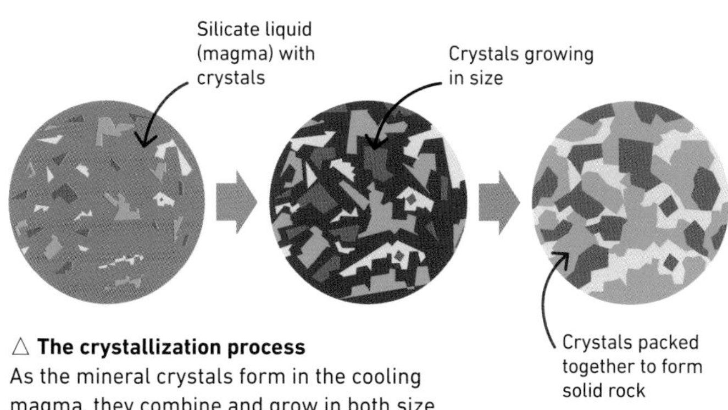
Silicate liquid (magma) with crystals
Crystals growing in size
Crystals packed together to form solid rock

△ **The crystallization process**
As the mineral crystals form in the cooling magma, they combine and grow in both size and number until they form a solid mass.

Rock chemistry

Apart from crystal size, igneous rocks vary in their chemistry. They are mixtures of different minerals, and the amount of each mineral in the mix alters the rocks' color, weight, and hardness. It also affects how they behave when they melt.

Basalt

◁ **Basic rocks**
Basic rocks are low in sand and rich in iron and magnesium. They are dark in color, and form from runny magma or lava.

Acid rocks ▷
Acid rocks are much sandier and made with minerals. They are lighter in color and form from sticky, runny magma or lava.

Granite

Weathering and erosion

WEATHERING IS THE BREAKDOWN OF EXPOSED ROCK; EROSION IS THE WEARING AWAY OF ROCK BY RIVERS, GLACIERS, THE SEA, AND WIND.

As soon as rock is exposed on the surface, it is battered constantly by wind, rain, and other physical and chemical effects. Below ground, rock can be eaten away by water seeping passed.

SEE ALSO
Mountain streams 56–57 ❯
Glacial erosion 64–65 ❯
Glacial deposition 66–67 ❯
Coastal erosion 68–69 ❯
Coastal deposition 70–71 ❯

Weathering

Rocks are broken down over thousands, or even millions, of years by a combination of mechanical, chemical and organic processes. They can be shattered by frost, or cracked by heat and cold. They can be dissolved by chemicals in rainwater, and they can be attacked by chemicals from microorganisms or split by plant roots.

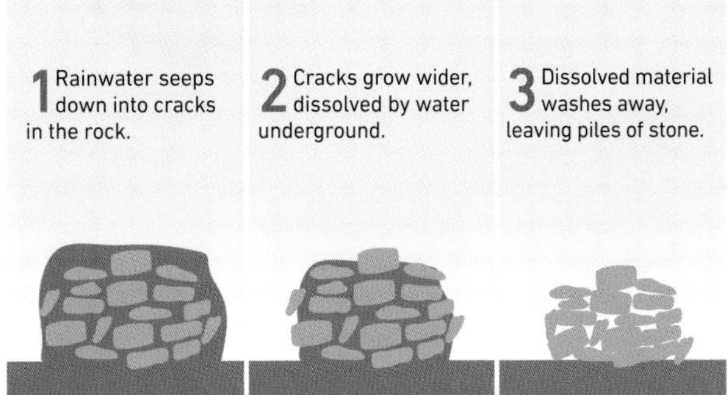
1 Rainwater seeps down into cracks in the rock.
2 Cracks grow wider, dissolved by water underground.
3 Dissolved material washes away, leaving piles of stone.

△ **Chemical weathering**
Rainwater is slightly acidic, especially in warmer conditions. Where it seeps through cracks in rocks, it can begin to slowly eat out the cracks.

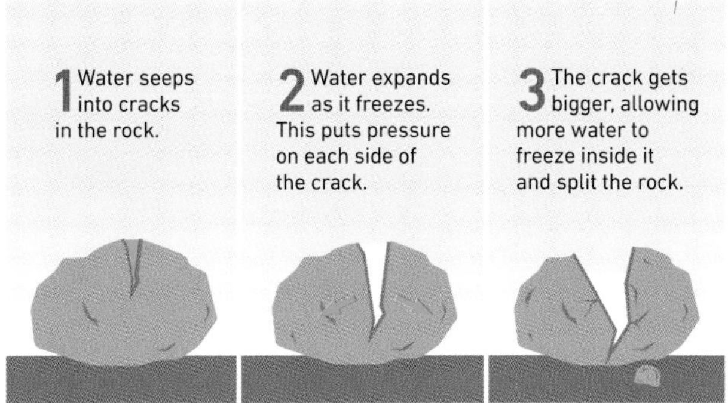
1 Water seeps into cracks in the rock.
2 Water expands as it freezes. This puts pressure on each side of the crack.
3 The crack gets bigger, allowing more water to freeze inside it and split the rock.

△ **Freeze-thaw weathering**
Cracks in rocks fill with water. As the water freezes to ice, it expands, forcing the cracks farther apart. Eventually rocks split and fall apart.

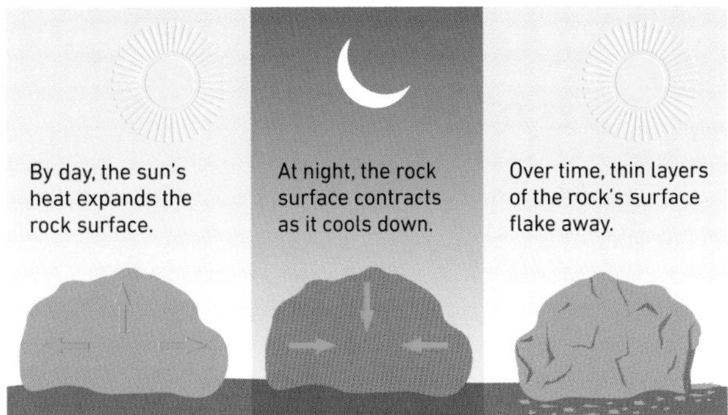
By day, the sun's heat expands the rock surface.
At night, the rock surface contracts as it cools down.
Over time, thin layers of the rock's surface flake away.

△ **Thermal weathering**
Intense sunshine bakes desert rocks by day and clear skies chill them at night causing the rocks to expand and then contract. The surface of the rock begins to flake away.

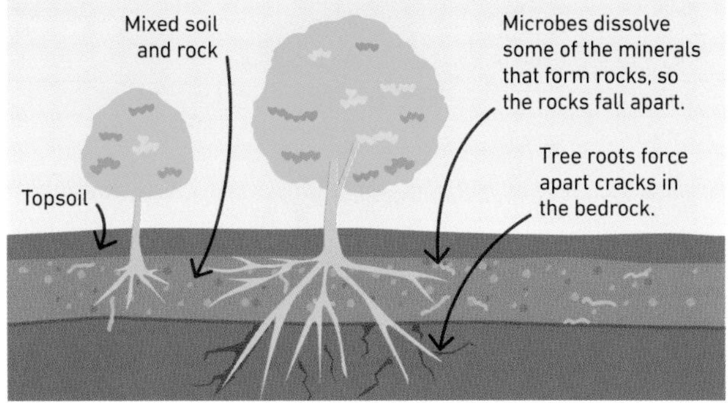
Mixed soil and rock
Microbes dissolve some of the minerals that form rocks, so the rocks fall apart.
Tree roots force apart cracks in the bedrock.
Topsoil

△ **Biological weathering**
As tree roots grow, they force their way into cracks in the rock and break it up to form soil. Microscopic life in the soil also breaks down rocks to get at their minerals.

Erosion

Weathering breaks rocks into fragments. Erosion—by forces such as rivers, waves, and moving ice—wears rocks down and transports fragments away. Deposition is when these transported fragments are dropped and build up. There are five major agents of erosion: water, waves, wind, moving ice, and landslides or "mass movement."

The **Mississippi River** carries **500 million tons** of sediment into the sea **every year.**

Worn down

Sometimes dramatic events such as storms can cause cataclysmic erosion. But mostly it is a slow, almost invisible process happening over millions of years. Some hard rocks resist erosion for a long time, standing proud as other rocks are worn away around them. Many soft sediments are broken down very quickly.

Erosion by ice
Their massive weight means glaciers and ice sheets can wear away huge amounts of rock as they move.

Water erosion
Water shapes hills and cuts out valleys. Sometimes water erosion is simply drops of rain falling on the land. Sometimes it is mighty rivers surging through a gorge.

Wave erosion
Waves can batter coastlines with tremendous force. Sea cliffs are dramatic evidence of wave power, showing how waves have sliced into a hill.

Wind erosion
Wind or "aeolian" erosion is important in dry landscapes where the wind can pick up loose dust and sand and hurl it against rocks.

REAL WORLD

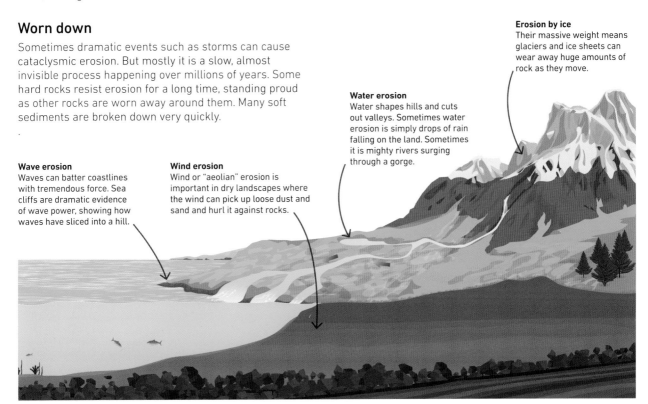

Rising Appalachians

The Appalachians have been wearing down ever since they were formed 290 million years ago. The loss of rock makes them lighter—so they are floating a few millimeters higher every decade. Even the highest mountain ranges are worn flat in time. The Appalachians may once have been as high as the Himalayas. But just as mountains are worn down, so new ones are raised—only to be worn down. This is called the cycle of erosion.

Erosion on slopes

Weathered fragments and soil are always on the move whenever there is a slope. This can happen slowly causing "soil creep." However, when water is added, slopes can collapse in a more catastrophic way with large amounts of material suddenly tumbling downhill.

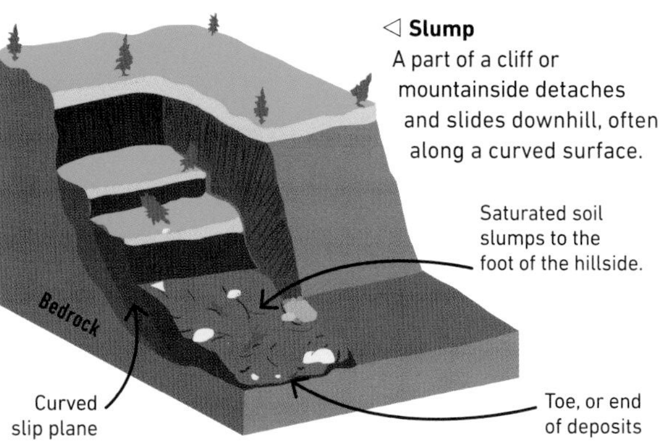
Saturated soil slumps to the foot of the hillside.
Bedrock
Curved slip plane
Toe, or end of deposits

◁ **Slump**
A part of a cliff or mountainside detaches and slides downhill, often along a curved surface.

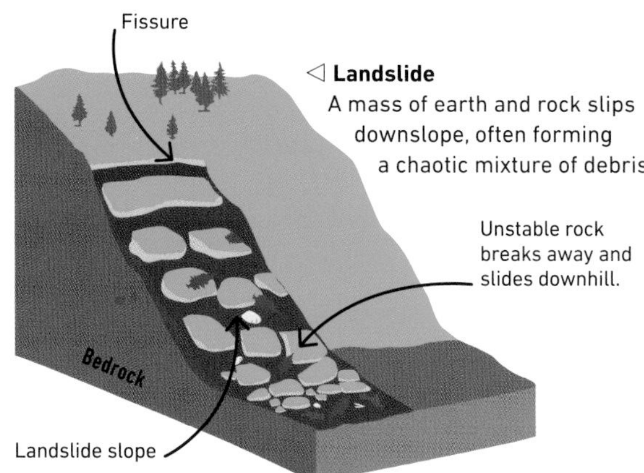
Fissure
Unstable rock breaks away and slides downhill.
Bedrock
Landslide slope

◁ **Landslide**
A mass of earth and rock slips downslope, often forming a chaotic mixture of debris.

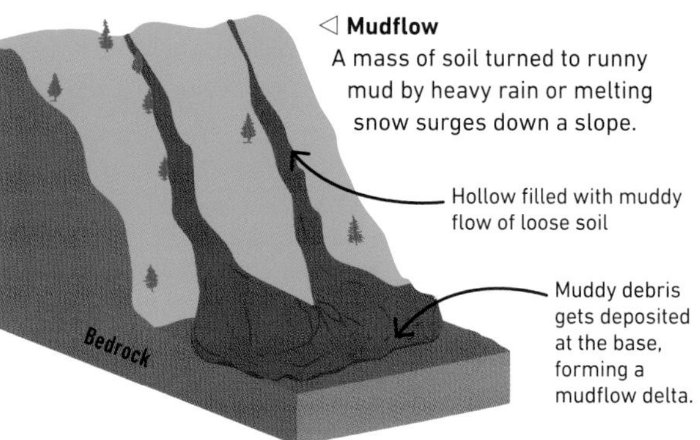
Hollow filled with muddy flow of loose soil
Muddy debris gets deposited at the base, forming a mudflow delta.
Bedrock

◁ **Mudflow**
A mass of soil turned to runny mud by heavy rain or melting snow surges down a slope.

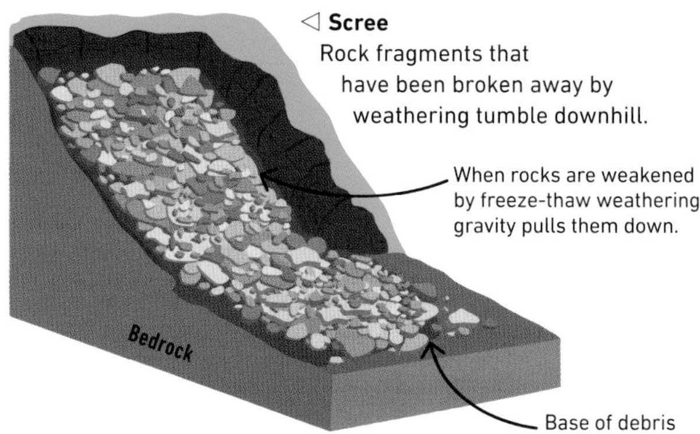
When rocks are weakened by freeze-thaw weathering, gravity pulls them down.
Bedrock
Base of debris

◁ **Scree**
Rock fragments that have been broken away by weathering tumble downhill.

REAL WORLD

Sculpting natural wonders

Weathering and erosion can create spectacular landforms like the Grand Canyon in southwestern US. Carved out of the desert by flowing water and frost action over the last five million years, the canyon is up to 6,092 ft (1,857 m) deep, 277 miles (446 km) long, and more than 18 miles (29 km) wide at its broadest point.

Sedimentary rocks and fossils

SEE ALSO
❮ 24–25 Moving plates and boundaries
❮ 40–41 Rocks and minerals
❮ 44–46 Weathering and erosion
The rock cycle 52–53 ❯

THREE-QUARTERS OF EARTH'S LAND IS COVERED IN THIN LAYERS OF SEDIMENTARY ROCK.

Sedimentary rocks are made from rock fragments, or organic or chemical deposits, that settled long ago on the beds of oceans, rivers, and lakes. These deposits were gradually buried, compacted, and cemented over time to form hard rock.

Sediment size

Many sedimentary rocks are "clastic"—that is, they are made from fragments or grains of other weathered and eroded rocks. Geologists classify these rocks according to the size of grain they are made of, from fine silt to stones.

▽ **Sorted sediments**
When water transports sediments, the heaviest drop first, while the lighter ones are carried farther. This sorts them into different sizes.

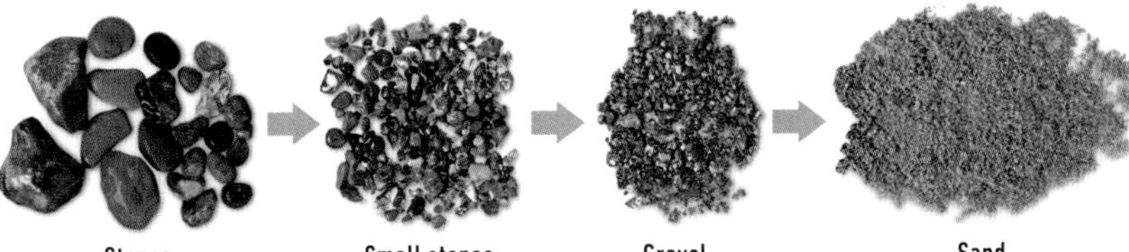
Stones
Small stones
Gravel
Sand

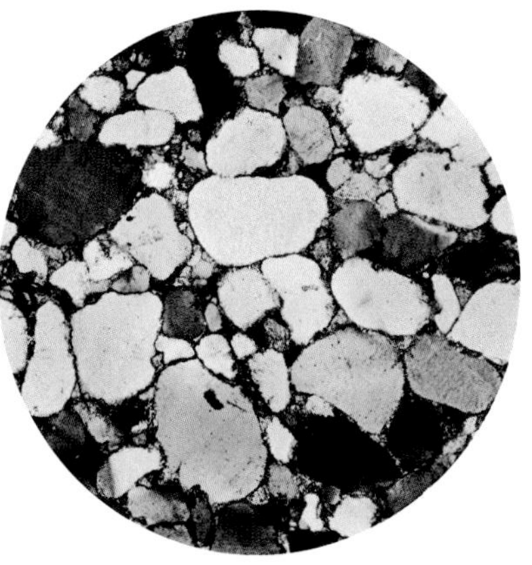

△ **Compression and cementation**
Loose sediments are laid down in layers compressed by the weight of sediments on top. Minerals in water may then cement grains together to form solid rock. This view through a microscope shows the grains packed together in a rock.

Rock strata

Over millions of years, different types of sediment can settle on the same site and turn into different types of sedimentary rock. This creates layers called strata. When strata form, they are horizontal, but they can be buckled and snapped, or faulted, by the forces that build mountains.

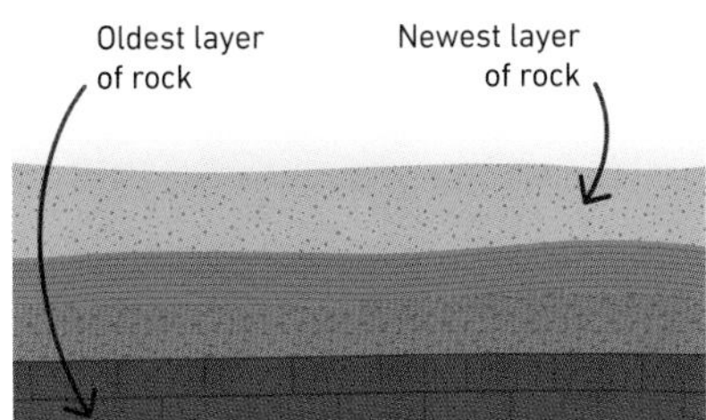
Oldest layer of rock
Newest layer of rock

△ **Undeformed strata**
Sedimentary rock strata nearly always form horizontally, from sediments laid down on lake or seabeds. Older rocks lie beneath strata that formed more recently.

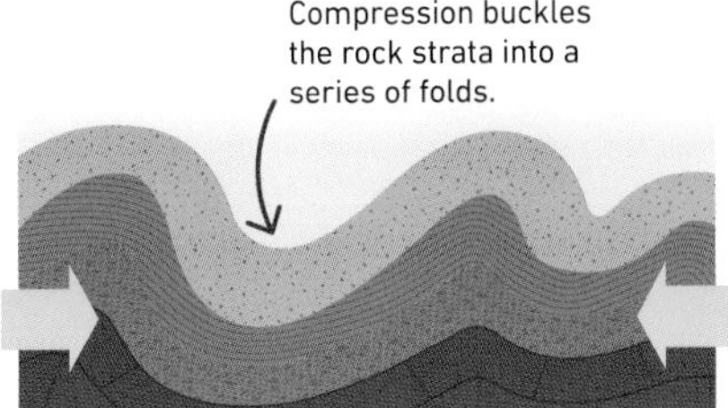
Compression buckles the rock strata into a series of folds.

△ **Folded strata**
When moving plates of the Earth's crust push against each other, rock strata are squeezed and folded. They can be thrust up on end or even overturned.

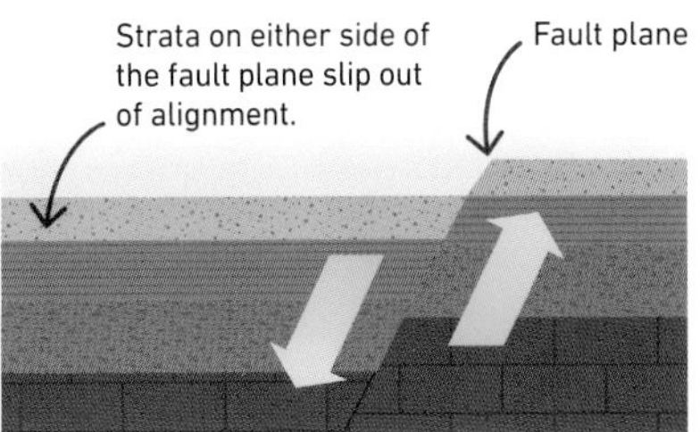
Strata on either side of the fault plane slip out of alignment.
Fault plane

△ **Faulted strata**
When strata are stretched, the rock layers break at a point called a fault plane. Similar faults can form if compression occurs too fast for the rock to adjust to it by folding.

Sedimentary rock types

Layers of sediment that have been sorted naturally by grain size lock together into sedimentary rocks. Beds of pebbles may be cemented together into conglomerate, sand becomes sandstone, and clay turns to very fine-grained shale. Some rocks such as chert (which is made of silica) have a glasslike texture, with no visible grains at all.

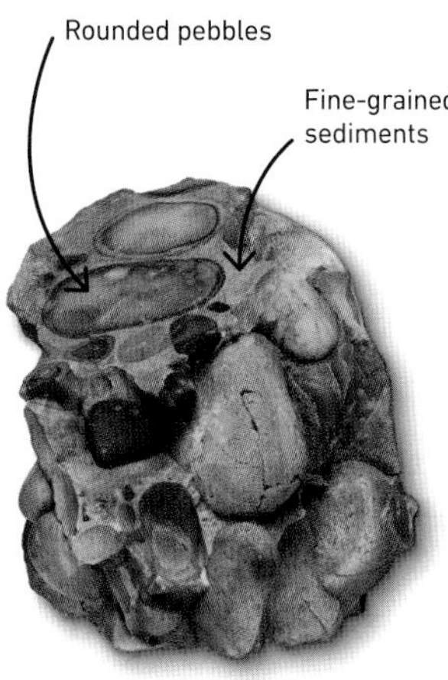
Rounded pebbles
Fine-grained sediments

Conglomerate
This rock is formed by rounded, water-worn pebbles cemented together by finer sediments.

Fine matrix of sediments surround angular fragments

Breccia
This is made of sharp-edged stones, held together by sand and mud.

Folds created due to movement of original deposit

Sandstone
Most sandstones are made of quartz grains once carried by water or swept into dunes.

The dark color is due to the carbon-rich matter in the rock.

Siltstone
Silt consists of much smaller particles than sand and forms siltstone.

Visible layers that split into sheets

Shale
Clay can harden into shale, which has many thin layers that often split apart.

Compact silica in its fine-grained structure

Chert
Made of silica—the same mineral as glass—chert is very hard. Flint is a type of chert.

Iron-deficient chert
Iron-rich hematite

Banded ironstone
These are ancient rocks formed in the sea with layers of chert and iron-rich rocks.

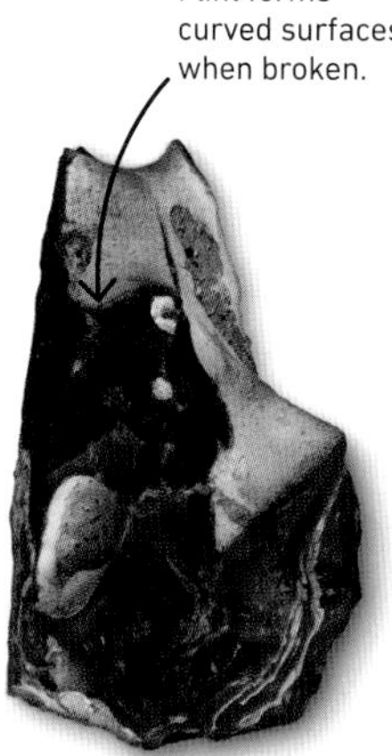
Flint forms curved surfaces when broken.

Flint
Found as lumps in chalk rock, flint forms very sharp edges where it breaks.

Fossils

Many sedimentary rocks contain remains of animals and plants that fell into the soft sediments before they turned to rock. Their buried tissue and bones are gradually replaced by minerals to form stony fossils preserved in the rock. The remains tell us about ancient life and also help geologists figure out how old the rocks are.

1 Animal dies
Some 67 million years ago, this *Triceratops* died of old age and its body slipped into a lake.

2 Skeleton is buried
Mud settling through the water buried the body before it could be torn apart and eaten by other animals.

3 Bones turn to fossil
Over millions of years, the mud turned to rock and the buried bones became fossils.

Biogenic rocks

Some sedimentary rocks consist almost entirely of fossils. Shelly limestone is full of fossil seashells, chalk is made up of the microscopic skeletons of marine plankton, and coal is made up of the remains of compressed plants.

The **hard parts** of the animal—the **bones, teeth**, or **shell**—are the **most likely to fossilize**.

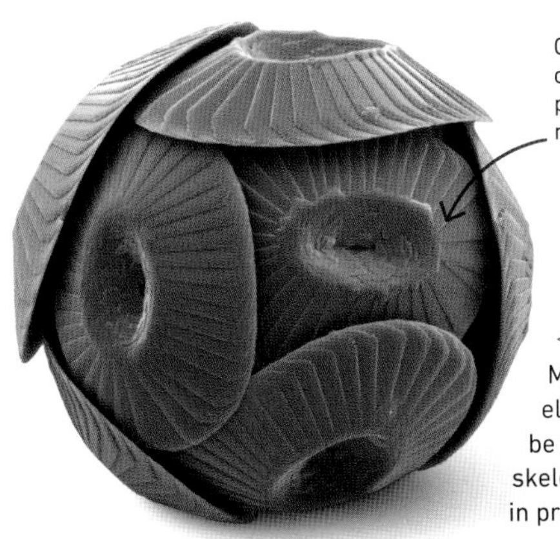

Overlapping plates called coccoliths protected the plankton from harm. They are made of calcium carbonate.

◁ **Microscopic structure of chalk**
Magnified thousands of times by an electron microscope, chalk is seen to be made up of coccolithophores—the skeletal remains of plankton that lived in prehistoric oceans.

REAL WORLD

Set in stone

Some sedimentary rocks are so fine-grained that they show every detail of the fossils in them. This stone slab from the Green River formation, Wyoming, has preserved a rare fossil of the Cretaceous-period turtle *Trionyx*, conserving the minutest details of the animal.

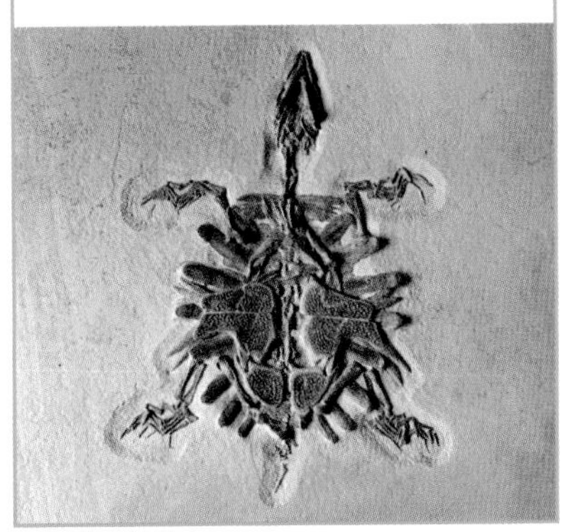

Metamorphic rocks

UNDER EXTREME CONDITIONS, SOME ROCKS ARE TRANSFORMED INTO OTHER TYPES OF ROCK.

SEE ALSO
❮ 22–23 Earth's structure
The rock cycle 52–53 ❯

Extreme pressure or heat can compress or cook rocks and change their character, turning them into metamorphic rocks. These are usually much harder than the rocks that formed them, and they often glitter with crystals.

Types of metamorphism

Heat and pressure create new metamorphic rocks in various ways by changing a rock's mineral content and the arrangement of the crystals. Each kind of metamorphic rock begins with a particular rock or "protolith." Marble, for instance, is formed from pure calcite limestone and sandstone is baked into quartzite.

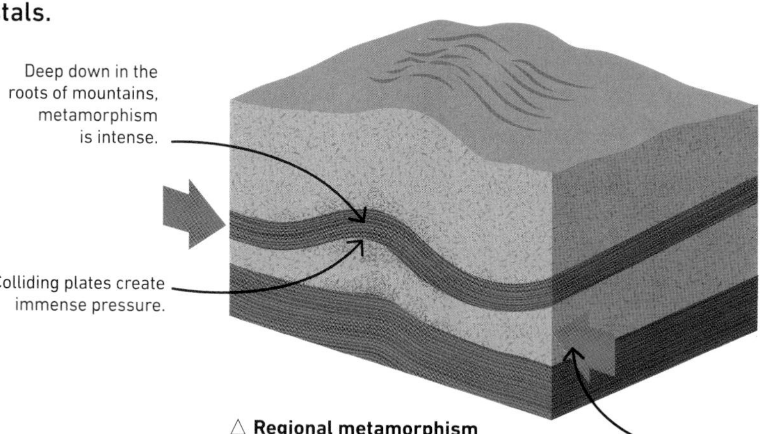
Deep down in the roots of mountains, metamorphism is intense.
Colliding plates create immense pressure.
Crystals reform at right angles to the pressure.

△ **Regional metamorphism**
The forces of plate tectonics cook and crush rock over vast areas, creating bands of different metamorphic rock.

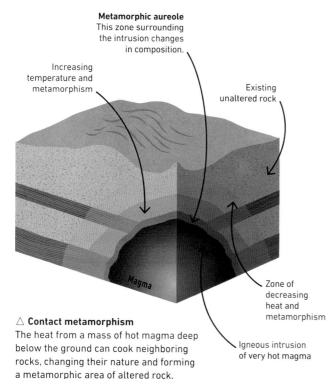

Metamorphic aureole
This zone surrounding the intrusion changes in composition.

Increasing temperature and metamorphism

Existing unaltered rock

Magma

Zone of decreasing heat and metamorphism

Igneous intrusion of very hot magma

△ **Contact metamorphism**
The heat from a mass of hot magma deep below the ground can cook neighboring rocks, changing their nature and forming a metamorphic area of altered rock.

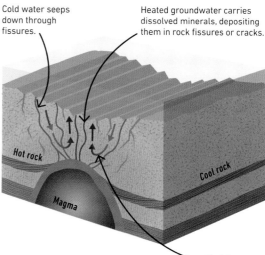

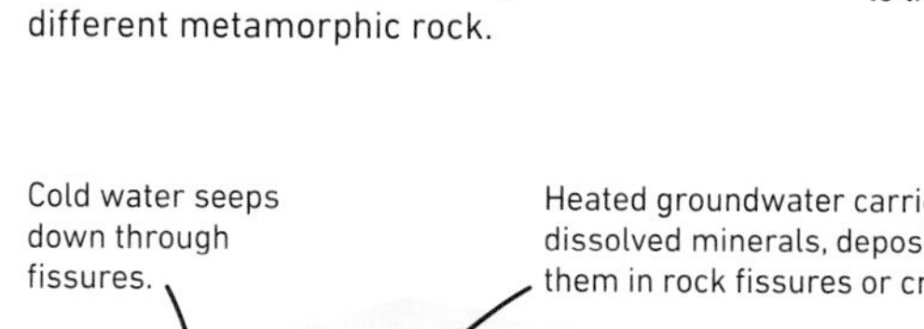
Cold water seeps down through fissures.
Heated groundwater carries dissolved minerals, depositing them in rock fissures or cracks.

Hot rock

Cool rock

Magma

Heat from magma causes groundwater to warm up, expand, and rise toward the surface.

△ **Hydrothermal metamorphism**
Water superheated by hot rock flows up through cracks carrying dissolved minerals, depositing them as veins of crystals and metals, including gold.

Grades of metamorphism

In regional metamorphism, increasing heat and pressure creates grades of metamorphism as rocks are changed progressively. Shale changes to slate, to schist, then gneiss. As rocks are pressed, they become increasingly banded or "foliated," starting with sheets in slate, changing to stripes or "schistosity" in schist, then swirling bands in gneiss.

Shale ▷
Compressed mud or clay deposited in layers on a lake or seabed forms the sedimentary rock shale. The shale rock retains the layers, which may flake apart.

Slate ▷
If shale is heated and compressed, it forms dark gray, tough, and slightly shiny slate. The clay grains regrow in flat sheets that can be easily split apart.

Schist ▷
In this medium-grade metamorphic rock, heat and pressure have turned the original clay minerals to flat, shiny mica crystals that glitter like tiny mirrors.

Gneiss ▷
More heat and pressure can convert mica into harder feldspar, turning schist into gneiss. This high-grade metamorphic rock can also form from other rocks like granite.

REAL WORLD

Earth's oldest rocks

So far, the oldest rocks found on the Earth's surface are metamorphic gneisses, which are almost 4 billion years old. These ancient rocks form the foundations of the continents. They are mostly buried deep beneath younger rocks, but in a few places, the forces of plate tectonics have pushed them to the surface. The Amitsoq gneiss rock outcrop on the western fringe of Greenland dates back 3.8 billion years and shows the banded structure typical of gneiss.

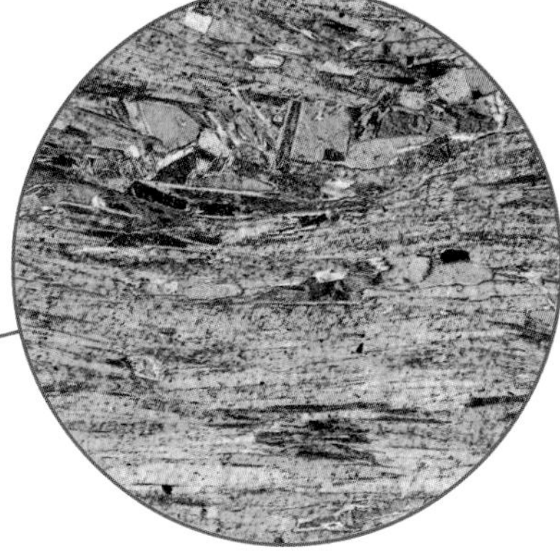

△ **Vertical section of schist**
A microscopic view of a thin slice of schist shows how the mica crystals have been forced into aligned layers by pressure.

Metamorphic rocks can be **formed from igneous** or **sedimentary** rocks.

The rock cycle

ROCKS IN THE EARTH'S CRUST ARE PART OF A CYCLE OF MELTING AND WEATHERING.

SEE ALSO
❮ 40–41 Rocks and minerals
❮ 42–43 Igneous rocks
❮ 47–49 Sedimentary rocks and fossils
❮ 50–51 Metamorphic rocks

The rocks that form the Earth's surface layers are constantly turning into different types of rock. Over millions of years, they can be transformed completely in a cycle of weathering, erosion, sedimentation, hardening, melting, and crystallization.

Recycled rocks

Rocks are mixtures of minerals broken down and recycled to form other rocks. The igneous rocks formed when molten magma, or lava, cools and crystallizes are weathered into soft clays and sand. These may become sedimentary shale or sandstone, which under heat and compression, can form harder metamorphic rocks. These may then melt and recrystallize as igneous rocks.

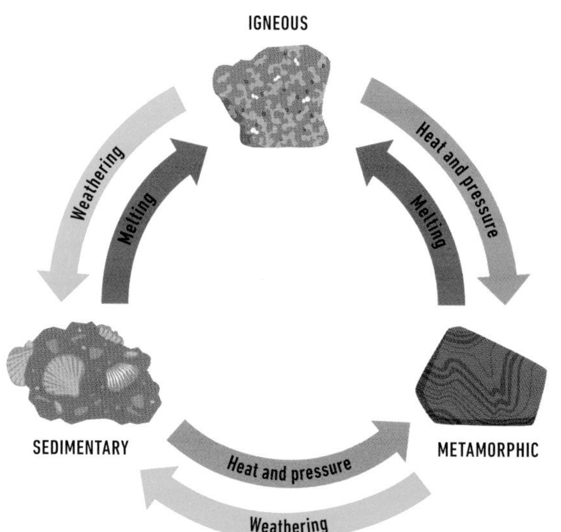
IGNEOUS
Weathering
Melting
Heat and pressure
Melting
SEDIMENTARY
METAMORPHIC
Heat and pressure
Weathering

◁ **How rocks change**
The rock cycle is not a one-way process. Each type of rock can be recycled in either direction to form a completely different rock.

2 Lithification
Over millions of years, soft sediments are compressed and cemented together, or lithified, to form much harder sedimentary rocks.

1 Deposition
Fragments of weathered rock are carried off by moving ice, water, or wind. They are deposited in lakes and seas as soft sediments.

Sedimentary rock

7 Melted rock
Many rocks are dragged down toward the Earth's mantle by subduction. The extreme heat and pressure melts them into magma, which may rise to form new igneous rock.

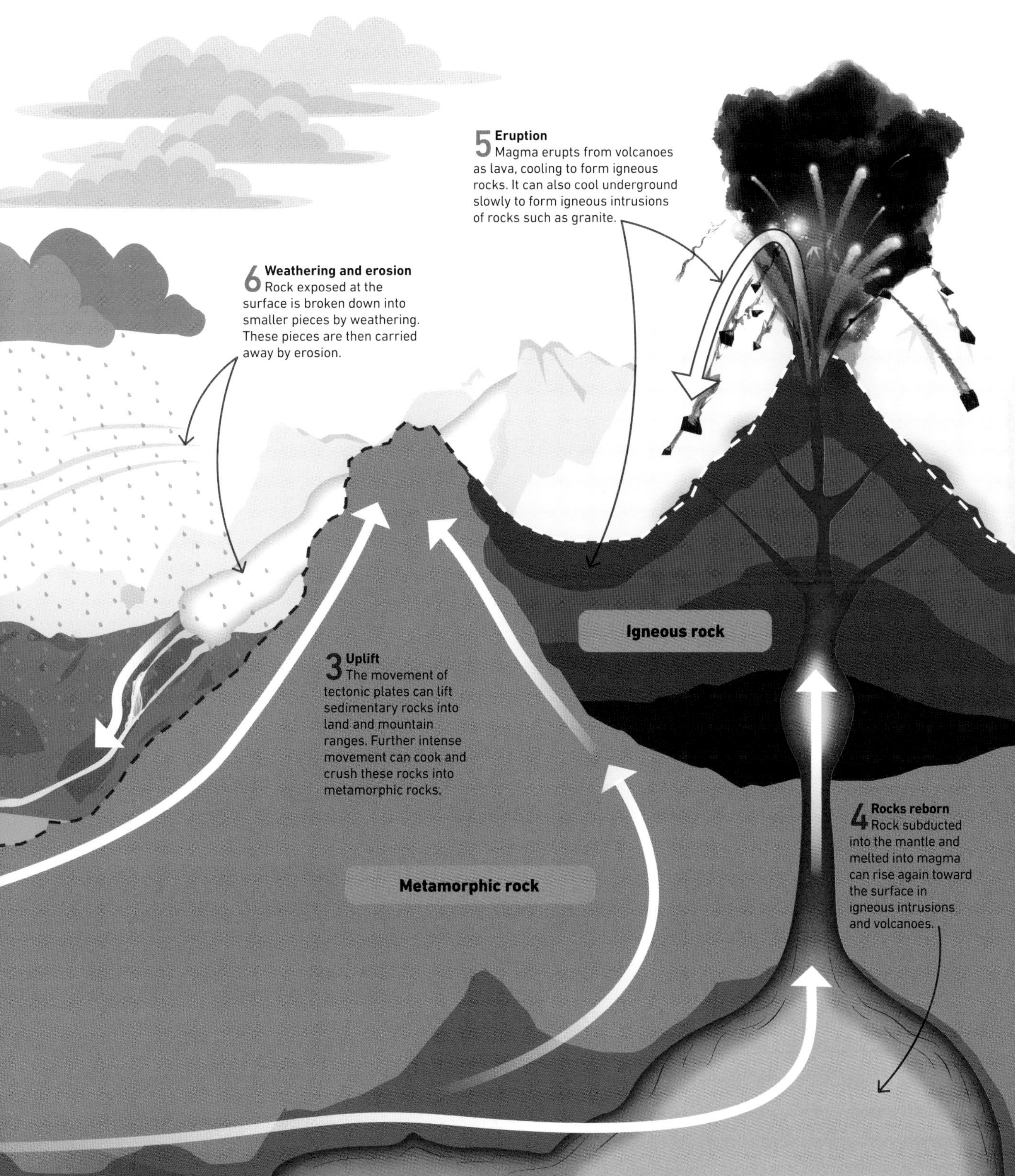
5 Eruption
Magma erupts from volcanoes as lava, cooling to form igneous rocks. It can also cool underground slowly to form igneous intrusions of rocks such as granite.
6 Weathering and erosion
Rock exposed at the surface is broken down into smaller pieces by weathering. These pieces are then carried away by erosion.
Igneous rock
3 Uplift
The movement of tectonic plates can lift sedimentary rocks into land and mountain ranges. Further intense movement can cook and crush these rocks into metamorphic rocks.
4 Rocks reborn
Rock subducted into the mantle and melted into magma can rise again toward the surface in igneous intrusions and volcanoes.
Metamorphic rock

Soil

ROCK FRAGMENTS MIX WITH THE DECAYED REMAINS OF LIVING THINGS TO FORM SOIL.

SEE ALSO
❮ 44–46 Weathering and erosion
❮ 52–53 The rock cycle
Biomes 98–99 ❯
Deforestation 172–174 ❯

Soil provides plants with food, water, and an anchor for their roots. Without soil, plants could not grow. There are many types of soil, depending on the nature of the local rock, climate, and geography.

Every handful of healthy soil contains **millions of living things** that create the nutrients needed for **plant growth**.

Rock and humus

Most soils are based on fragments of solid rock that are broken down by the process of weathering into sand, silt, and clay. These mineral ingredients are mixed with a dark, crumbly substance called humus, which is made of dead plant and animal material broken down by fungi, bacteria, and other living things, and may become plant nutrients.

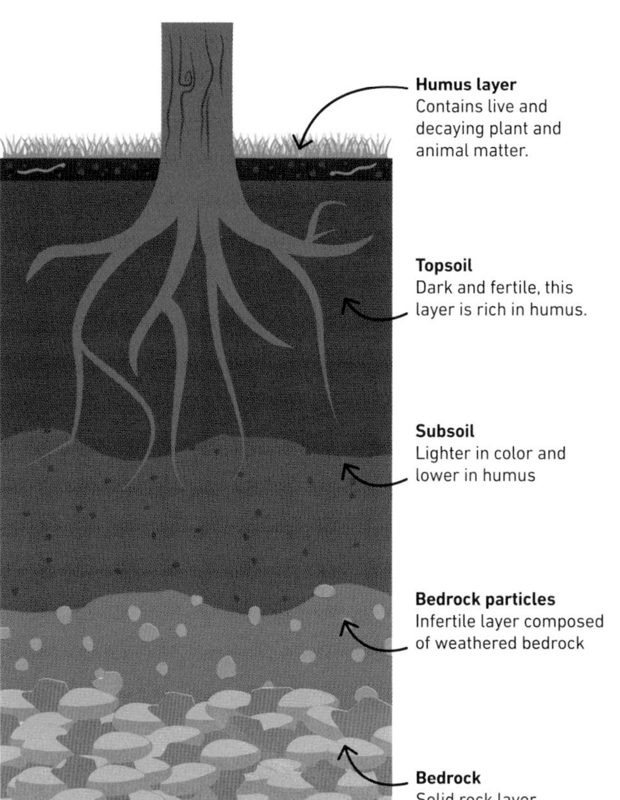
Humus layer
Contains live and decaying plant and animal matter.
Topsoil
Dark and fertile, this layer is rich in humus.
Subsoil
Lighter in color and lower in humus
Bedrock particles
Infertile layer composed of weathered bedrock
Bedrock
Solid rock layer

Soil life

The upper layers of healthy soil are full of decaying plant matter and microscopic life, such as bacteria. These provide food for animals that live in the soil, such as earthworms, slugs, and insects, which are preyed upon by burrowing moles. This teeming ecosystem creates the humus in the soil and helps mix and aerate the soil layers.

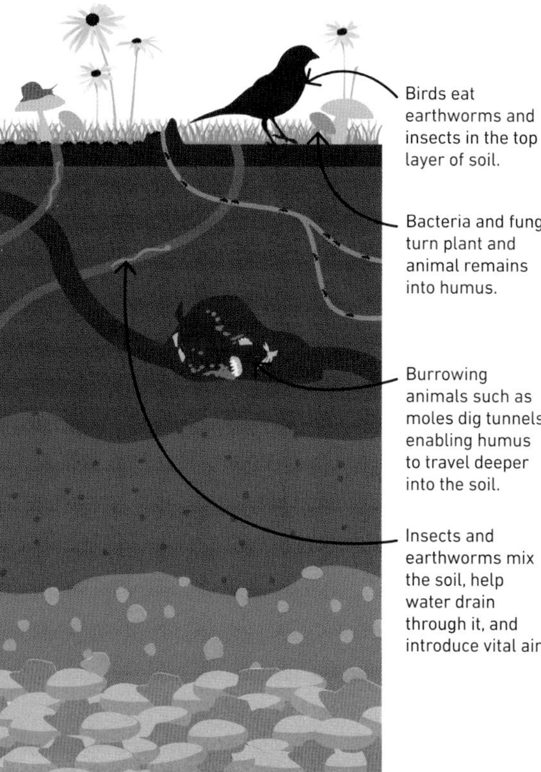
Birds eat earthworms and insects in the top layer of soil.
Bacteria and fungi turn plant and animal remains into humus.
Burrowing animals such as moles dig tunnels, enabling humus to travel deeper into the soil.
Insects and earthworms mix the soil, help water drain through it, and introduce vital air.

Soil formation

Most soils are made of similar basic ingredients, but in different proportions depending on where and how they form. The soil on a sandy hillside will be unlike one formed where fine silt and mud have been deposited by floods in a river valley. Several other factors play a part in soil formation, including temperature and rainfall.

Time
Soils develop over time, often getting deeper but also forming different layers as rainwater drains through them.

Climate
Soils form more quickly in warm climates. But heavy rain can waterlog the ground, stopping the plant decay that creates humus.

Topography
The shape of the land affects how water drains through the soil, or across it. Water runs faster down steep slopes, causing erosion.

Soil life
A healthy mix of lifeforms in the soil ensures that plant and animal remains break down quickly, forming the humus that makes soil fertile.

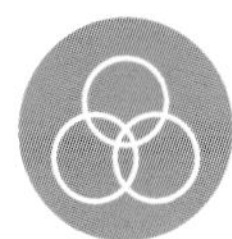

Parent material
The rocks, minerals, and living material that make up the soil have a major effect on its physical qualities and chemical nature.

Fertile soil

Most of the humus that makes the soil fertile is near the surface, and the deeper soil contains more mineral material (sand and rock). The depth of the humus-rich layer can vary a lot, which is why some soils make better farmland than others.

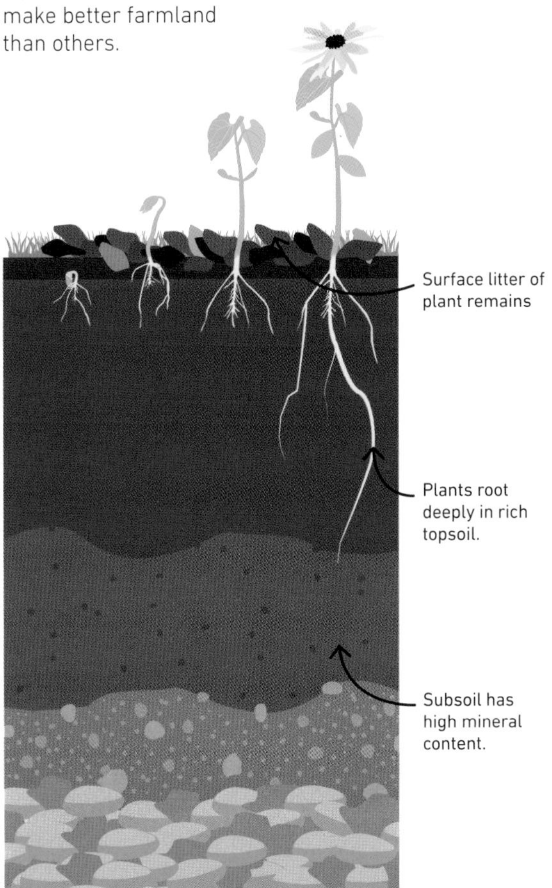
Surface litter of plant remains
Plants root deeply in rich topsoil.
Subsoil has high mineral content.

Infertile soil

Rainwater draining through the soil dissolves plant nutrients and some minerals, and carries them to a deeper level. This leaching process creates distinct layers in the soil. In areas of fast-draining sandy soil or high rainfall, the layer just below the surface becomes increasingly acidic and infertile, and turns ashy gray as iron is washed away.

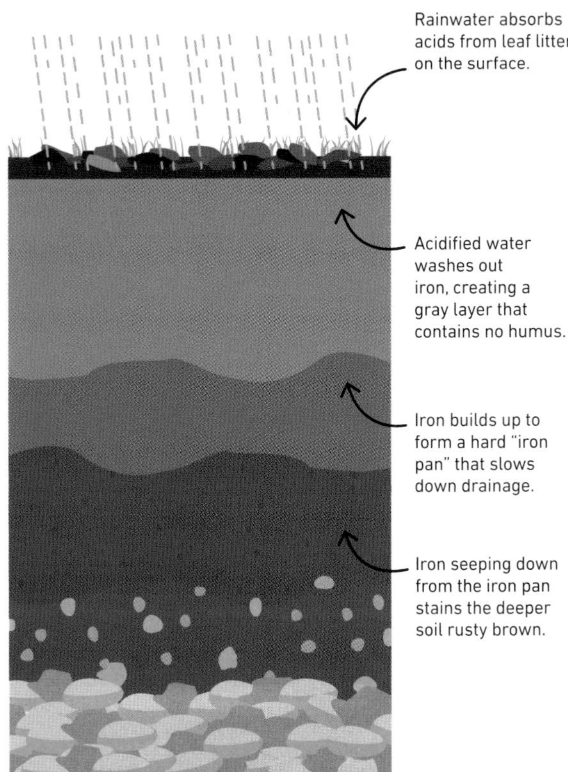
Rainwater absorbs acids from leaf litter on the surface.
Acidified water washes out iron, creating a gray layer that contains no humus.
Iron builds up to form a hard "iron pan" that slows down drainage.
Iron seeping down from the iron pan stains the deeper soil rusty brown.

Mountain streams

WATER FLOWING RAPIDLY DOWN STEEP SLOPES IN THE UPPER COURSE OF A RIVER ERODES AWAY ROCK AND SOIL.

SEE ALSO
‹ 44–46 Weathering and erosion
Rivers 58–61 ›
Coastal erosion 68–69 ›
The hydrological cycle 80–81 ›

The fast-moving water of an upland stream flows turbulently, carving out the landscape and removing soil and rock. It carries debris downstream, dropping some on the way and carrying some to the sea.

Rapids and waterfalls

When rain falls, some soaks into the ground, but some runs off downhill into streams. Over time, mountain streams cut down into the rock to create a narrow, V-shaped valley. These streams have steep, often rocky beds, and tumble over boulders and stones, creating "rapids." In places, they run over rock ledges in waterfalls.

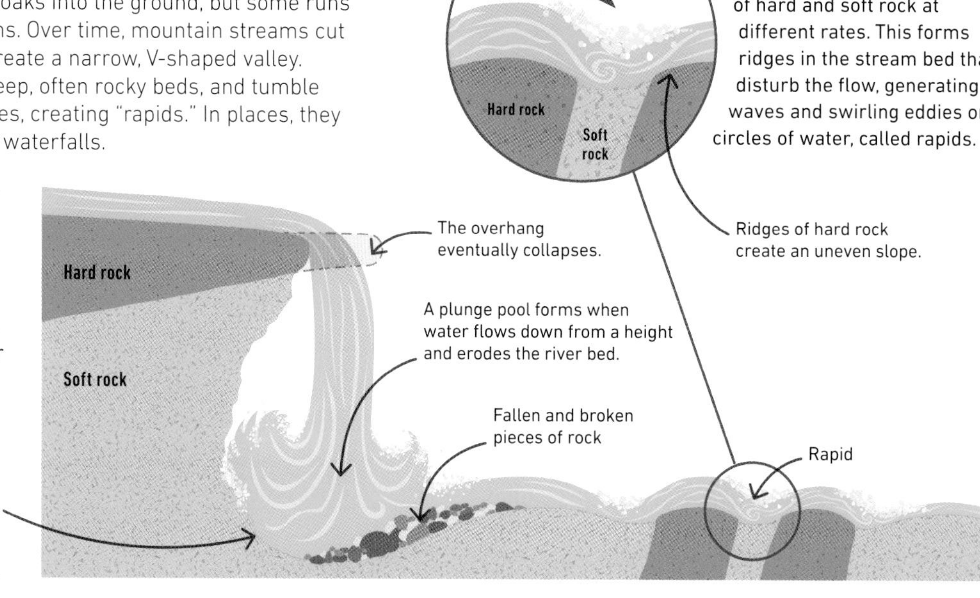
Hard rock
Soft rock
The overhang eventually collapses.
Ridges of hard rock create an uneven slope.
Hard rock
Soft rock
A plunge pool forms when water flows down from a height and erodes the river bed.
Fallen and broken pieces of rock
Rapid
As the waterfall retreats, a steep-sided gorge is formed.

◁ **River rapid**
Flowing water erodes layers of hard and soft rock at different rates. This forms ridges in the stream bed that disturb the flow, generating waves and swirling eddies or circles of water, called rapids.

How a waterfall forms ▷
Many waterfalls cascade over ledges of hard rock that lie above soft rocks. The falling water carves away the softer rock until the hard layer is no longer supported, and its edge collapses into the plunge pool below.

Carrying the load

Streams and rivers carry minerals and rock debris downstream in different ways. Soluble minerals, such as lime, dissolve in the water. Small particles are suspended in the water, while larger particles bounce along the riverbed. Heavier rocks may also roll down the river bed, especially when the flow of the water is at its most powerful. All this transported material is called the river's load.

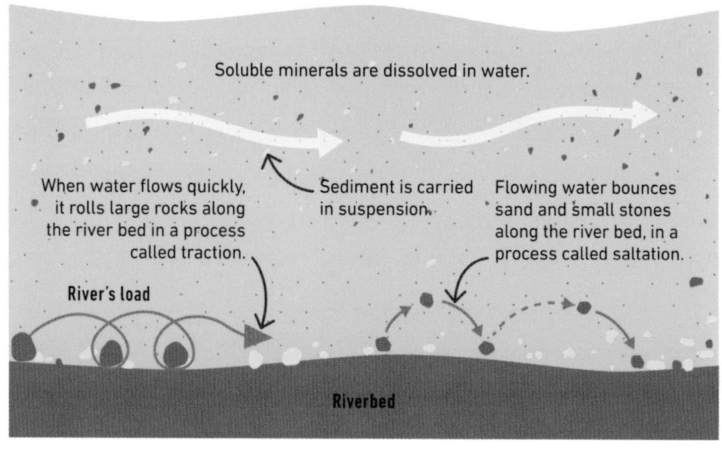
Soluble minerals are dissolved in water.
When water flows quickly, it rolls large rocks along the river bed in a process called traction.
Sediment is carried in suspension.
Flowing water bounces sand and small stones along the river bed, in a process called saltation.
River's load
Riverbed

Transporting sediments ▷
The four types of sediment transportation are called solution, suspension, saltation, and traction. They can all help carry the load downstream to the lowlands.

Sinkholes, caves, and gorges

Rainwater is slightly acidic. If it soaks into certain kinds of pervious (permeable) rock, such as limestone, the acidic water slowly dissolves the rock. This enlarges cracks and fissures into broad, open sinkholes. Streams plunge into these and flow underground through chains of limestone caves, which can include huge caverns that may eventually collapse, forming steep-sided gorges open to the sky.

Some **limestone caves contain** underground **rivers** and **lakes**.

Limestone cave systems ▽
Acidic water dissolves the alkaline limestone and carries the lime away. When the water flows into a cave, some of the dissolved lime is deposited to form features such as stalactites.

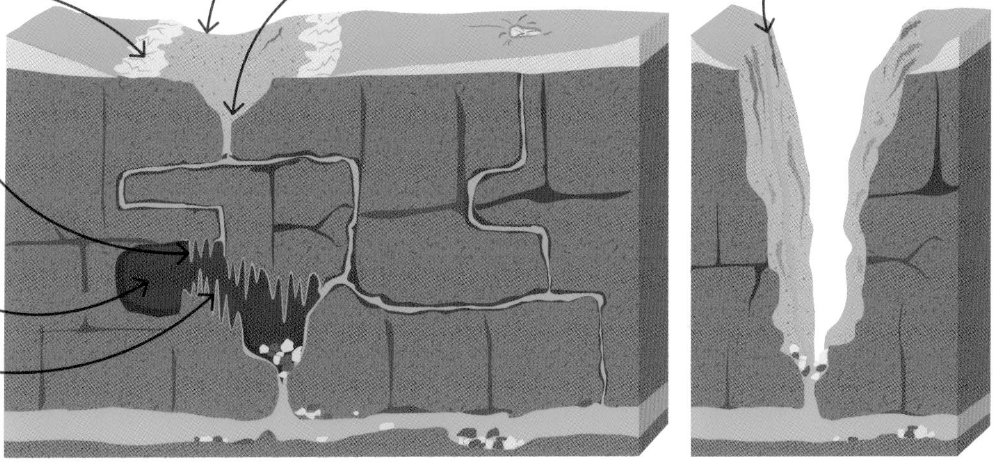

Sinkhole
Acidic rainwater flows into a deep limestone fissure, enlarging it to create a sinkhole, or doline.

Gorge
A narrow valley with steep sides forms when the roof of a cave collapses into an underground stream.

Limestone pavement
Rainwater makes a network of fissures in the exposed limestone.

The surface stream disappears down the sinkhole.

Stalactite
When lime-saturated water drips from the cave ceiling, the lime is deposited as hanging spikes.

Cave
Rainwater flowing through a crack erodes the rock in its path to form a cave.

Stalagmite
Water dripping onto the cave floor deposits lime that builds up into spiky stalagmites.

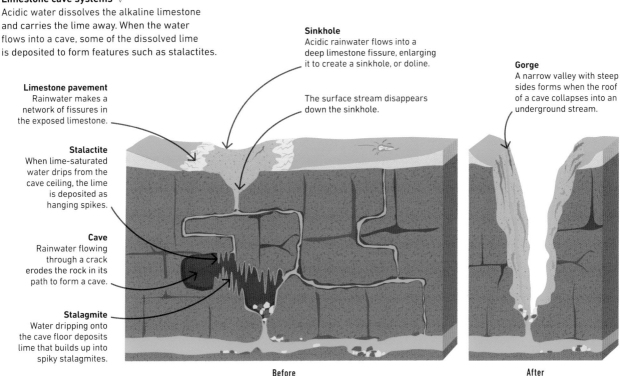

Before **After**

Aquifers and springs

If the rock below the ground is porous, rainwater soaks into it and fills it up like a wet sponge. This saturated rock forms a reservoir of groundwater called an aquifer. Water from an aquifer can be tapped by sinking a well or a borehole into the saturated zone.

How a spring forms ▷
When saturated rock and an aquifer lie above impermeable rock, water seeps out from the base of the aquifer through the ground and emerges as a spring.

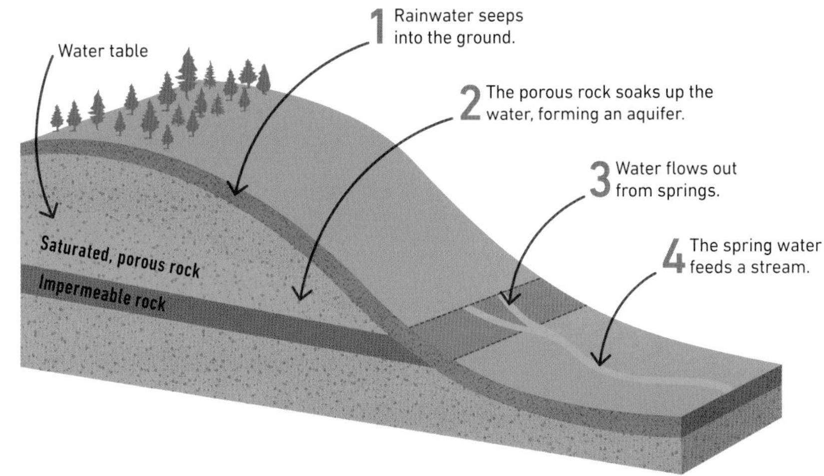
Water table
1 Rainwater seeps into the ground.
2 The porous rock soaks up the water, forming an aquifer.
3 Water flows out from springs.
4 The spring water feeds a stream.
Saturated, porous rock
Impermeable rock

Rivers

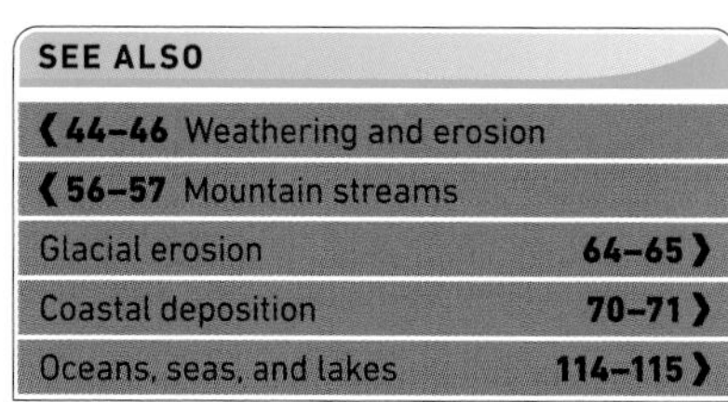
SEE ALSO
❮ 44–46 Weathering and erosion
❮ 56–57 Mountain streams
Glacial erosion 64–65 ❯
Coastal deposition 70–71 ❯
Oceans, seas, and lakes 114–115 ❯

WATER FALLING AS RAIN AND SNOW DRAINS OFF THE LAND TO FORM RIVERS THAT FLOW DOWNHILL TO THE SEA.

Landscapes are shaped by the relentless processes of weathering, erosion, and deposition. The flowing water of rivers plays a crucial part by moving material from the uplands to the lowlands.

From the source to the mouth

Where it begins on high land, a river's course is very steep and the flow very rough. But as it is joined by other "tributary" rivers lower down, it begins to flatten out, and become deep and smooth flowing, often winding over a wide plain of the debris it has washed down. As it nears the sea, it may widen into a muddy estuary or broad delta.

Vertical erosion carves out a narrow V-shaped valley with steep sides.

Fast-flowing water scours the river bed, carrying away the rock.

Shallow channel

△ **Upper course**
The river is relatively narrow in its upper course but tumbles down steep gradients. The fast-flowing water cuts a deep valley through the landscape.

Source
Water draining off hills forms a lake that feeds the river.

Rain and snow

Tributary
This is a smaller stream that flows into the main river.

Upper course

Middle course

△ **The course of a river**
A typical river is divided into three courses. The upper course is a powerful agent of erosion. The middle course starts depositing sediment. The lower course carries sediment to the sea.

Valley

Channel

River

◁ **Valley and channel**
The river flows in a channel that can be full of water. The channel lies in a broader valley carved by the river over time.

Floodplain
When high rainfall makes the river overflow, water floods the valley, depositing sediment that causes the river to form a flat plain.

Interlocking spurs

In the uplands, rivers carve deep, narrow valleys. The winding course of the river makes the valley wind too, creating alternating ridges known as "interlocking spurs." If a glacier flows down the valley, it often cuts right across the interlocking spurs to create a broad, straight valley with truncated spurs.

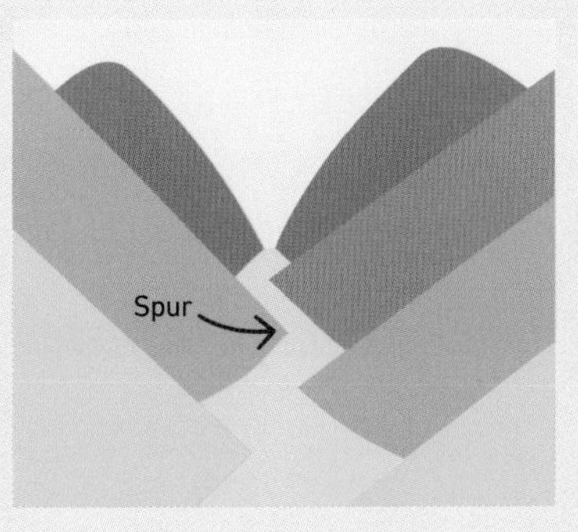
Spur

REAL WORLD

Catchment area

A river is fed by streams that form a pattern like the branches of a tree. The area it drains is called its catchment. This satellite image of a mountain landscape clearly shows the network of dark rivers and streams, divided by snow-covered rocky ridges.

On the inside of bends, the flow is less, and shoals of sediment build up.

Meanders steer the flow of water against the outside of bends, causing erosion.

Wider, deeper channel

Sediments

△ **Middle course**
Farther down, tributaries swell the river and it begins to wind in big bends, called meanders. As it flows from side to side, it carves out a broad valley.

Natural embankments called levees are created by sediments.

Very wide, almost flat valley

Sediments transported from uplands settle near the mouth of the river.

Wide, flat floodplain

Very wide, deep channel

△ **Lower course**
As the river flows toward the coast, it deposits most of its sediment load, forming a broad floodplain. The sediment may extend out to sea as a fan-shaped delta.

Meander
As the river flows over its floodplain, it forms large bends.

Delta
An area of sediment is left by the river as it flows out into the sea.

Sea

Oxbow lake
A curved lake is formed by a bend in the river.

Lower course

Mouth
Where the river flows into the sea, it forms either a delta or an estuary.

Floodplains

Heavy rain or melting snow upstream can dramatically increase the flow in lowland rivers shortly afterward. Sometimes, the river can swell so much that it spills over its banks and floods the surrounding land. The floodwater forms a temporary shallow lake with very little water flow. When floods recede, they leave layers of fertile sediment (called alluvium), which slowly build up to form a floodplain.

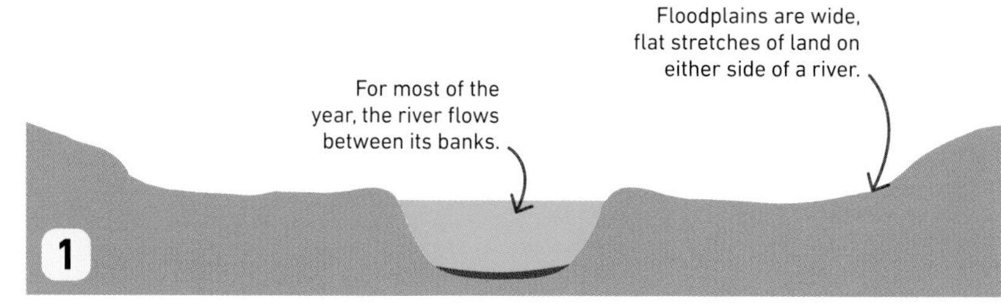
Floodplains are wide, flat stretches of land on either side of a river.
For most of the year, the river flows between its banks.
1

When in flood, the river overflows its banks.
Heavier sediments deposited along the river form a raised levee.
Floodwater carries lighter sediments further across the floodplain.
2

Natural floodbanks ▷ As floodwater spills over a river bank, heavy sediments are dropped first, forming ridges called levees along the bank. These allow the level of the river to rise above the floodplain.

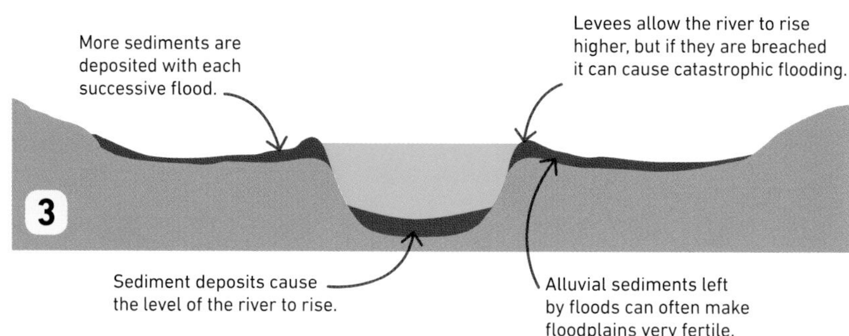
More sediments are deposited with each successive flood.
Levees allow the river to rise higher, but if they are breached it can cause catastrophic flooding.
3
Sediment deposits cause the level of the river to rise.
Alluvial sediments left by floods can often make floodplains very fertile.

Meanders

Bends in the course of a river begin with small shoals or riffles on the river bed, which divert the strongest flow from side to side. Where the flow hits the bank, the bank is worn away, creating ever bigger loops, or meanders. Sometimes, the loop breaks right through to the next loop down, cutting off a section of river to create an oxbow lake.

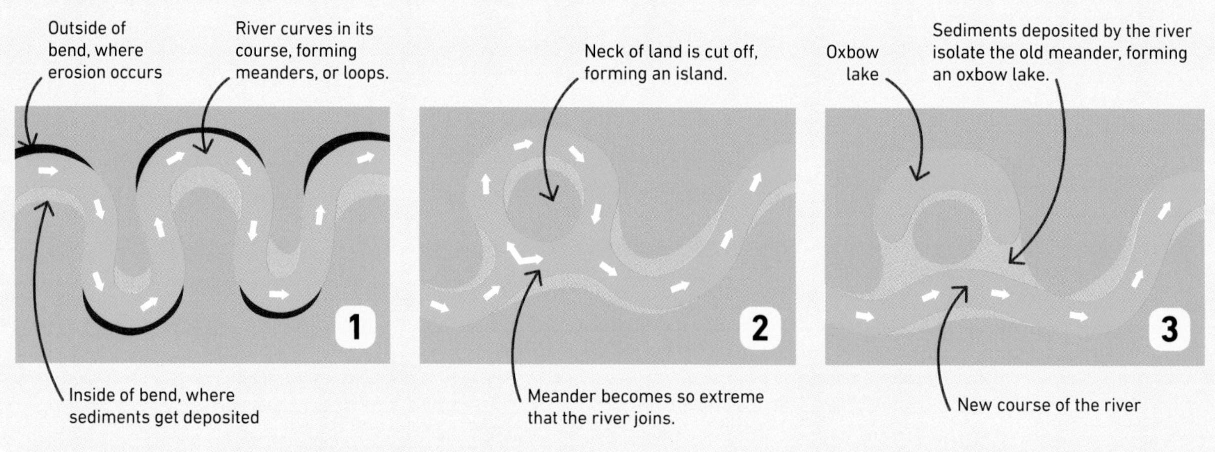
Outside of bend, where erosion occurs
River curves in its course, forming meanders, or loops.
Neck of land is cut off, forming an island.
Oxbow lake
Sediments deposited by the river isolate the old meander, forming an oxbow lake.
1
2
3
Inside of bend, where sediments get deposited
Meander becomes so extreme that the river joins.
New course of the river

Estuaries

When rivers approach the sea, they often meet tidal water at a widening area of the river called an estuary. Twice a day, the incoming tide stops the river flow, making it drop fine particles of suspended sediment. These settle to form extensive mud flats, which are revealed when the water level drops at low tide.

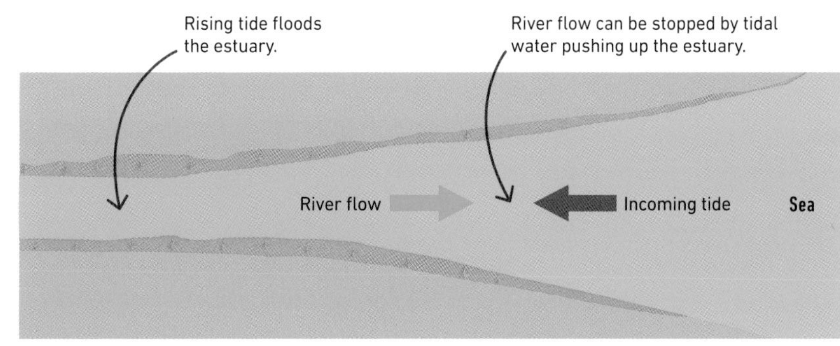
Rising tide floods the estuary.
River flow can be stopped by tidal water pushing up the estuary.
River flow
Incoming tide
Sea

△ **Estuary at high tide**
As the tide rises, seawater flows up the estuary, slowing the flow of the river. Salty seawater makes fine particles clump together to form heavier ones that sink easily and settle on the river bed.

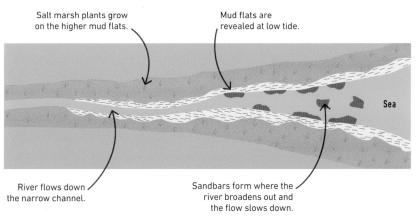

Salt marsh plants grow on the higher mud flats.

Mud flats are revealed at low tide.

Sea

River flows down the narrow channel.

Sandbars form where the river broadens out and the flow slows down.

◁ **Estuary at low tide**
As the tide falls, the river flows strongly over mud, carving a narrow but deep channel. It can also carry sand out to sea, depositing it as sandbars.

Deltas

Some rivers flow into lakes, or out to sea in places with a small tidal range, and deposit sediments in deep layers well beyond the original mouth of the river. This creates a triangular delta. The river typically fans out across the delta in a series of channels known as distributaries.

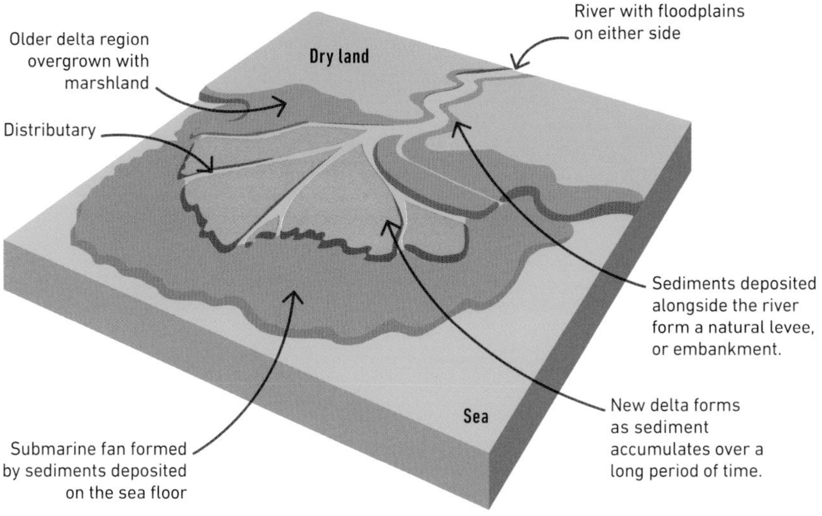
Older delta region overgrown with marshland
Dry land
River with floodplains on either side
Distributary
Sediments deposited alongside the river form a natural levee, or embankment.
Sea
New delta forms as sediment accumulates over a long period of time.
Submarine fan formed by sediments deposited on the sea floor

REAL WORLD

Taming a river

Many rivers flowing through farmland and cities, such as the Moskva in Moscow, are controlled by building artificial embankments that raise the height of natural levees. This can stop local flooding, but since it forces the floodwater to flow down the river, it may cause worse floods further downstream. Allowing a river to naturally flood an area can prevent such problems.

Ice ages

ICE AGES ARE LONG PERIODS WHEN MUCH OF THE EARTH IS COVERED IN VAST SHEETS OF ICE.

SEE ALSO
Seasons 76–77 ❯
The hydrological cycle 80–81 ❯
Climate change 175–177 ❯
Hemispheres and latitude 210–211 ❯

Geologists believe that the Earth has gone through at least five "ice ages." There are arguments about what causes them but one of the key factors is probably changes in the composition of the atmosphere.

The Earth is **now in** an **interglacial period** within **the Pleistocene period**.

Cycles of change

Each ice age lasts millions of years. When people talk about "The Ice Age," they mean just the most recent. Scientists call this the Pleistocene, and it began 2.6 million years ago. Within each ice age, there are especially cold phases called glacials and warmer phases called interglacials. We are now living in an interglacial period. The most recent glacial ended just 11,700 years ago.

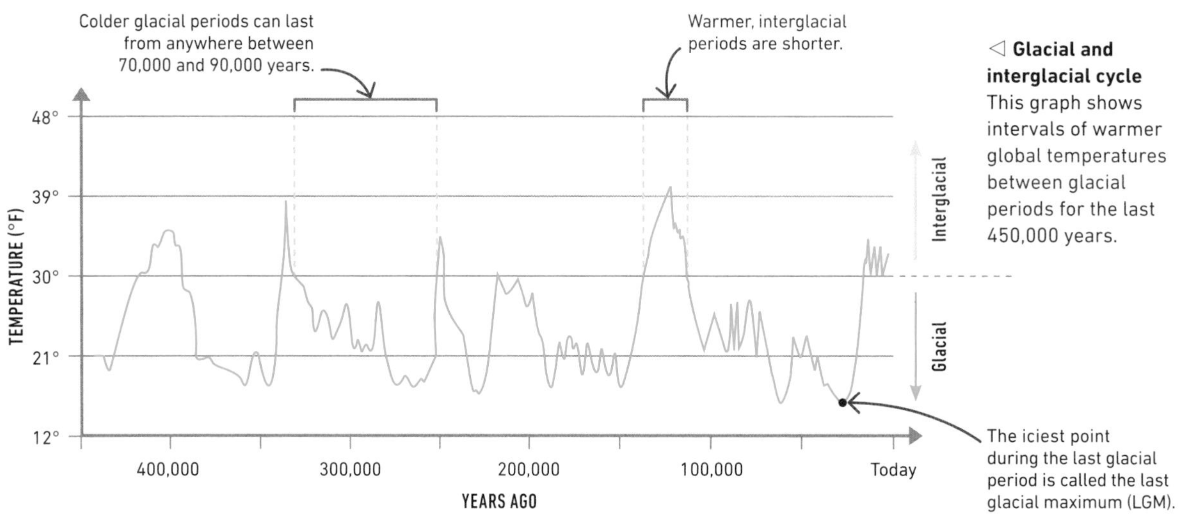
Colder glacial periods can last from anywhere between 70,000 and 90,000 years.
Warmer, interglacial periods are shorter.
48°
39°
30°
21°
12°
TEMPERATURE (°F)
400,000
300,000
200,000
100,000
Today
YEARS AGO
Interglacial
Glacial
The iciest point during the last glacial period is called the last glacial maximum (LGM).

◁ **Glacial and interglacial cycle** This graph shows intervals of warmer global temperatures between glacial periods for the last 450,000 years.

Shrinking ice sheets

The vast areas of ice that cover parts of the Earth are called ice sheets. These build up during glacial periods and recede during interglacials. In the northern hemisphere, the ice sheet is currently shrinking as the average global temperature rises.

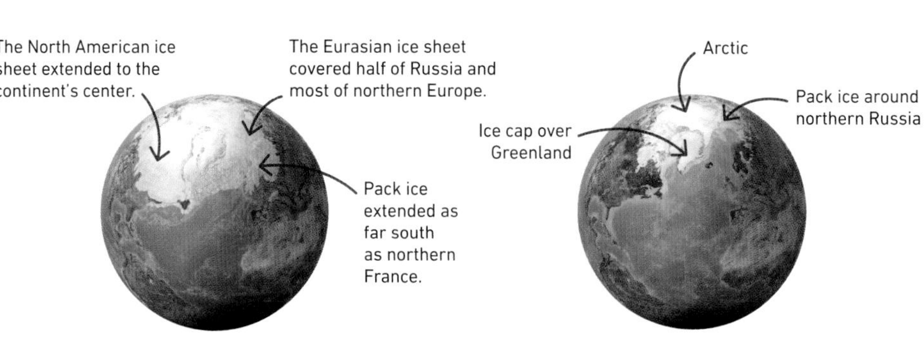
The North American ice sheet extended to the continent's center.
The Eurasian ice sheet covered half of Russia and most of northern Europe.
Pack ice extended as far south as northern France.
Arctic
Pack ice around northern Russia
Ice cap over Greenland

△ **Last glacial period** During the last glacial period, Arctic ice covered a huge area of the Earth.

△ **Current interglacial period** We are now in a warm interglacial period and the Arctic ice has shrunk to the far north.

The last glacial maximum (LGM)

During the last glacial maximum, vast areas of the Earth's oceans and land were covered by ice sheets. The land at the fringes of the ice sheets was treeless tundra, which froze in winter, as did parts of the sea. There were also cool grasslands grazed by bison, wild horses, and giant woolly mammoths.

▽ **A colder past**
During the LGM, which occurred 26,000 to 20,000 years ago, the Arctic ice sheets grew to cover most of Canada, much of northern Europe, and Scandinavia.

Laurentide ice sheet
Greenland ice sheet
Eurasian ice sheet
Himalayas
Patagonian ice sheet
Antarctic ice sheet

KEY

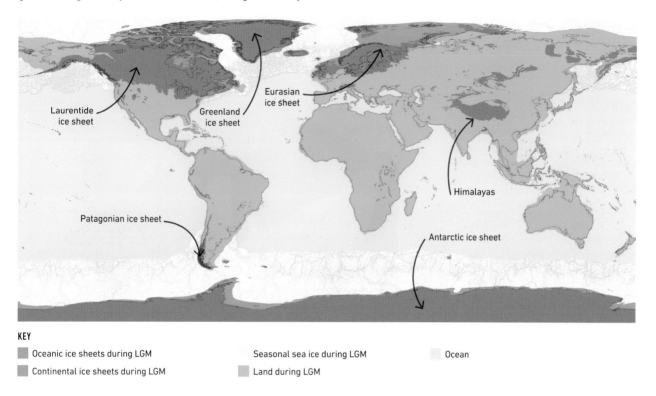

- Oceanic ice sheets during LGM
- Continental ice sheets during LGM
- Seasonal sea ice during LGM
- Land during LGM
- Ocean

Changing sea levels

During ice ages, so much water may turn to ice that sea levels drop. At the LGM, sea levels around the world were over 300 ft (100 m) lower. However, the weight of ice pushed the land down in some places. After the ice melted, as sea levels rose again, the land began to bounce back—a slow process continuing to this day.

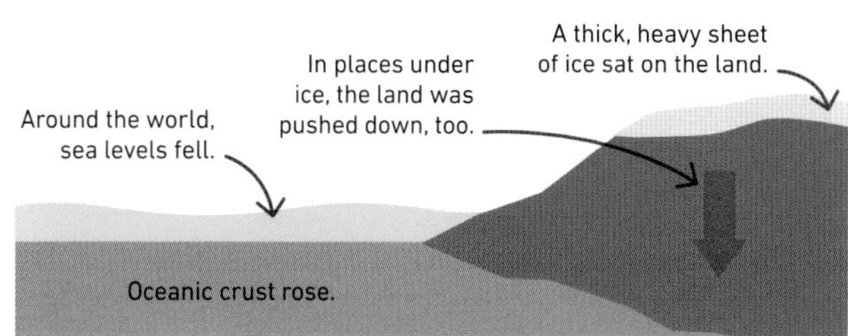
Around the world, sea levels fell.
In places under ice, the land was pushed down, too.
A thick, heavy sheet of ice sat on the land.
Oceanic crust rose.

△ **Under pressure**
Around the world, sea levels fell during the glacials. However, in some places, the immense weight of ice pushed the Earth's crust down into the mantle, making sea levels rise locally.

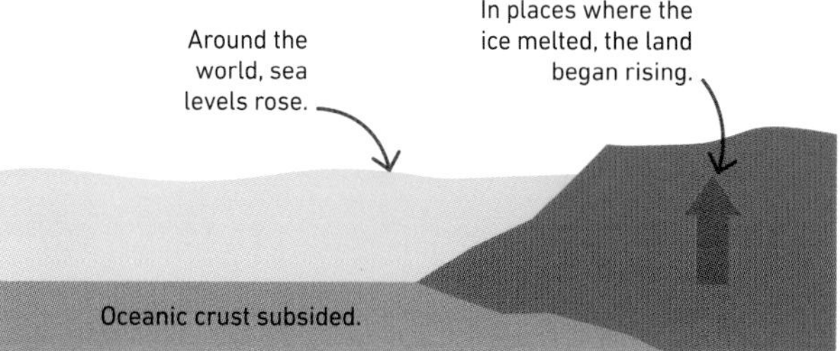
Around the world, sea levels rose.
In places where the ice melted, the land began rising.
Oceanic crust subsided.

△ **Slow release**
Melting ice sheets and increased precipitation caused the sea levels to rise. The weight of water started pushing the ocean floor down, while ice-free continents began to rise.

Glacial erosion

SEE ALSO
❮ 44–46 Weathering and erosion
Glacial deposition 66–67 ❯

MOUNTAIN GLACIERS MOVE DOWNHILL BY GRAVITY AND HAVE ENORMOUS EROSIVE POWER, CREATING DISTINCTIVE LANDFORMS.

A glacier is a large mass of ice moving slowly downhill. The moving ice causes a lot of erosion, carving valleys and removing rock. When the ice melts, it reveals a transformed landscape.

How glaciers form

Snow that falls in regions where the air temperature usually stays below freezing point, such as high mountains, never fully melts away. Instead, it builds up over centuries, compressing the snow beneath it into ice. Under the effect of gravity, the massive weight of the ice forms a glacier that flows very slowly downhill.

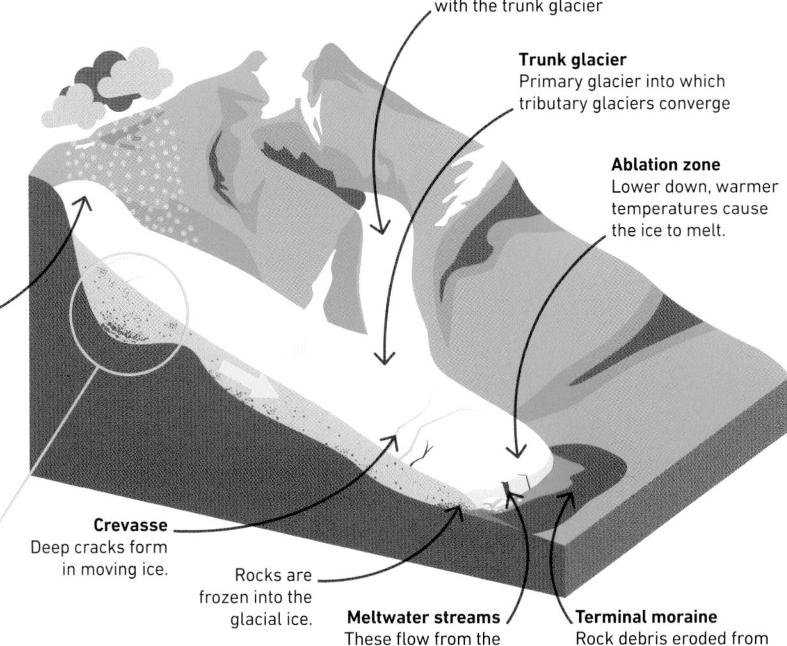
Snow builds up in the accumulation zone.
Tributary glacier A smaller glacier that merges with the trunk glacier
Trunk glacier Primary glacier into which tributary glaciers converge
Ablation zone Lower down, warmer temperatures cause the ice to melt.
Crevasse Deep cracks form in moving ice.
Rocks are frozen into the glacial ice.
Meltwater streams These flow from the ablation zone, where ice starts to melt.
Terminal moraine Rock debris eroded from the valley is deposited at this end of a glacier.

River of ice ▷
Heavy ice that forms on slopes creeps downhill through existing valleys. On reaching warmer, lower levels, it melts, forming streams.

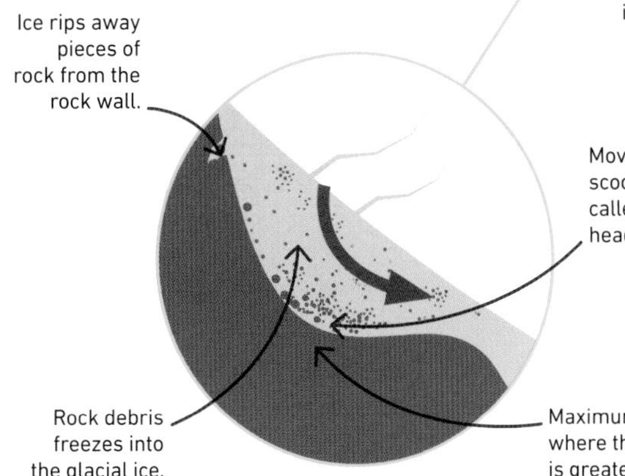
Ice rips away pieces of rock from the rock wall.
Moving ice and rock scoop out a hollow called a cirque at the head of the glacier.
Rock debris freezes into the glacial ice.
Maximum erosion occurs where the weight of ice is greatest.

Scoured by ice

Plucking and scouring

Moving ice freezes onto nearby rock and rips pieces away as water in rock fissures freezes, expanding and wedging the rock apart in a process called "plucking." Any broken rock fragments are frozen into the moving ice and, like sandpaper, "scour" any rock the glacier passes over.

Rounded rocks

Rock outcrops in the path of the moving ice are worn smooth by abrasion at the upstream end but have rock ripped away by plucking at the downstream end. The result is a "rôche moutonnée," which means "sheep-shaped rock"—although they can be much bigger than sheep.

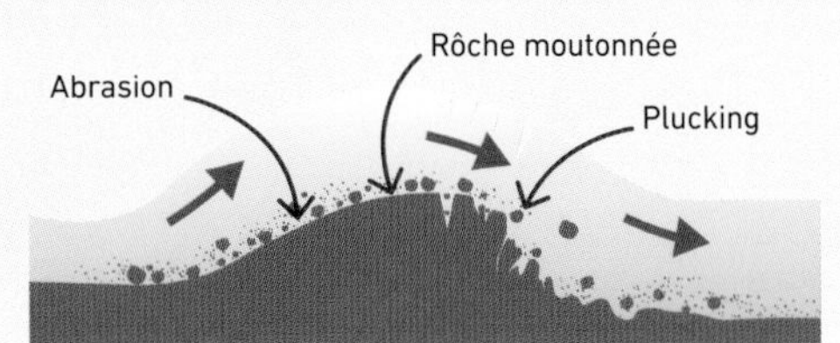
Abrasion
Rôche moutonnée
Plucking

Glacial landforms

Where accumulating snow turns to ice at mountain tops, erosion creates cirques (armchair-shaped curves). As more ice forms, it spills over the lip of its cirque and a glacier flows downhill, carving a U-shaped valley. As it moves, it slices away spurs of higher land and scoops out deep hollows in the valley floor.

REAL WORLD

The Norwegian fjords

Glacial erosion during the last ice age, which began around 2.5 million years ago, had a massive impact on landscapes in the northern hemisphere. In Norway, glaciers flowing down to the sea from an ice sheet covering the mountains gouged many deep, U-shaped valleys. When the ice melted, seawater flooded the valleys, creating the Norwegian fjords.

Arête
Sharp narrow ridges may be left between adjoining cirques.

Side glacier flowing into main glacier

Horn
Three or more cirques forming back to back create a pyramidal peak.

Cirque

◁ **Active glacier**
Small side glaciers (tributary glaciers) erode valleys into deep hollows, which are known as cirques (or corries). Intense glacial erosion creates arêtes between valleys and may also form horns.

Glacier

During glaciation

U-shaped valley
The glacier erodes the existing valley into a U-shaped one.

Truncated spur
Moving ice carves away projecting spurs of land, leaving cliffs.

Hanging valleys and waterfalls
Tributary glaciers that once flowed into the main valley form hanging valleys. Rivers flowing down these hanging valleys create waterfalls.

Horn

Arête

Ribbon lake
Hollows scooped out of the valley floor by ice fill with water to form long, narrow lakes.

Tarn
Small lakes that form in empty cirques are called tarns.

Transformed landscape ▷
When a glacier melts, it leaves a deep U-shaped valley dotted with lakes that form in the hollows. The cirques fill with water, creating tarns.

After glaciation

U-shaped valley
Wide U-shaped valleys are left behind in the landscape.

Glacial deposition

SEE ALSO
❮ 64–65 Glacial erosion
Climate change 175–177 ❯

GLACIAL ICE PULLED DOWNHILL BY GRAVITY CARRIES HUGE AMOUNTS OF ROCKY DEBRIS, DEPOSITED WHEREVER THE ICE MELTS.

The rocky material that the glacier ice tears away from the landscape is deposited as heaps of rubble and clay where the glacier ends, or where it existed before melting away.

Valley glaciers

Moving glaciers carry both frost-shattered rock falling from the mountains above, and rock torn from the valley walls by the glacier itself. As a glacier melts, it leaves behind piles of debris, known as moraine, that form distinctive patterns over the valley floor—including lateral moraines along the sides of the valley, and terminal moraines across the valley.

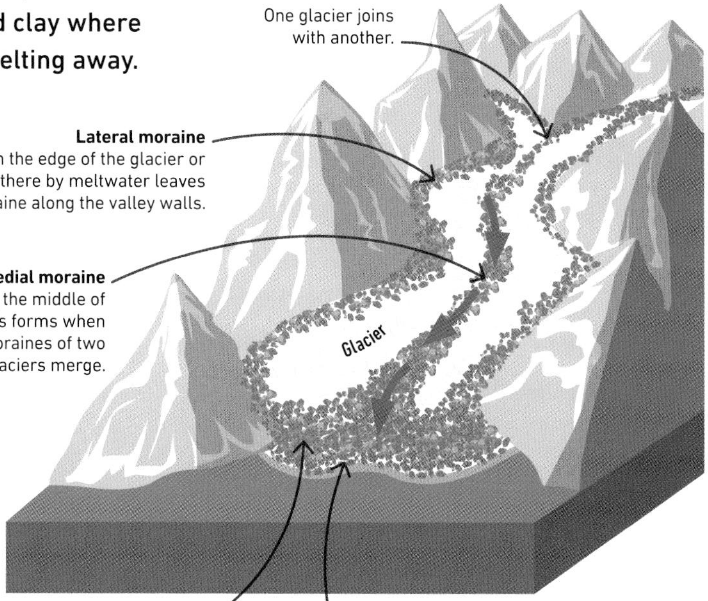

One glacier joins with another.
Lateral moraine Debris falling on the edge of the glacier or washed there by meltwater leaves shelves of moraine along the valley walls.
Medial moraine Lying along the middle of a glacier, this forms when the lateral moraines of two glaciers merge.
Glacier
Terminal moraine The rock debris deposited at the end of the glacier, in semicircular mounds, forms terminal moraines.
Ground moraine Glacial till or debris is washed by streams under the glacier into a hummocky landscape known as ground moraine.

Types of moraine ▷ An active glacier creates several types of moraine. Where the glacier melts and retreats, the moraines are left behind on the landscape.

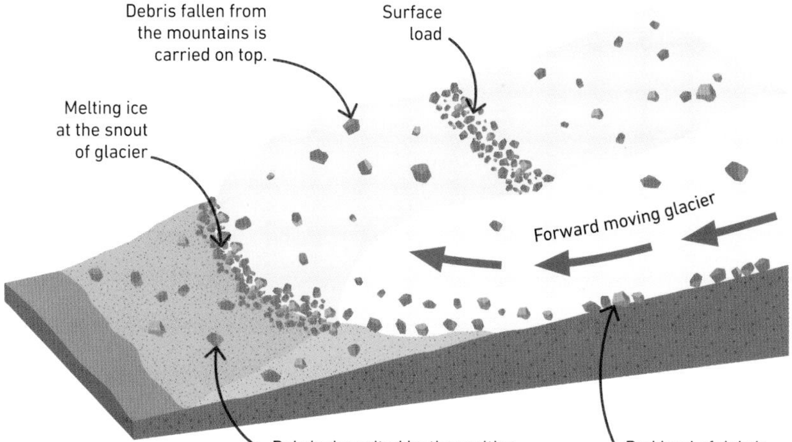

How moraines form

The moving glacier carries debris in various ways. Some is carried on top, some embedded within it, and some dragged along underneath. Finer debris may be carried by streams flowing within and under the glacier.

Debris fallen from the mountains is carried on top.
Surface load
Melting ice at the snout of glacier
Forward moving glacier
Debris deposited by the melting ice forms the terminal moraine.
Bed load of debris inside the ice

◁ Glacial conveyor belt Rocks and finer material frozen in the ice are carried by the glacier. The glacier moves slowly downhill to the point where the ice melts away.

Ice sheets retreat

As the vast sheets of the Ice Ages melted, they left behind landscapes covered in glacial deposits or "drift." Some dropped directly by the ice, but much washed into plains by floods of meltwater. The drift often covers the landscape with a thick blanket, burying earlier hills and valleys. But here and there are distinctive landforms such as eskers and drumlins.

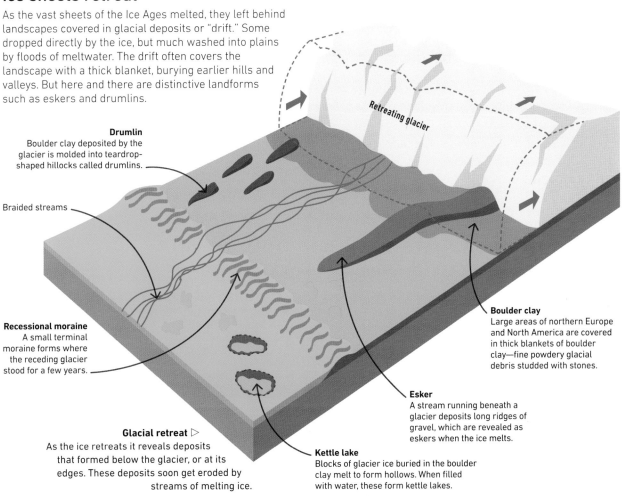

Retreating glacier

Drumlin
Boulder clay deposited by the glacier is molded into teardrop-shaped hillocks called drumlins.

Braided streams

Recessional moraine
A small terminal moraine forms where the receding glacier stood for a few years.

Boulder clay
Large areas of northern Europe and North America are covered in thick blankets of boulder clay—fine powdery glacial debris studded with stones.

Esker
A stream running beneath a glacier deposits long ridges of gravel, which are revealed as eskers when the ice melts.

Glacial retreat ▷
As the ice retreats it reveals deposits that formed below the glacier, or at its edges. These deposits soon get eroded by streams of melting ice.

Kettle lake
Blocks of glacier ice buried in the boulder clay melt to form hollows. When filled with water, these form kettle lakes.

Glacial erratics

Moving ice sheets and glaciers can transport big rocks far from their place of origin. These are eventually deposited where the ice melts beneath them. The rock of these glacial boulders may be quite different from the local rock of the region where they come to rest, making them easy to identify.

Out of place ▷
Perched on the local limestone of a hillside in northern England, this huge sandstone boulder was carried up to 2 miles (3 km) by a moving ice sheet that melted about 18,000 years ago.

Coastal erosion

COASTLINES ARE CONSTANTLY CHANGING, WORN AWAY BY THE ACTION OF BREAKING WAVES.

SEE ALSO
❮ 52–53 The rock cycle
Coastal deposition 70–71 ❯
Continents and oceans 202–203 ❯

Storm waves crashing onto rocky shores have enormous power. They can shatter and grind away solid rock to create sheer cliffs, caves, and rock arches.

Wearing away the coast

Waves are almost constantly at work along the coast, but storm waves carry huge power and can be enormously destructive. Big waves force seawater into cracks in the cliffs at high pressure, breaking the rock apart. Loosened blocks fall away, undercutting the rock above them so it collapses, creating more rubble to be picked up by the waves.

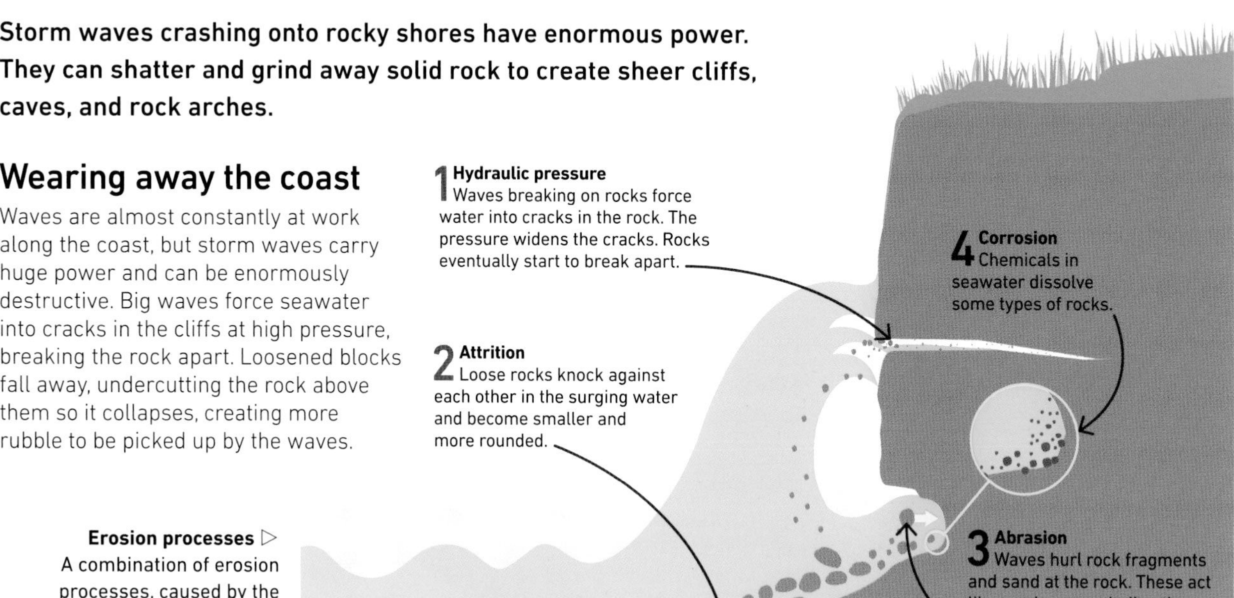
1 Hydraulic pressure
Waves breaking on rocks force water into cracks in the rock. The pressure widens the cracks. Rocks eventually start to break apart.
2 Attrition
Loose rocks knock against each other in the surging water and become smaller and more rounded.
3 Abrasion
Waves hurl rock fragments and sand at the rock. These act like sandpaper, grinding the rock away.
4 Corrosion
Chemicals in seawater dissolve some types of rocks.

Erosion processes ▷
A combination of erosion processes, caused by the force of the breaking waves, cuts into rocky shores.

Cliffs and rock platforms

In areas of hard rock, coastal erosion creates steep cliffs as it wears hills away. Continual attack by waves undercuts the cliffs, creating a notch at the high-tide mark, where wave energy is greatest. As the undercut cliffs collapse and retreat, they leave behind a gently shelving platform of rock, uncovered at low tide to reveal rock pools.

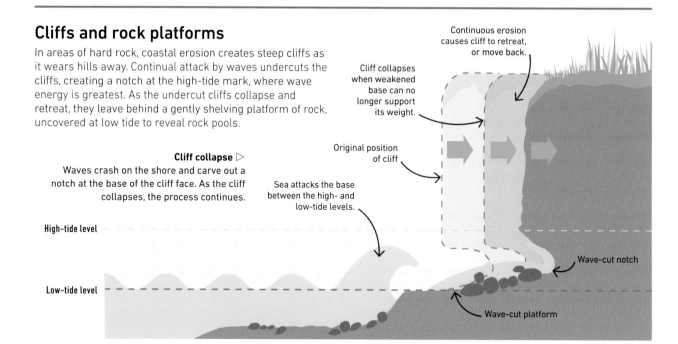

Cliff collapse ▷
Waves crash on the shore and carve out a notch at the base of the cliff face. As the cliff collapses, the process continues.

Continuous erosion causes cliff to retreat, or move back.

Cliff collapses when weakened base can no longer support its weight.

Original position of cliff

Sea attacks the base between the high- and low-tide levels.

High-tide level

Low-tide level

Wave-cut notch

Wave-cut platform

Headlands and bays

Some coasts consist of both hard and soft rocks. The softer rocks are eroded faster, creating bays that lie between headlands of harder rock. Headlands absorb most of the impact of waves sweeping in from the open sea, sheltering the bays from further erosion. This allows stones and sand to build up as beaches in the bays.

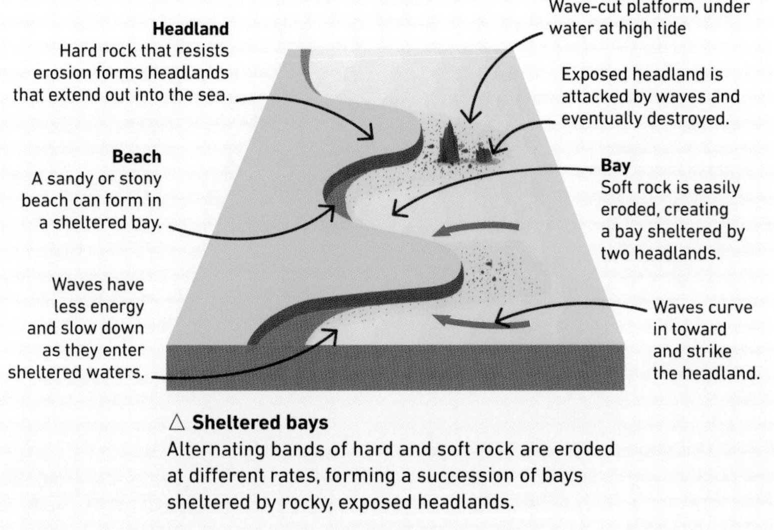
Headland
Hard rock that resists erosion forms headlands that extend out into the sea.
Wave-cut platform, under water at high tide
Exposed headland is attacked by waves and eventually destroyed.
Beach
A sandy or stony beach can form in a sheltered bay.
Bay
Soft rock is easily eroded, creating a bay sheltered by two headlands.
Waves have less energy and slow down as they enter sheltered waters.
Waves curve in toward and strike the headland.

△ **Sheltered bays**
Alternating bands of hard and soft rock are eroded at different rates, forming a succession of bays sheltered by rocky, exposed headlands.

REAL WORLD

Erosion threat

Shores made of soft rock, such as the Holderness coast in the UK, are very vulnerable to coastal erosion. Waves driven by winter storms undercut and weaken the cliffs so they collapse, along with anything on the land above. This can make houses fall into the sea, and many coastal communities are abandoned each year because they are no longer safe.

Caves, arches, and stacks

Waves breaking on an exposed rocky shore may enlarge a crack or joint to hollow out a cave in the rock. When this happens on a headland, a cave may also form on the other side, and the sea will eventually break through to form a rock arch. This may collapse to leave an isolated stack, which finally crumbles to a stump.

▽ **Pounding waves**
Although headlands are made of hard rock, the sea wears away any weak points to create caves, arches, and stacks.

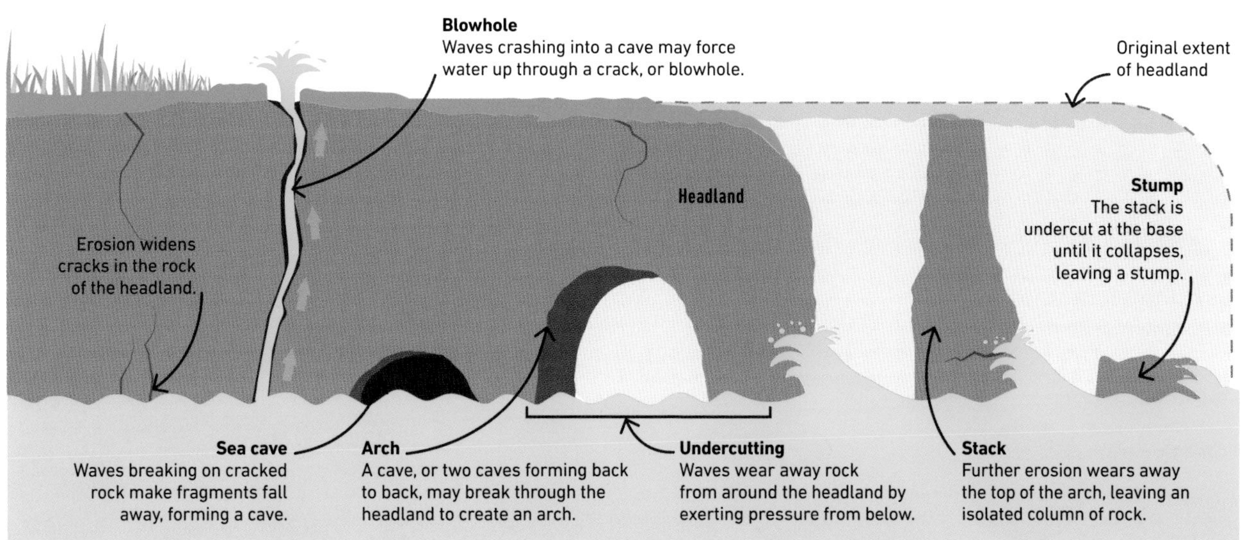
Blowhole
Waves crashing into a cave may force water up through a crack, or blowhole.
Original extent of headland
Headland
Stump
The stack is undercut at the base until it collapses, leaving a stump.
Erosion widens cracks in the rock of the headland.
Sea cave
Waves breaking on cracked rock make fragments fall away, forming a cave.
Arch
A cave, or two caves forming back to back, may break through the headland to create an arch.
Undercutting
Waves wear away rock from around the headland by exerting pressure from below.
Stack
Further erosion wears away the top of the arch, leaving an isolated column of rock.

Coastal deposition

EXPOSED SHORES ARE CUT AWAY BY EROSION AND SHELTERED SHORES BUILT UP BY DEPOSITION.

SEE ALSO
❮ 64–65 Glacial erosion
❮ 66–67 Glacial deposition
❮ 68–69 Coastal erosion
Erosion in deserts 72–73 ❯

Waves erode rocks and break them up into shingle, sand, and fine silt. These are then swept away and deposited on shores that are less exposed to violent storms.

Waves show **energy moving through** the **water.**

The work of waves

On relatively sheltered coasts, breaking waves transport materials, such as stones and sand, onto the shore and deposit them there to form beaches. During storms, bigger and more violent waves crash on the shore, eroding the beach and carrying stones and sand away into deeper water.

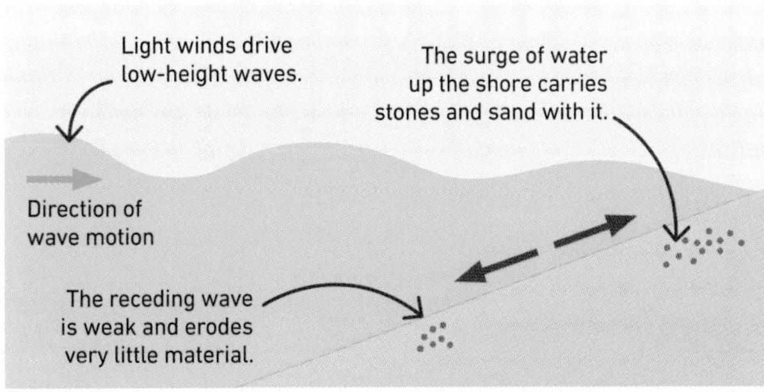
Light winds drive low-height waves.
The surge of water up the shore carries stones and sand with it.
Direction of wave motion
The receding wave is weak and erodes very little material.

△ **Constructive waves**
A cross section of the beach during calm weather shows that the waves have a lower height and deposit more material than they carry away, gradually building up the beach.

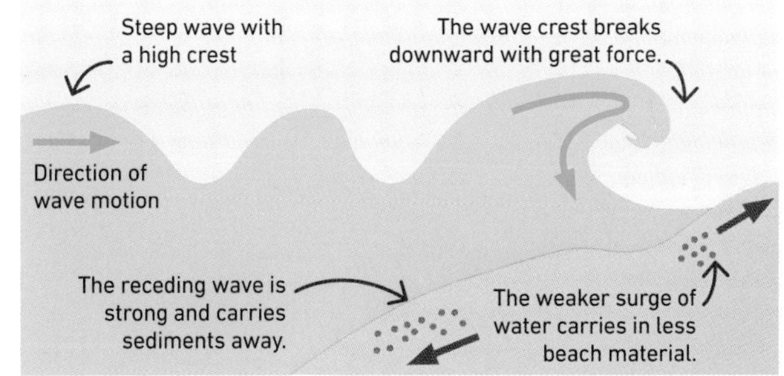
Steep wave with a high crest
The wave crest breaks downward with great force.
Direction of wave motion
The receding wave is strong and carries sediments away.
The weaker surge of water carries in less beach material.

△ **Destructive waves**
A cross section of the beach during a storm shows the waves rising higher and breaking violently on the shore, sweeping beach material away. The beach is built up again in calm weather.

Deposited material

Waves dislodge rocks, toss them around, and erode their corners, creating boulders and shingle. Smaller fragments turn into sand, silt, and mud. More wave energy is needed to move heavy stones, so they are left higher up the shore than lighter sediments.

▽ **Sorted sediments**
Waves move smaller particles more easily than bigger ones, so different types of sediments are deposited in different places.

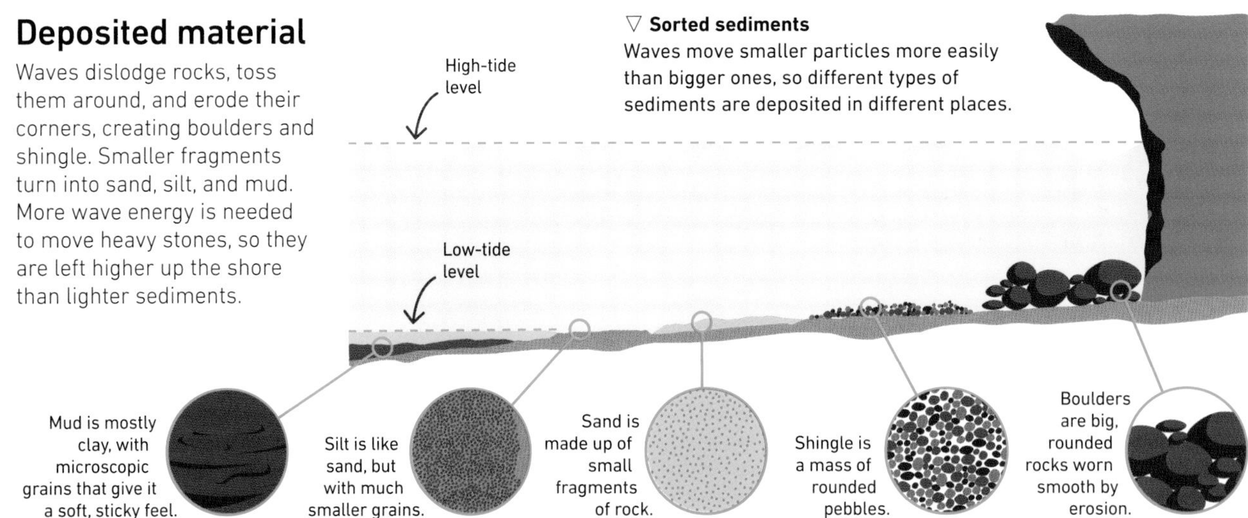
High-tide level
Low-tide level
Mud is mostly clay, with microscopic grains that give it a soft, sticky feel.
Silt is like sand, but with much smaller grains.
Sand is made up of small fragments of rock.
Shingle is a mass of rounded pebbles.
Boulders are big, rounded rocks worn smooth by erosion.

Longshore drift

When waves strike the beach at an angle, the power of the incoming wave carries sand up the beach at an angle too. But when the wave runs out of energy, it falls back down the beach at right angles pulling the sand with it. And so sand is gradually shifted along the beach in an in-out zigzag. This is called longshore drift.

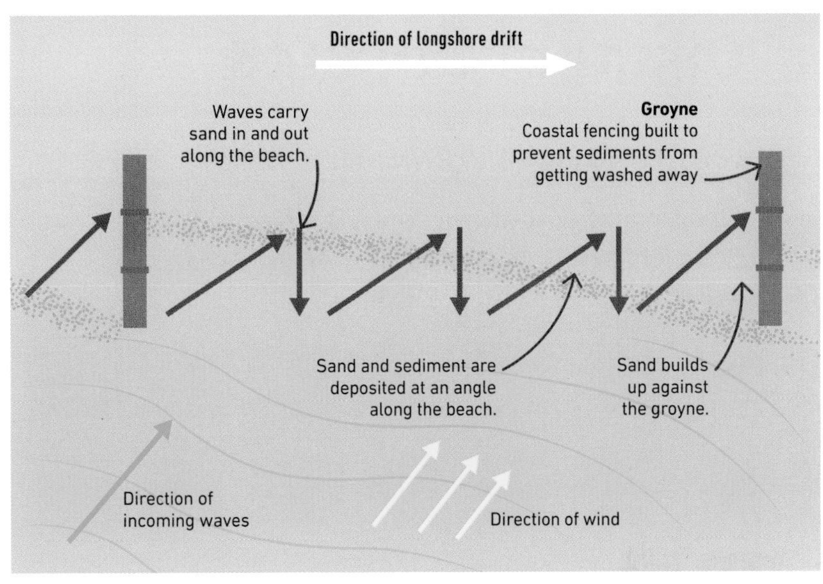
Direction of longshore drift
Waves carry sand in and out along the beach.
Groyne
Coastal fencing built to prevent sediments from getting washed away
Sand and sediment are deposited at an angle along the beach.
Sand builds up against the groyne.
Direction of incoming waves
Direction of wind

Beaches on the move ▷
An overhead view of the beach shows how particles of sand roll down toward the sea and are carried back up by the waves at an angle as they move in and out of the beach. The particles gather at a barrier or groyne.

Sand spits and sandbars

Longshore drift can create long beaches that extend for many miles and may project from headlands as spits. They can form bars and barrier islands that cross bays and inlets, sometimes isolating lagoons of calm water. A similar feature called a tombolo may also extend from the shore to join up with an island.

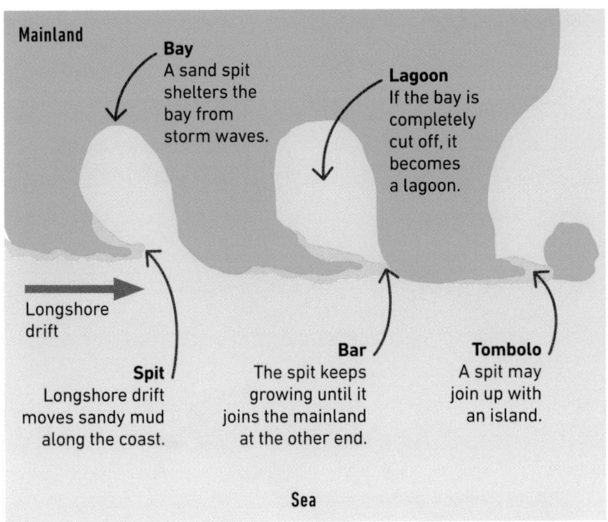
Mainland
Bay
A sand spit shelters the bay from storm waves.
Lagoon
If the bay is completely cut off, it becomes a lagoon.
Longshore drift
Spit
Longshore drift moves sandy mud along the coast.
Bar
The spit keeps growing until it joins the mainland at the other end.
Tombolo
A spit may join up with an island.
Sea

△ Sandy barriers
Beach material can extend into open water to create spits, bars, and tombolos. These act as barriers that protect the shore from storm waves.

Salt marshes and mangroves

Where spits and bars form at the mouth of a river, they create very sheltered conditions on the side nearest to the shore. This allows fine silt and mud to settle and form mudbanks that are flooded at high tide. Either salt marsh plants or tropical mangrove trees, which can adapt to salty, waterlogged mud, grow there.

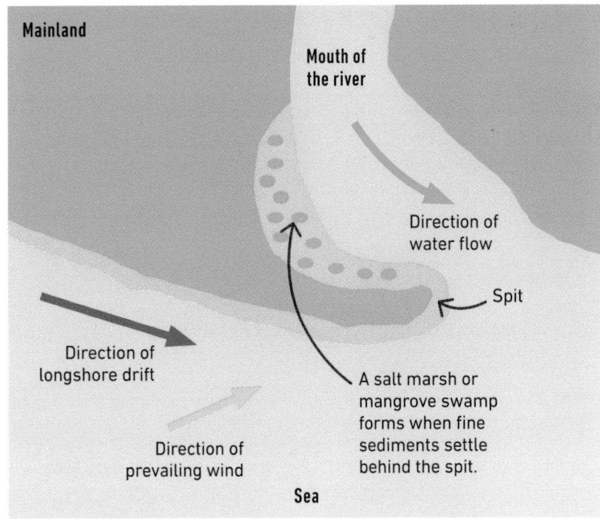
Mainland
Mouth of the river
Direction of water flow
Spit
Direction of longshore drift
A salt marsh or mangrove swamp forms when fine sediments settle behind the spit.
Direction of prevailing wind
Sea

△ Coastal wetlands
The smallest sediments form mudbanks only in very calm water. The spit provides shelter from the prevailing wind and big waves, and marshes or mangrove swamps form behind it.

Erosion in deserts

SEE ALSO
❮ 44–46 Weathering and erosion
❮ 64–65 Glacial erosion
❮ 68–69 Coastal erosion
Deserts 106–107 ❯

IN DESERTS, THE ARIDITY (DRYNESS) IS A POWERFUL FORCE OF WEATHERING AND EROSION.

In very dry climates, wind-blown sand and torrential flash floods, triggered by intense tropical rainstorms, can attack and sculpt rock into a variety of forms.

Barren ground

Sand and loose rock are common in deserts, and there is little vegetation due to low rainfall. With few plant roots to hold the soil together, wind sweeps away topsoil easily, exposing the rock underneath it to erosion by wind and water.

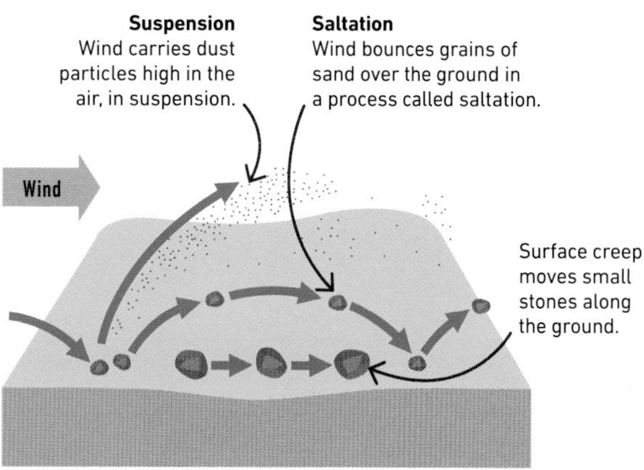
Suspension
Wind carries dust particles high in the air, in suspension.
Saltation
Wind bounces grains of sand over the ground in a process called saltation.
Wind
Surface creep moves small stones along the ground.

△ **Suspension and saltation**
The wind picks up and carries small particles of dust in suspension. It also bounces grains of sand along the ground (saltation), and makes stones creep slowly along the surface.

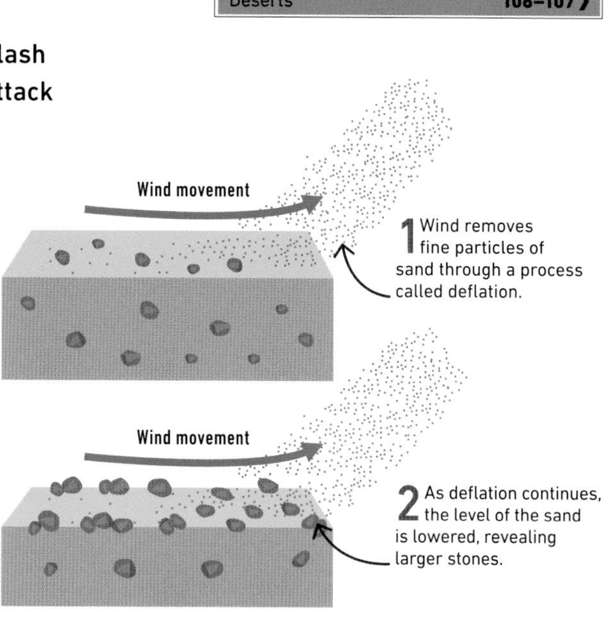
Wind movement
1 Wind removes fine particles of sand through a process called deflation.
Wind movement
2 As deflation continues, the level of the sand is lowered, revealing larger stones.
Wind movement
3 Over time, deflation exposes more stones, which are densely packed and form a hard desert pavement.

△ **How a desert pavement forms**
Wind moves dust and sand more easily than heavier stones and strips away all the fine material over time, forming a desert pavement – a more solid, pebbly ground, containing little sand.

Wind sculptures

Dry sand picked up by the desert wind acts like a sand-blasting machine, scouring away the surface of rocks. The rocks are worn away mostly at the base, where the process of saltation bounces along grains of sand. Over time, this creates spectacular wind-eroded rocks called ventifacts.

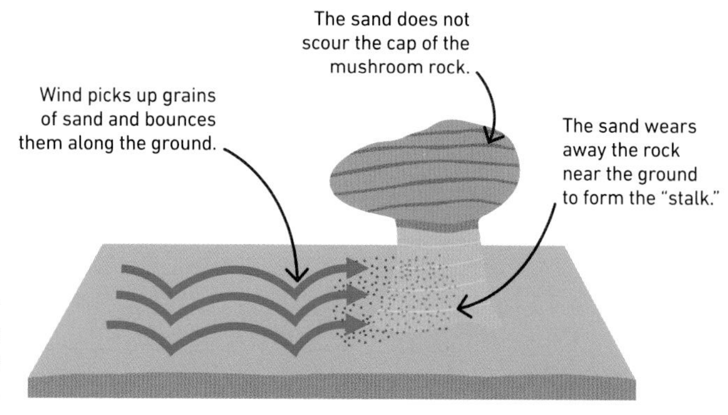
The sand does not scour the cap of the mushroom rock.
Wind picks up grains of sand and bounces them along the ground.
The sand wears away the rock near the ground to form the "stalk."

Mushroom rocks ▷
The desert wind can carve rocks that rise more than 3 ft (1 m) above the ground into mushroom shapes.

Sand dunes

In desert landscapes, grains of sand moved by the wind constantly collide with each other, and the repeated impact gives each grain a rounded, "frosted" look. The wind easily sweeps these grains into heaps called dunes, whose shape depends on the direction and nature of the wind.

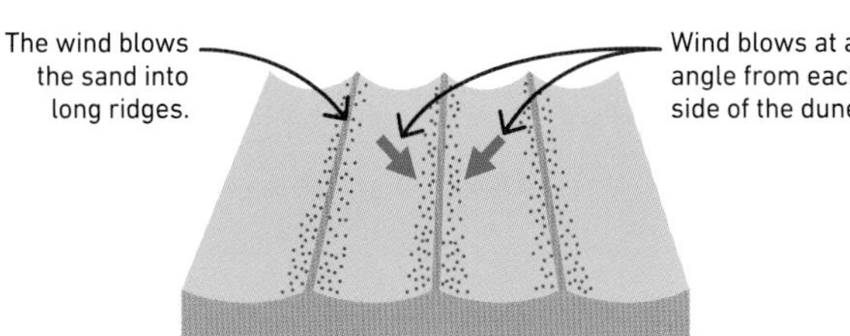
The wind blows the sand into long ridges.
Wind blows at an angle from each side of the dune.

△ **Longitudinal dunes**
When the direction of the wind varies slightly, parallel sand ridges build up, extending in the average direction of the wind. The ground between the dunes may have very little sand.

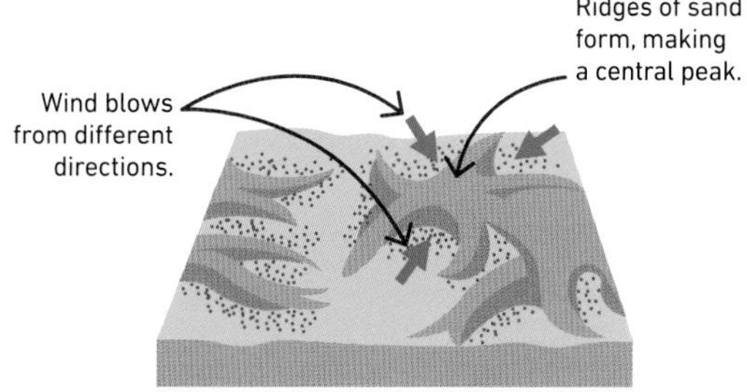
Ridges of sand form, making a central peak.
Wind blows from different directions.

△ **Star dunes**
Desert winds that blow from many different directions can sweep the sand into star-shaped dunes. Each dune has many ridges that meet to form a high central peak.

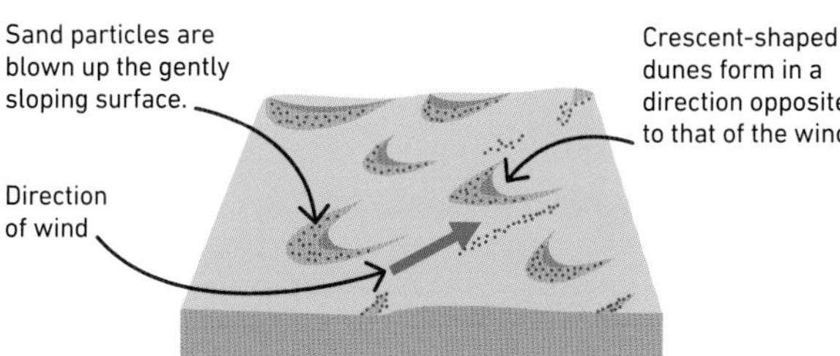
Sand particles are blown up the gently sloping surface.
Crescent-shaped dunes form in a direction opposite to that of the wind.
Direction of wind

△ **Barchan dunes**
Wind blowing mostly from one direction can form crescent-shaped dunes. The wind blows over the outer curve of the dune toward the pointed ends.

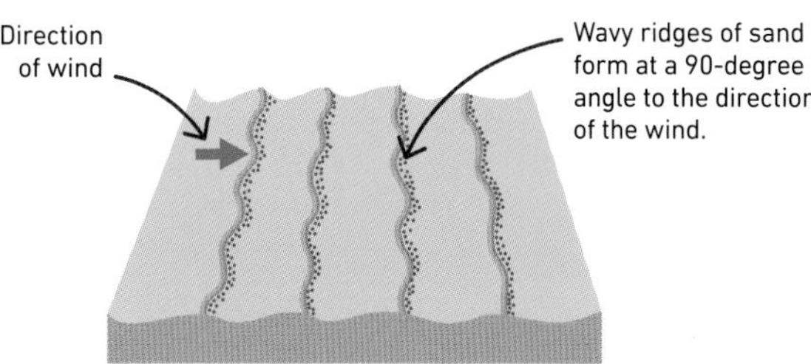
Direction of wind
Wavy ridges of sand form at a 90-degree angle to the direction of the wind.

△ **Transverse dunes**
In deserts with a lot of sand, steady winds form broad transverse dunes perpendicular to the direction of the wind. These break up later to form barchan dunes.

Flash floods

Rainstorms can occur even in deserts. These storms are intense, with rainwater flowing downhill in surging torrents in the absence of any soil to soak it up. The torrents, or flash floods, carry rocky debris that erodes deep canyons, leaving isolated rocky towers called mesas and buttes.

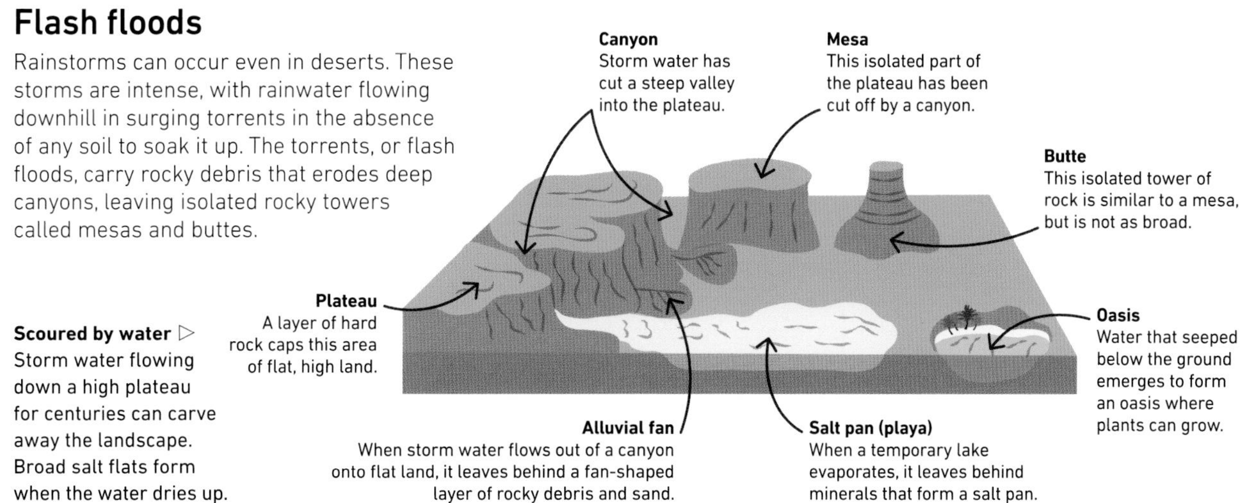
Canyon
Storm water has cut a steep valley into the plateau.
Mesa
This isolated part of the plateau has been cut off by a canyon.
Butte
This isolated tower of rock is similar to a mesa, but is not as broad.
Plateau
A layer of hard rock caps this area of flat, high land.
Oasis
Water that seeped below the ground emerges to form an oasis where plants can grow.
Alluvial fan
When storm water flows out of a canyon onto flat land, it leaves behind a fan-shaped layer of rocky debris and sand.
Salt pan (playa)
When a temporary lake evaporates, it leaves behind minerals that form a salt pan.

Scoured by water ▷
Storm water flowing down a high plateau for centuries can carve away the landscape. Broad salt flats form when the water dries up.

The atmosphere

SEE ALSO
❮ 22–23 Earth's structure
Weather systems 87–89 ❯
Weather forecasting 90–91 ❯
Climate change 175–177 ❯

THE EARTH IS SURROUNDED BY A BLANKET OF GASES 430 MILES (700 KM) DEEP, CALLED THE ATMOSPHERE.

The atmosphere provides air and water, making life on Earth possible. It regulates the temperature and protects living things from the sun's harmful rays.

Composition of the atmosphere

Our atmosphere, or air, is a mixture of different gases and tiny particles. Two gases, nitrogen (78 percent) and oxygen (21 percent), make up 99 percent of the atmosphere. The remaining 1 percent includes carbon dioxide and water vapor, plus traces of the gases argon, helium, methane, ozone, and neon. Oxygen is vital for supporting life on Earth.

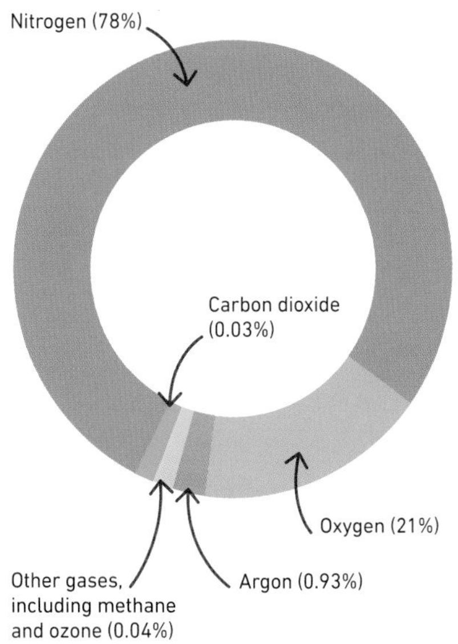
Nitrogen (78%)
Carbon dioxide (0.03%)
Oxygen (21%)
Other gases, including methane and ozone (0.04%)
Argon (0.93%)

△ **Component gases**
The average composition of the atmosphere stays stable, but slight variations in trace gases such as carbon dioxide, methane, and ozone can have a significant impact on the Earth's climate.

Layers of the atmosphere

The Earth's atmosphere has five layers, each of a different thickness, from the troposphere at the bottom to the exosphere at the top. The thermosphere includes a thin layer called the ionosphere, packed with ions (electrically charged particles), while the "ozone layer" in the stratosphere is made up of a thin concentration of ozone gas.

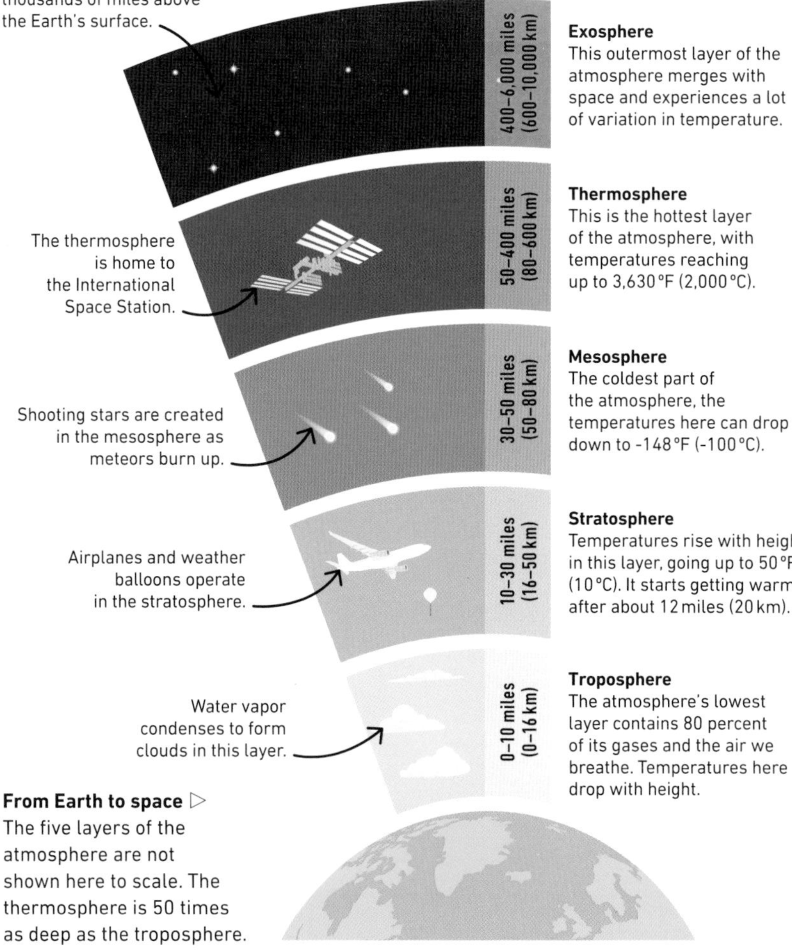
This outer layer extends to thousands of miles above the Earth's surface.
400–6,000 miles (600–10,000 km)
Exosphere
This outermost layer of the atmosphere merges with space and experiences a lot of variation in temperature.
The thermosphere is home to the International Space Station.
50–400 miles (80–600 km)
Thermosphere
This is the hottest layer of the atmosphere, with temperatures reaching up to 3,630°F (2,000°C).
Shooting stars are created in the mesosphere as meteors burn up.
30–50 miles (50–80 km)
Mesosphere
The coldest part of the atmosphere, the temperatures here can drop down to -148°F (-100°C).
Airplanes and weather balloons operate in the stratosphere.
10–30 miles (16–50 km)
Stratosphere
Temperatures rise with height in this layer, going up to 50°F (10°C). It starts getting warmer after about 12 miles (20 km).
Water vapor condenses to form clouds in this layer.
0–10 miles (0–16 km)
Troposphere
The atmosphere's lowest layer contains 80 percent of its gases and the air we breathe. Temperatures here drop with height.

From Earth to space ▷
The five layers of the atmosphere are not shown here to scale. The thermosphere is 50 times as deep as the troposphere.

The greenhouse effect

Most of the sun's energy passes through the atmosphere and heats up the land. Energy radiated from the warming land is then absorbed by the atmosphere with the help of greenhouse gases, such as carbon dioxide (CO_2). Burning carbon in the form of coal, oil, and gas releases CO_2 into the atmosphere, enhancing the greenhouse effect.

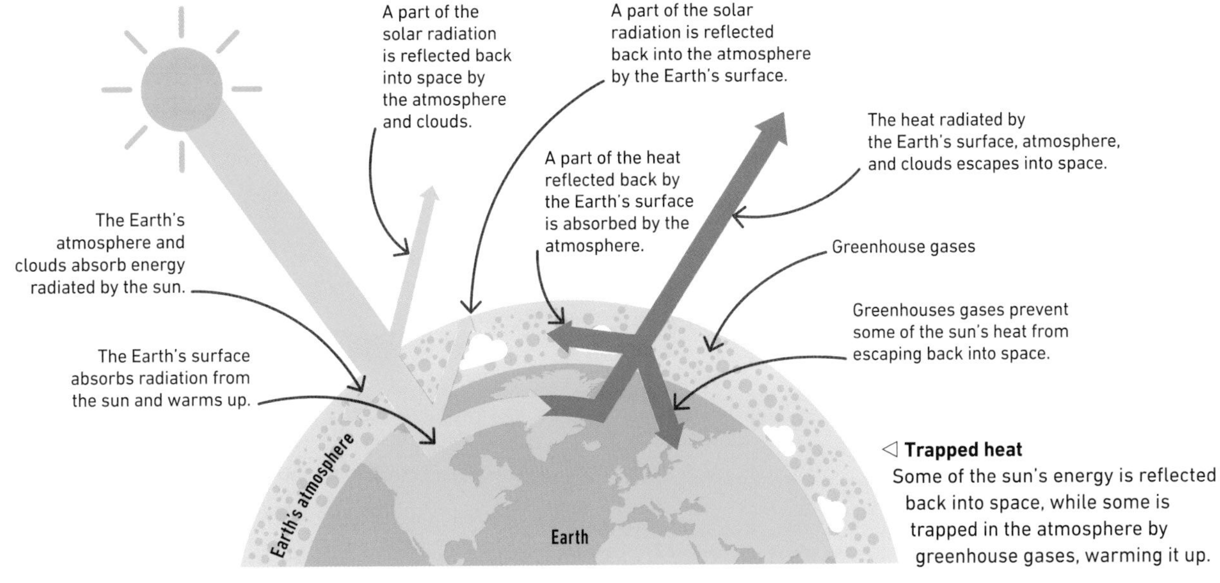
A part of the solar radiation is reflected back into space by the atmosphere and clouds.
A part of the solar radiation is reflected back into the atmosphere by the Earth's surface.
The heat radiated by the Earth's surface, atmosphere, and clouds escapes into space.
A part of the heat reflected back by the Earth's surface is absorbed by the atmosphere.
The Earth's atmosphere and clouds absorb energy radiated by the sun.
Greenhouse gases
Greenhouses gases prevent some of the sun's heat from escaping back into space.
The Earth's surface absorbs radiation from the sun and warms up.
Earth's atmosphere
Earth

◁ **Trapped heat**
Some of the sun's energy is reflected back into space, while some is trapped in the atmosphere by greenhouse gases, warming it up.

Atmospheric pressure

Air is packed with countless molecules that are constantly moving. These movements create atmospheric pressure, which is measured in units called bars and millibars. The higher the number of molecules, the greater the pressure. Atmospheric pressure changes in different weather conditions and at different heights.

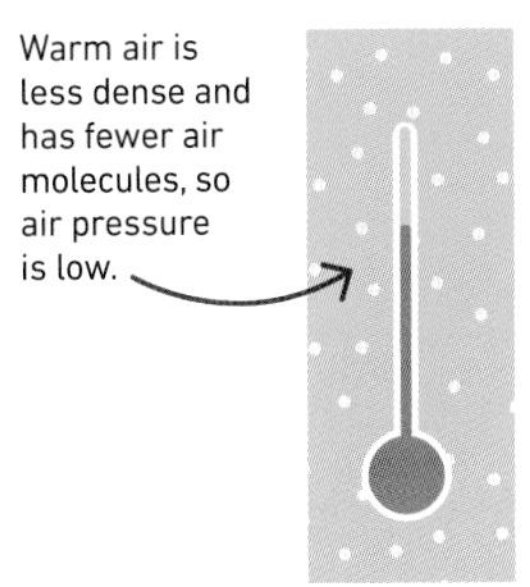
Warm air is less dense and has fewer air molecules, so air pressure is low.

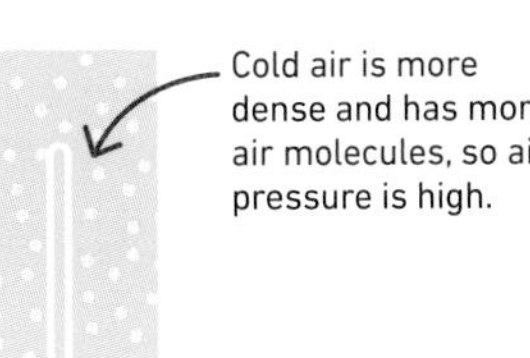
Cold air is more dense and has more air molecules, so air pressure is high.

△ **Effect of temperature**
Warmth expands air and reduces air pressure, creating a "low pressure" area. Cold makes air contract and increases pressure, creating a "high pressure" area.

Effect of height ▷
Pressure is highest at sea levels, where the air is dense and contains the most air molecules. It drops with increasing height as the air thins out.

REAL WORLD

The ozone layer

The ozone layer is a concentration of ozone gas in the stratosphere and protects life on Earth from the sun's ultraviolet rays. Release of harmful chlorofluorocarbon (CFC) gases from aerosols and refrigerators has made the ozone layer thin and caused holes to appear, though these are reducing.

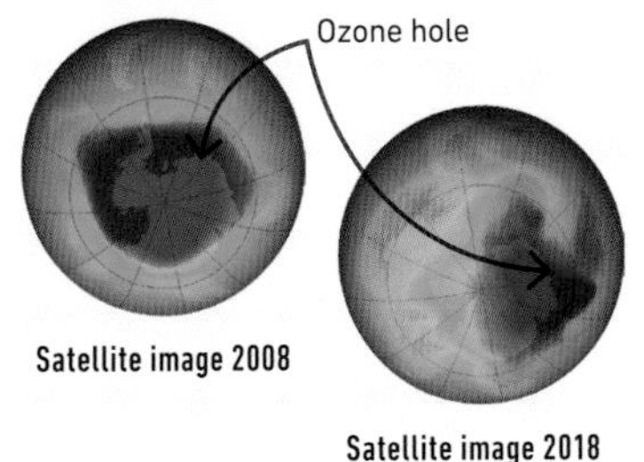
Ozone hole
Satellite image 2008
Satellite image 2018

Seasons

SEE ALSO
Climate zones 78–79 ❯
Hemispheres and latitude 210–211 ❯

WE EXPERIENCE SEASONS AS A RESULT OF THE EARTH'S ORBIT AROUND THE SUN. THE SEASONS ARE LESS OBVIOUS AT THE EQUATOR AND MOST EXTREME IN HIGHER LATITUDES.

Places in the tropics mostly have two seasons, a wet season when it rains a lot and a dry season when it doesn't. The temperate regions see four distinct seasons—winter, spring, summer, and fall.

Why do we have seasons?

As the Earth travels around the sun, it also spins on its slightly tilted axis. It is this angle toward the sun that creates the seasons. Different parts of the world lean toward the sun at different times. When a part of the world leans toward the sun, it experiences summer, and when it leans away, it experiences winter.

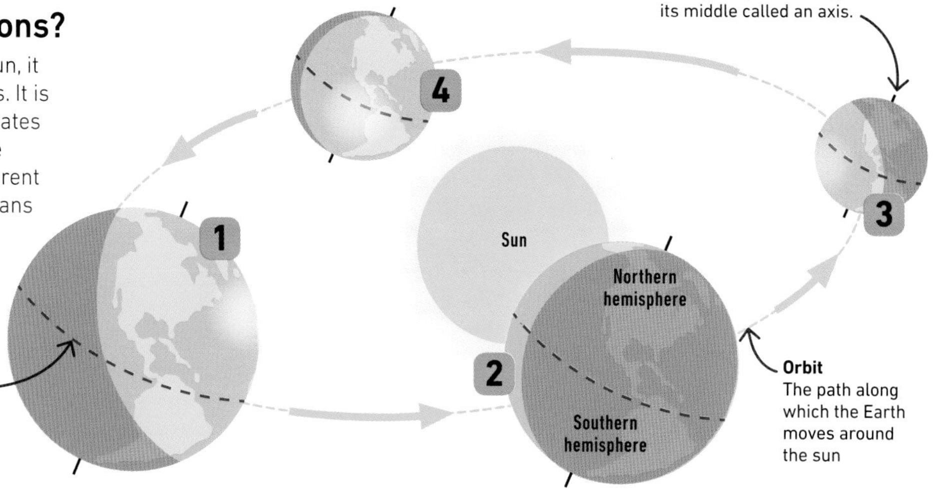

Axis
The Earth rotates around an imaginary line through its middle called an axis.
4
3
1
Sun
Northern hemisphere
2
Southern hemisphere
Equator
An imaginary line that divides the Earth in two equal halves
Orbit
The path along which the Earth moves around the sun

1 June
At noon, the sun is overhead in the northern hemisphere due to the Earth's tilt. This brings long, warm summer days. It is winter in the southern hemisphere.

2 September
As the north tilts away, the overhead sun is now moving south to the equator. Days and nights even out, bringing fall in the north of the world and spring in the south.

3 December
The overhead sun is now even further south into the southern hemisphere, bringing summer there. Meanwhile, in the north, nights get longer and winter sets in.

4 March
The overhead sun moves back north to the equator. Days and nights even out again. Winter warms into spring in the north, and summer cools into fall in the south.

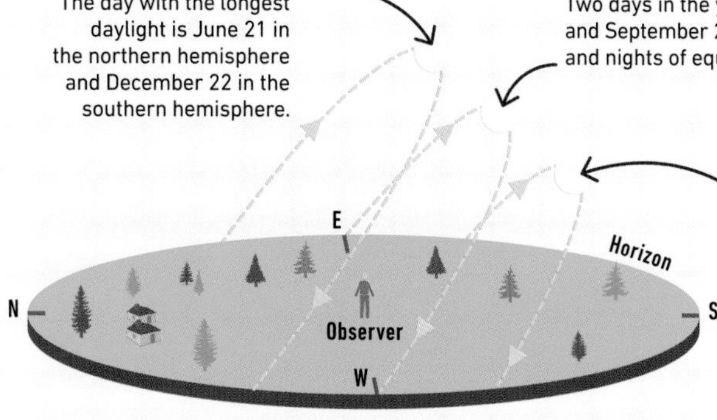

Summer solstice
The day with the longest daylight is June 21 in the northern hemisphere and December 22 in the southern hemisphere.
Equinox
Two days in the year—March 21 and September 23—have days and nights of equal length.
Winter solstice
The shortest day in the year is December 22 in the northern hemisphere and June 21 in the southern hemisphere.
E
Horizon
N
S
Observer
W
Northern hemisphere

Solstices and equinoxes

As the seasons change, the sun's path through the sky appears to shift. The sun is at its highest point at the summer solstice and at its lowest at the winter solstice. In between, there are two equinoxes, or midpoints, in spring and fall.

The word **"equinox"** means **"equal night."**

The four seasons

At middle and higher latitudes, seasons bring different weather because the sun's intensity and the length of the day varies through the year. When the sun is high and hot, it is summertime. When the sun is low and the tilt of the Earth is away from the sun, it is winter. Fall and spring come in between.

Winter
Days are short, cold, and dark. There may be frost, snow, and blizzards.

Spring
Days start to warm, though nights stay cool. There may be brief showers.

Summer
Days are hot and long. Rain falls in afternoon thunderstorms or not at all.

Fall
Days are cooler and damp with misty mornings. The weather is often stormy.

Wet and dry seasons

Many parts of the world, especially the tropical regions, have just two seasons, a wet season when it rains a lot and a dry season. Some equatorial regions experience no seasons at all—it is either hot and dry or warm and wet throughout the year.

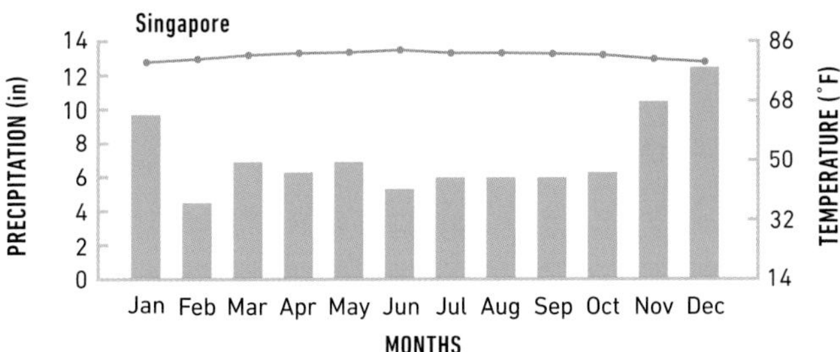
Singapore
PRECIPITATION (in)
14
12
10
8
6
4
2
0
TEMPERATURE (°F)
86
68
50
32
14
Jan Feb Mar Apr May Jun Jul Aug Sep Oct Nov Dec
MONTHS

△ **Equatorial climate**
Moist equatorial climates, such as Singapore in Asia, have the least seasonal variation. Rain falls throughout the year, and the air stays warm.

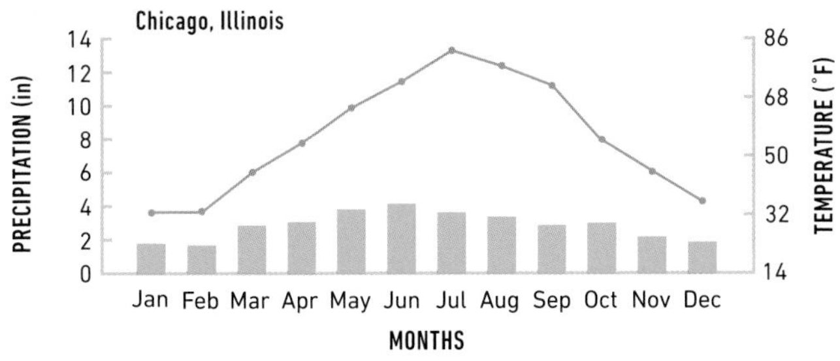
Chicago, Illinois
PRECIPITATION (in)
14
12
10
8
6
4
2
0
TEMPERATURE (°F)
86
68
50
32
14
Jan Feb Mar Apr May Jun Jul Aug Sep Oct Nov Dec
MONTHS

△ **Continental climate**
The interiors of large continents, such as the states in the middle of the US, have the most extreme variations between seasons. Winters may be bitterly cold and summers scorchingly hot.

KEY
Precipitation
Temperature

Monsoon

Much of southern Asia has a monsoon climate, with one dry season and another of torrential rain. In India, the dry season occurs from December to May, and the wet season from June to November. It is the change in the wind direction that brings the monsoon.

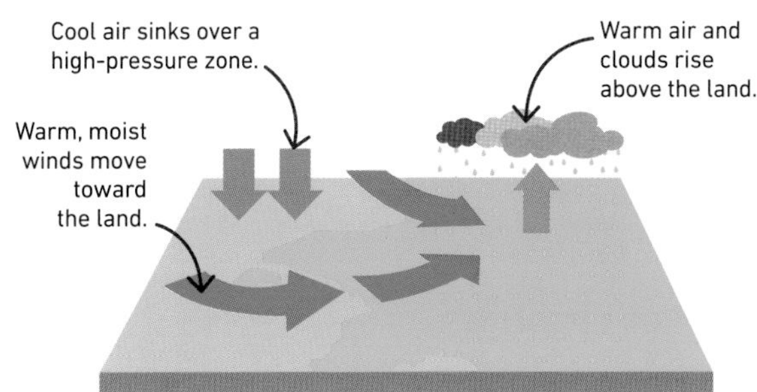
Cool air sinks over a high-pressure zone.
Warm air and clouds rise above the land.
Warm, moist winds move toward the land.

△ **Monsoon season**
The wet season starts when the interior of the continent warms up, drawing in the moisture-laden monsoon winds from the ocean.

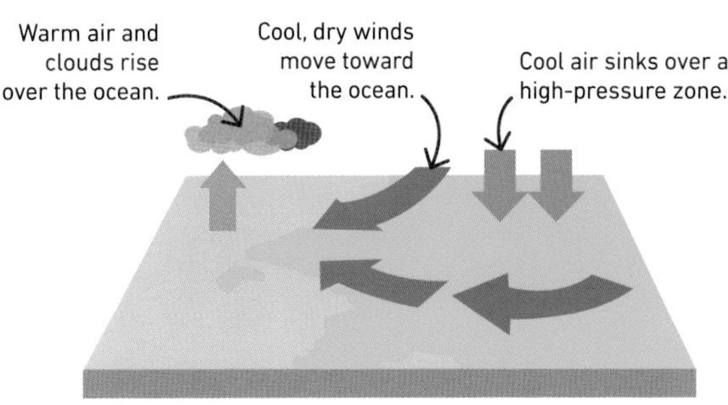
Warm air and clouds rise over the ocean.
Cool, dry winds move toward the ocean.
Cool air sinks over a high-pressure zone.

△ **Dry season**
The dry season starts when the interior of the continent cools, pushing the winds back to the ocean and drying the land.

Climate zones

SEE ALSO
Global winds 82–83 ❯
Weather systems 87–89 ❯
Biomes 98–99 ❯
Deserts 106–107 ❯

CLIMATE IS THE AVERAGE WEATHER IN A REGION MEASURED OVER 30 YEARS.

Every place has days of extreme weather, but the climate is how warm or cold and wet or dry it is on average for the time of year. Climate affects the plants and animals found in a region and the way people live there.

Climate bands

The world has three climate zones on each side of the equator. Either side of the equator is the warm tropical region that receives strong sunlight. At both ends of the world are the icy polar regions where the sunlight is the weakest. In between are the temperate zones with their distinct seasons.

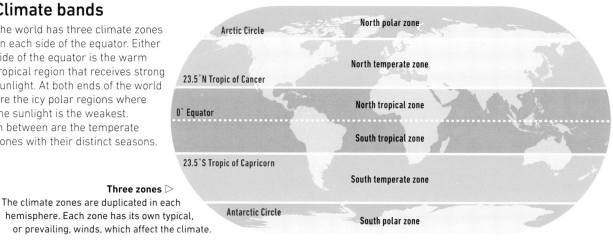

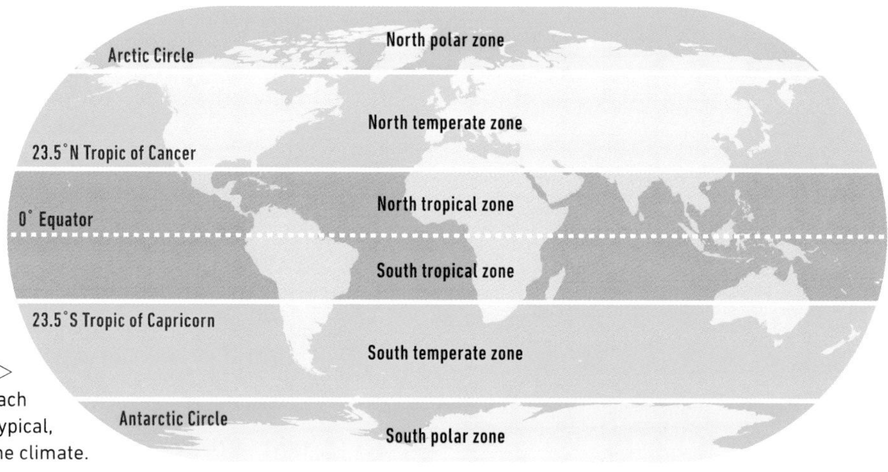
Arctic Circle
North polar zone
North temperate zone
23.5° N Tropic of Cancer
North tropical zone
0° Equator
South tropical zone
23.5° S Tropic of Capricorn
South temperate zone
Antarctic Circle
South polar zone

Three zones ▷
The climate zones are duplicated in each hemisphere. Each zone has its own typical, or prevailing, winds, which affect the climate.

Global climates

The climate of a region depends on three main factors: the climate zone it is in (tropical, temperate, or polar), its distance from the sea, and the side of the continent it is at in relation to the prevailing winds.

KEY

Subarctic
This region has long, cold winters and short, cool to mild summers.

Tundra
The climate is dry with very cold winters and warmish summers.

Ice cap
The climate here is very cold and no month averages above freezing.

Temperate
These regions have cool winters, warm summers, and distinct seasons.

Wet tropical
These regions have an extremely warm and very wet climate.

Continental
The interiors of continents are dry with hot summers and cold winters.

Dry tropical
The climate is warm, with a long dry season and a rainy season.

Mediterranean
These regions have warm, dry summers and mild winters.

Arid
These deserts receive less than 10 in (25.4 cm) of rainfall a year.

Semi-arid
The semi-arid, or steppe, regions have very long and dry summers.

Wet subtropical
This climate has mild winters and warm summers, with plenty of rain.

Monsoon
Regions in south Asia have one dry and one very wet season.

Climatic bands ▷
There are various ways of classifying climates. This map shows a few combinations across the world.

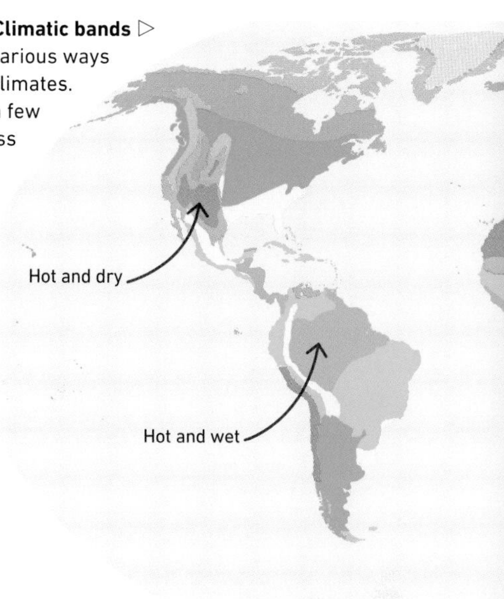
Hot and dry
Hot and wet

Where do deserts form?

Deserts cover about a fifth of the world's land. Many deserts are hot, but one of the largest deserts, Antarctica, is frozen. Big deserts occur in the subtropics where dense, descending air is very dry. Deserts are found either on the west of continents, such as the Namib desert in Africa, or far from the sea, such as the Arabian and Gobi deserts in the Eurasian interior.

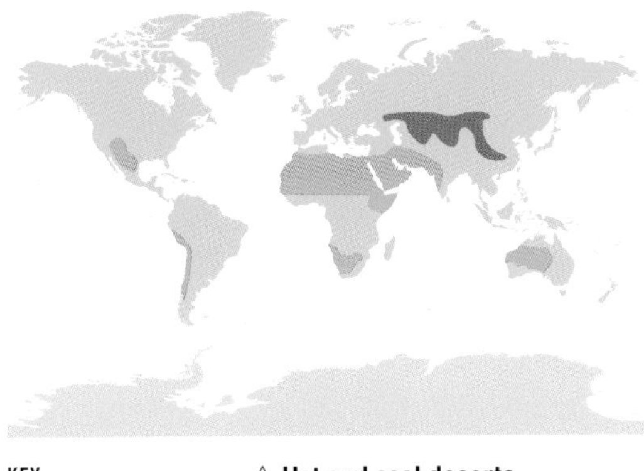

KEY
- Hot desert
- Polar desert
- Temperate desert

△ **Hot and cool deserts**
Deserts in tropical regions, such as the Sahara, are the hottest places on the Earth. Deserts in temperate regions can be bitterly cold in winter.

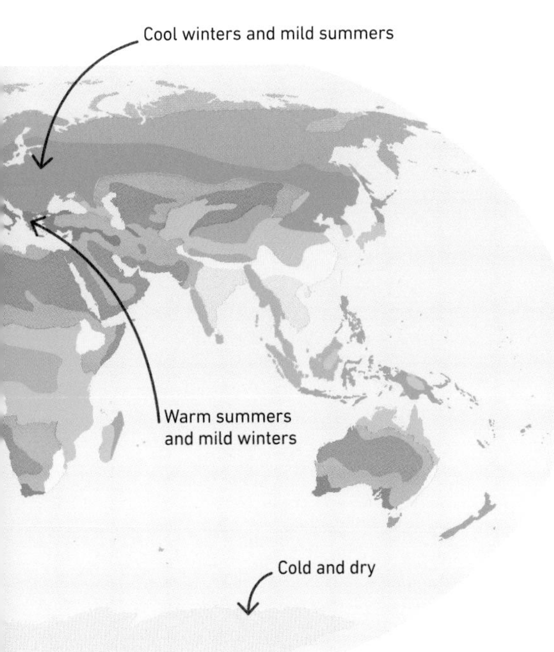

Cool winters and mild summers
Warm summers and mild winters
Cold and dry

Climate graphs

The simplest way to compare climates is to look at the average monthly temperatures and rainfall. Plotting these averages on a graph makes the differences clearer.

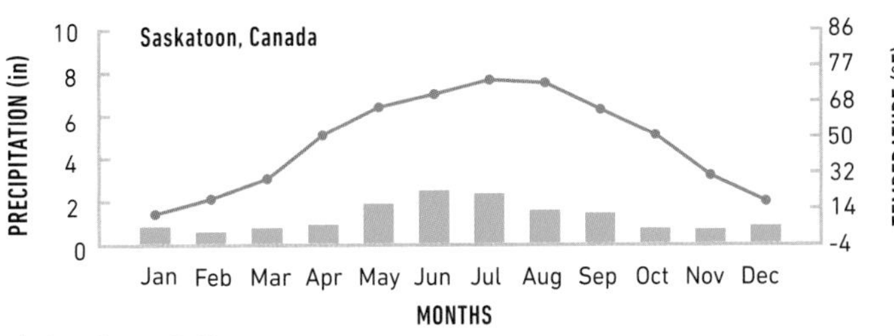

Saskatoon, Canada
PRECIPITATION (in)
10
8
6
4
2
0
TEMPERATURE (°F)
86
77
68
50
32
14
-4
Jan Feb Mar Apr May Jun Jul Aug Sep Oct Nov Dec
MONTHS

△ **Continental climate**
The climate in continental regions is extreme. Average temperatures in winter are much lower than in summer, and rainfall peaks in summer.

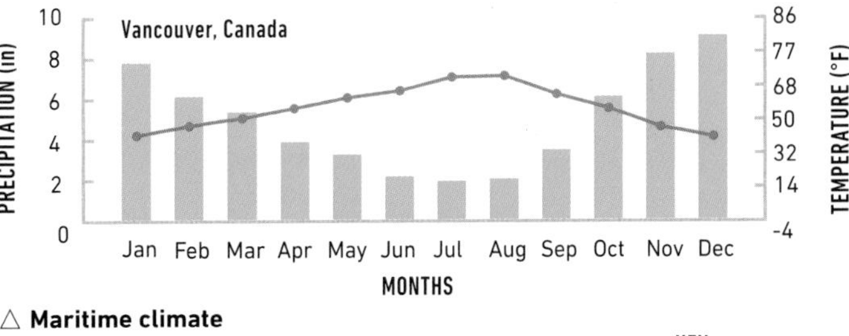

Vancouver, Canada
PRECIPITATION (in)
10
8
6
4
2
0
TEMPERATURE (°F)
86
77
68
50
32
14
-4
Jan Feb Mar Apr May Jun Jul Aug Sep Oct Nov Dec
MONTHS

△ **Maritime climate**
The climate in maritime, or coastal, regions is moderate. Summers are not too hot and winters not too cold. Rainfall peaks in winter while summers are dry.

KEY

- Precipitation
- Temperature

Microclimates

The climates of small areas, such as yards, cities, lakes, valleys, and forests, are called microclimates. Local factors, such as hills that block winds, trees that provide shade, and buildings that trap heat, have a significant effect on the local weather.

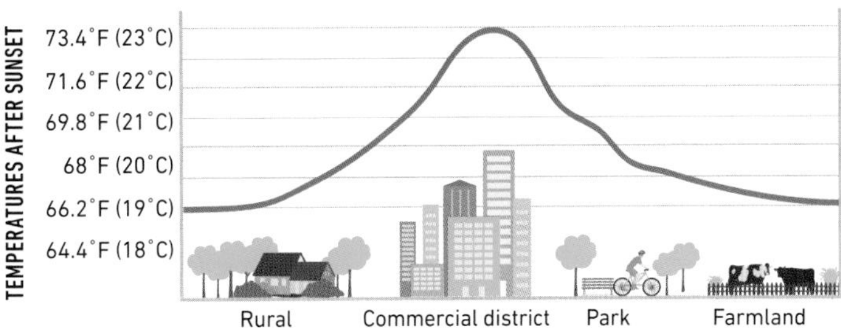

TEMPERATURES AFTER SUNSET
73.4°F (23°C)
71.6°F (22°C)
69.8°F (21°C)
68°F (20°C)
66.2°F (19°C)
64.4°F (18°C)
Rural
Commercial district
Park
Farmland

△ **Urban heat island**
Big cities tend to be significantly warmer than the surrounding countryside, especially at night, because of heat generated by buildings.

The hydrological cycle

WATER MOVES BETWEEN OCEANS AND LANDMASSES IN A NEVER-ENDING CYCLE THAT SUSTAINS LIFE ON EARTH.

SEE ALSO
❮ 58–61 Rivers
Cloud and fog 92–93 ❯
Precipitation 94–95 ❯
Water security 194–195 ❯

All the water on the Earth formed very early in the planet's history and has been circulating constantly between the atmosphere, oceans, and land ever since. This ongoing circulation is called the hydrological or water cycle.

Movement of water

More than 97 percent of the Earth's water is salty and is stored in the oceans. The remaining 3 percent is freshwater and is crucial for life on the land. When the sun's heat evaporates water from lakes, rivers, and oceans, it forms water vapor. This cools and condenses to form clouds from which rain or snow fall, returning the water to oceans, lakes, and rivers.

▽ **From ocean to ocean**
The heat of the sun powers the hydrological cycle. The cycle both starts from and ends at the ocean.

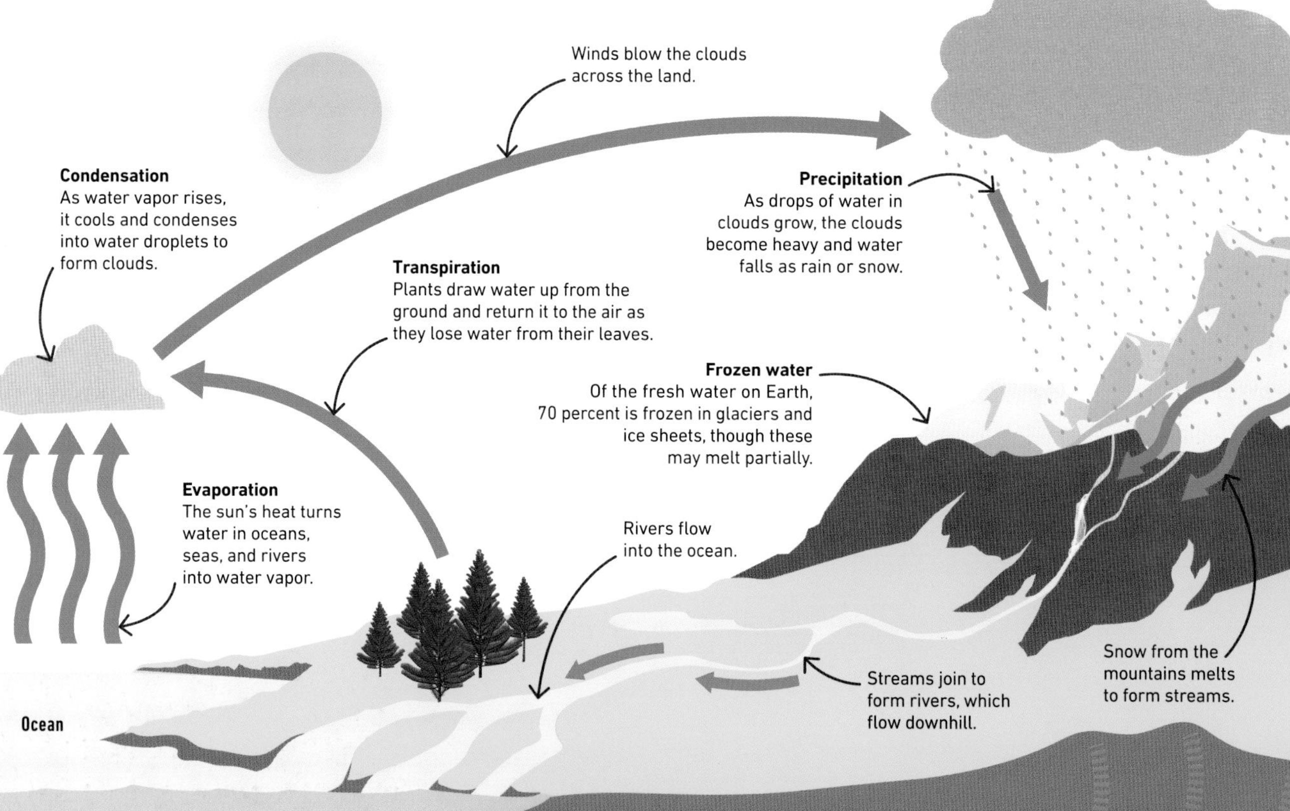
Winds blow the clouds across the land.
Condensation As water vapor rises, it cools and condenses into water droplets to form clouds.
Precipitation As drops of water in clouds grow, the clouds become heavy and water falls as rain or snow.
Transpiration Plants draw water up from the ground and return it to the air as they lose water from their leaves.
Frozen water Of the fresh water on Earth, 70 percent is frozen in glaciers and ice sheets, though these may melt partially.
Evaporation The sun's heat turns water in oceans, seas, and rivers into water vapor.
Rivers flow into the ocean.
Snow from the mountains melts to form streams.
Streams join to form rivers, which flow downhill.
Ocean
Infiltration Some rain and melting snow seep into the ground forming groundwater, which creeps toward the oceans underground.

Water vapor in the air

Humidity is the amount of water vapor that is held in the air. The weight of water vapor in a certain volume of air is its absolute humidity. Warm air holds more water vapor than cold air. Relative humidity, expressed in percentages, is the water vapor that is actually there in the air compared to the maximum amount of water vapor the air can hold at that temperature.

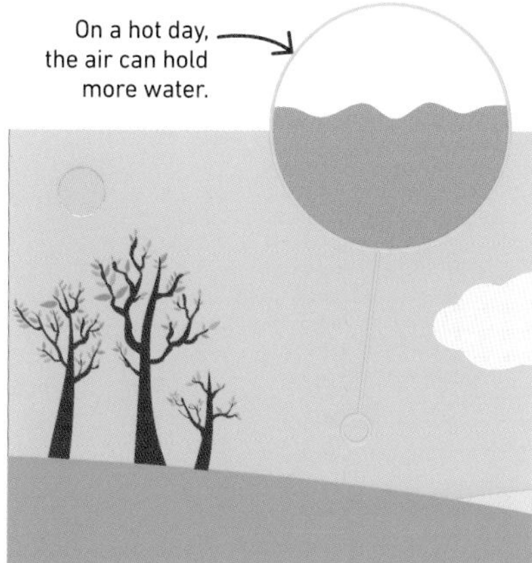
On a hot day, the air can hold more water.

△ **Hot and dry**
Hot summer air may have high absolute humidity, or contain a lot of water vapor, yet appear clear and feel dry because its relative humidity is low.

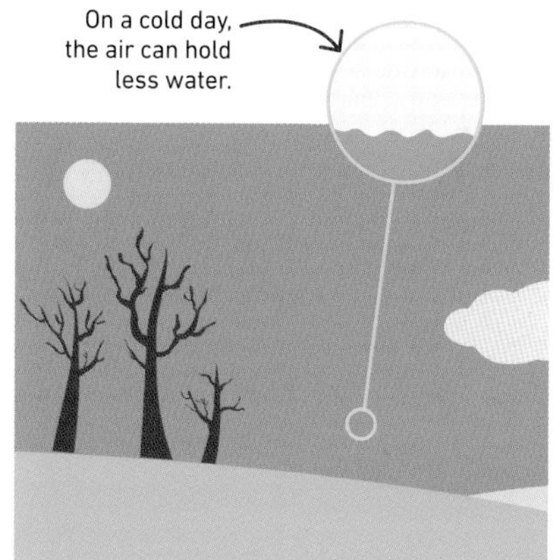
On a cold day, the air can hold less water.

△ **Cold and moist**
Cold winter air may have low absolute humidity, or contain much less water vapor, yet appear misty and feel damp because its relative humidity is high.

Dew point

Air holds water vapor in the space between its molecules. When the space fills up with water, the air is saturated. As air cools, the space contracts, squeezing the water vapor out and finally making it condense into dew, or droplets of water. The temperature at which dew forms is called the dew point.

△ **Dew drops**
When the ground cools at night, it may cool the air above it until water vapor condenses into tiny drops of water, or dew, on surfaces such as leaves.

Human impact on the hydrological cycle

By altering the flow of rivers, humans affect the hydrological cycle. Industry and homes require a large amount of water drawn from rivers, and sometimes the sea. Giant dams built to store water in order to generate hydroelectric power also affect the cycle.

△ **Irrigation**
When demand for food grows, so does the need for more water to irrigate food crops. Excessive irrigation can run rivers dry and damage the chemical composition of the soil.

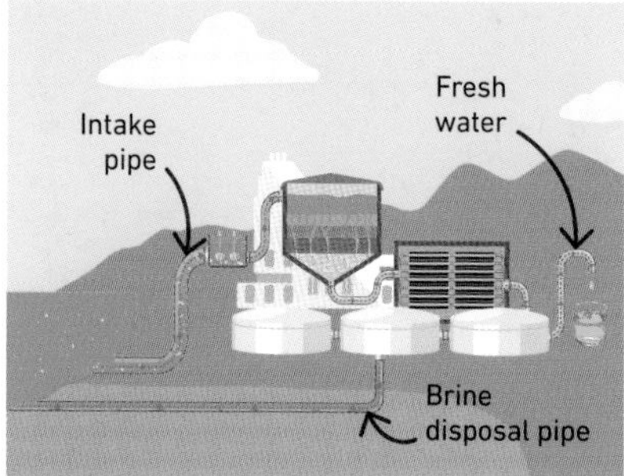
Intake pipe
Fresh water
Brine disposal pipe

△ **Desalination**
Sea water is abundant but is too salty to be used. Desalination takes salt out of sea water but uses a lot of energy and can harm local marine life. The process provides only 1 percent of the current water supply.

△ **Dams and reservoirs**
Built along the course of rivers, dams force water to accumulate in lakes or reservoirs behind them. Dams can trap sediments and lead to the loss of habitats and local species.

Global winds

THE AIR IN THE ATMOSPHERE IS CONSTANTLY ON THE MOVE, CIRCULATING HEAT AROUND THE GLOBE.

SEE ALSO
❮ 74–75 The atmosphere
❮ 78–79 Climate zones
Ocean currents 84–86 ❯
Weather systems 87–89 ❯
Weather forecasting 90–91 ❯

A wind that blows in a particular area and from a predictable direction is called a "prevailing" wind. Prevailing winds carry warm air from the equator to the poles and cold air toward the equator.

Why do winds blow?

The sun warms some places more than others, creating differences in air pressure. Where air is warmer and lighter, it rises and pressure is low. In other places, cool, dense, sinking air creates high pressure. When warm air rises, cooler air moves in to replace it, so wind often moves from colder to warmer areas.

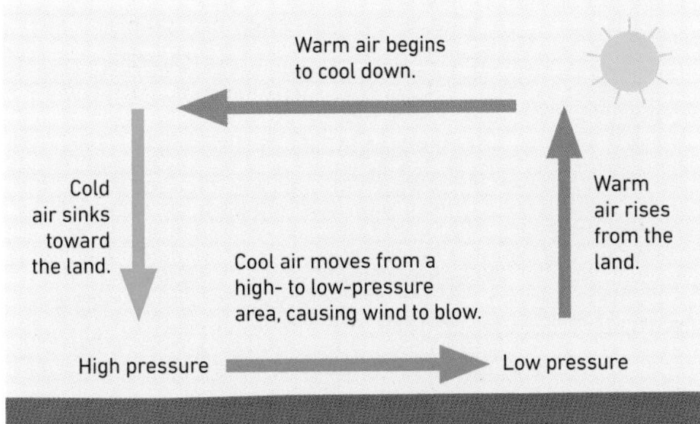
Warm air begins to cool down.
Cold air sinks toward the land.
Warm air rises from the land.
Cool air moves from a high- to low-pressure area, causing wind to blow.
High pressure
Low pressure

△ **Continuous cycle**
Winds circulate continually around the Earth, moving from high-pressure areas to low-pressure areas at ground level. Winds blow wherever there are differences in air temperature.

Global winds

Winds move around the Earth in bands of giant "cells," following a distinctive pattern called global atmospheric circulation. This continually circulates air from the tropics, where the sun is warmest, to the poles, where the sun is weakest. There are three distinct bands of cells in each hemisphere: around the tropics, in the temperate zones, and at the poles.

Trade winds
These winds emerge from the east in the tropics, blowing westward. They blow near the surface, in the lower parts of the atmosphere.

Prevailing westerlies
These originate in mid-latitudes and blow from west to east, toward the poles. They are also called antitrades because they blow in the opposite direction to trade winds.

Polar easterlies
These cold, dry winds emerge from the polar regions and blow westward.

The Coriolis effect

Winds never flow directly from areas of high pressure to areas of low pressure. Instead, the spinning motion of the Earth makes them curve around the equator. In the northern hemisphere, winds blowing south are bent to the west; and in the southern hemisphere, winds blowing north are also bent to the west. This is called the Coriolis effect.

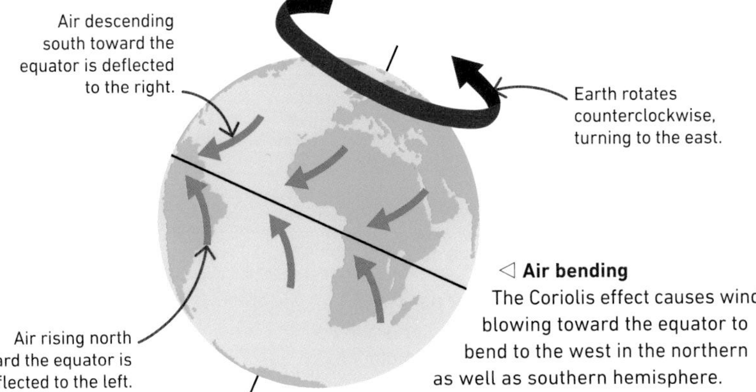
Air descending south toward the equator is deflected to the right.
Earth rotates counterclockwise, turning to the east.
Air rising north toward the equator is deflected to the left.

◁ **Air bending**
The Coriolis effect causes winds blowing toward the equator to bend to the west in the northern as well as southern hemisphere.

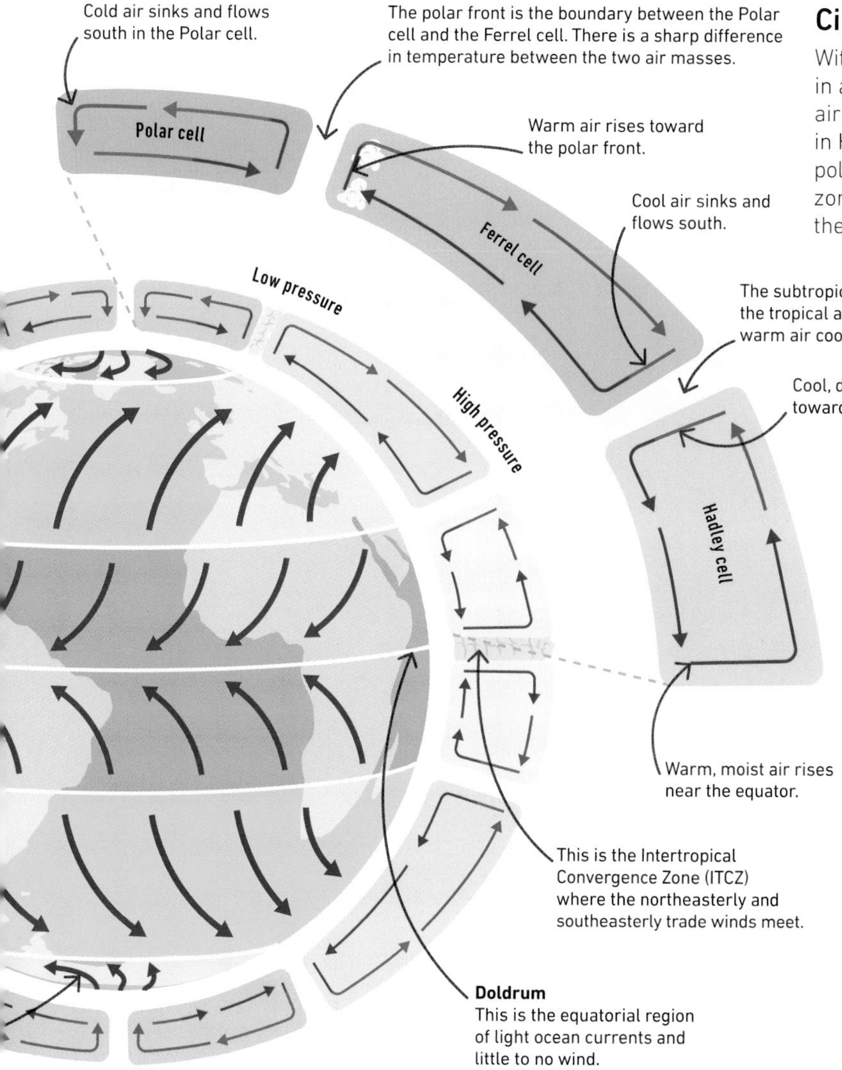

Cold air sinks and flows south in the Polar cell.

The polar front is the boundary between the Polar cell and the Ferrel cell. There is a sharp difference in temperature between the two air masses.

Polar cell

Warm air rises toward the polar front.

Cool air sinks and flows south.

Ferrel cell

Low pressure

High pressure

The subtropical zone is the region between the tropical and the temperate zones, where warm air cools down and starts sinking.

Cool, dry air sinks toward the equator.

Hadley cell

Warm, moist air rises near the equator.

This is the Intertropical Convergence Zone (ITCZ) where the northeasterly and southeasterly trade winds meet.

Doldrum
This is the equatorial region of light ocean currents and little to no wind.

Circulation cells

Within each of the bands of wind, the air circulates in a vertical circulation "cell." In the tropics, warm air rising near the equator flows north and south in Hadley cells. A similar circulation occurs in the polar regions, forming Polar cells. In the temperate zones, rising mid-latitude air divides, flowing to the poles and to the equator, inside Ferrel cells.

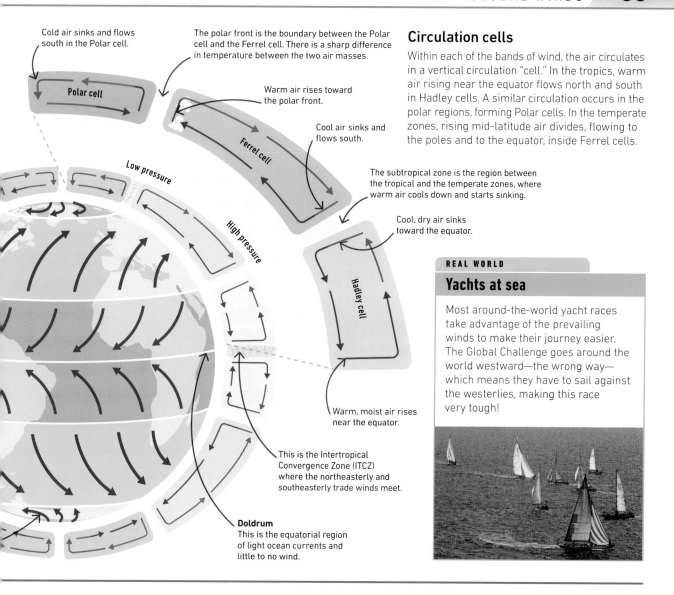

REAL WORLD

Yachts at sea

Most around-the-world yacht races take advantage of the prevailing winds to make their journey easier. The Global Challenge goes around the world westward—the wrong way—which means they have to sail against the westerlies, making this race very tough!

Major wind patterns

The world is split into three main zones or bands of prevailing winds that blow on either side of the equator. The three wind zones are linked to each of the three circulation cells. In each zone, the prevailing wind, the wind that blows most of the time, is the ground level wind created by each cell.

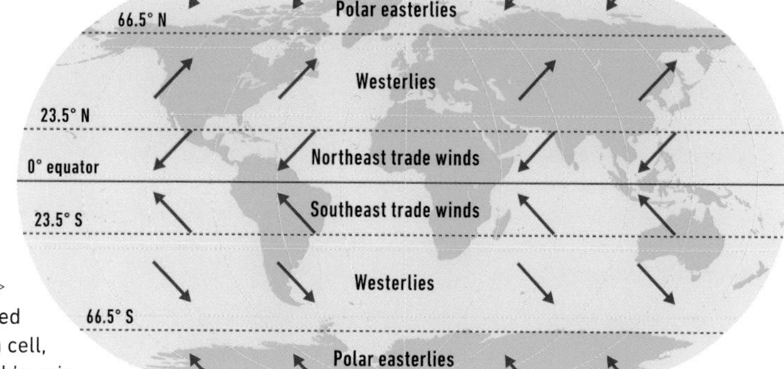

66.5° N
Polar easterlies
Westerlies
23.5° N
0° equator
Northeast trade winds
Southeast trade winds
23.5° S
Westerlies
66.5° S
Polar easterlies

Winds of the world ▷
In the three wind zones, air is moved north or south by the circulation cell, but bent east or west by the Earth's spin.

Ocean currents

VAST STREAMS OF WATER THAT FLOW THROUGH THE WORLD'S OCEANS ARE KNOWN AS THE OCEAN CURRENTS.

SEE ALSO
❮ 78–79 Climate zones
Oceans and seas map 112–113 ❯
Oceans, seas, and lakes 114–115 ❯
Continents and oceans 202–203 ❯

Ocean currents are huge masses of constantly moving water, both at the surface and in the depths of the oceans, caused by the need for water of different temperatures and densities to mix. The Earth's rotation also affects the pattern of ocean currents.

Cold deepwater currents and **warm surface currents** work together to **distribute heat** around the globe.

Surface currents and gyres

The movement of water at the ocean's surface is called a surface current. Warm water always mixes with cooler water, forming convection currents, and in the oceans, this happens on a global scale. Gyres flow around the major oceans and carry warm water toward the poles and cold water toward the tropics, on either side of the equator.

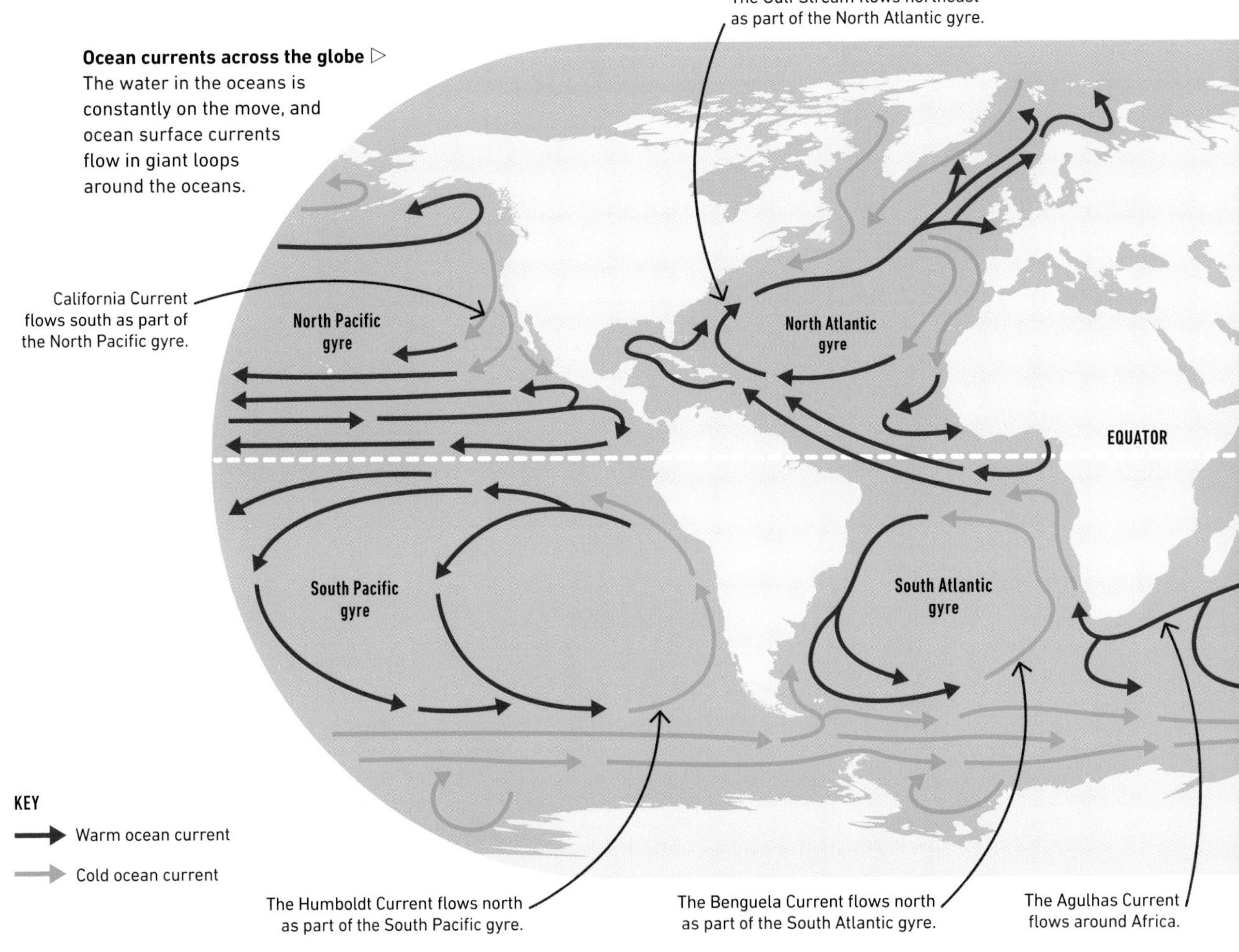
The Gulf Stream flows northeast as part of the North Atlantic gyre.
California Current flows south as part of the North Pacific gyre.
North Pacific gyre
North Atlantic gyre
EQUATOR
South Pacific gyre
South Atlantic gyre
KEY
Warm ocean current
Cold ocean current
The Humboldt Current flows north as part of the South Pacific gyre.
The Benguela Current flows north as part of the South Atlantic gyre.
The Agulhas Current flows around Africa.

Ocean currents across the globe ▷ The water in the oceans is constantly on the move, and ocean surface currents flow in giant loops around the oceans.

The movement of gyres

A combination of factors such as prevailing winds, the shape of oceans and continents, and the Earth's rotation makes surface currents flow clockwise in the northern hemisphere and counterclockwise in the southern hemisphere, creating vast, circular currents known as gyres. Trade winds can play a part in guiding currents, for example, in the northern hemisphere, toward the west in the tropics and then toward the east in temperate latitudes.

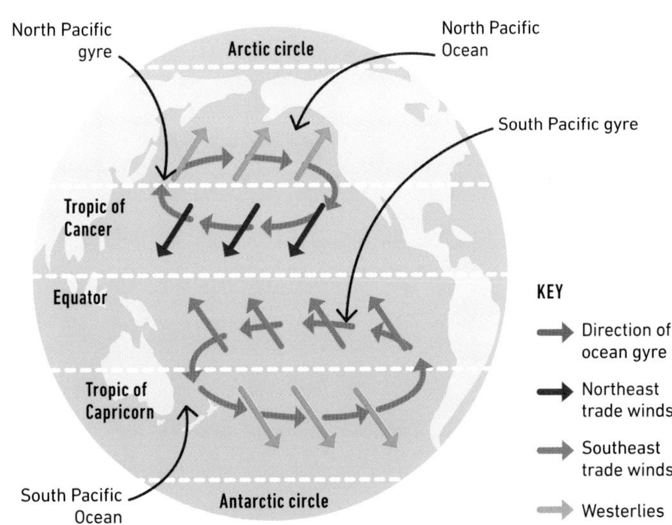
North Pacific gyre
Arctic circle
North Pacific Ocean
South Pacific gyre
Tropic of Cancer
Equator
Tropic of Capricorn
South Pacific Ocean
Antarctic circle
KEY
Direction of ocean gyre
Northeast trade winds
Southeast trade winds
Westerlies

Effects of the Earth's rotation ▷ Winds and currents are deflected in their paths through the air and the ocean by the spinning motion of the Earth, which causes their distinctive patterns.

The Kuroshio Current is part of the North Pacific gyre.

South Indian gyre

The Antarctic Circumpolar Current flows around Antarctica.

Boundary currents

The currents at the edges of gyres are called boundary currents. These currents link the westward and eastward flows of the gyres. Boundary currents on the western side of gyres are fast, deep, and narrow and move warm water to cooler regions, while eastern boundary currents are slow, shallow, and broad and move cold water toward the tropics.

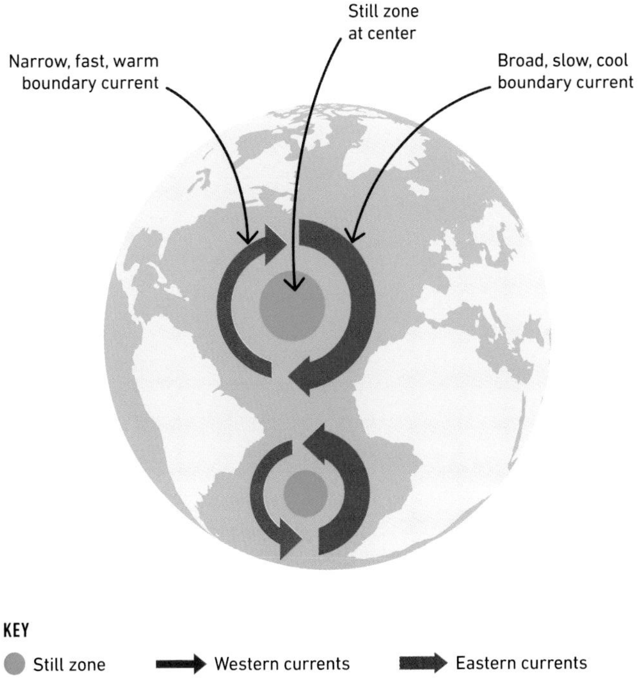
Still zone at center
Narrow, fast, warm boundary current
Broad, slow, cool boundary current
KEY
Still zone
Western currents
Eastern currents

Deepwater currents

In cold oceans in higher latitudes, the seawater beneath the ice is more salty. The salt and icy cold makes the water below the ice denser and heavier, so it sinks toward the ocean floor. This cold, dense water then flows beneath warmer, less-dense water as deepwater currents, eventually warming and coming back to the surface. Deepwater and surface currents carry ocean water all around the world as though on a giant conveyor belt.

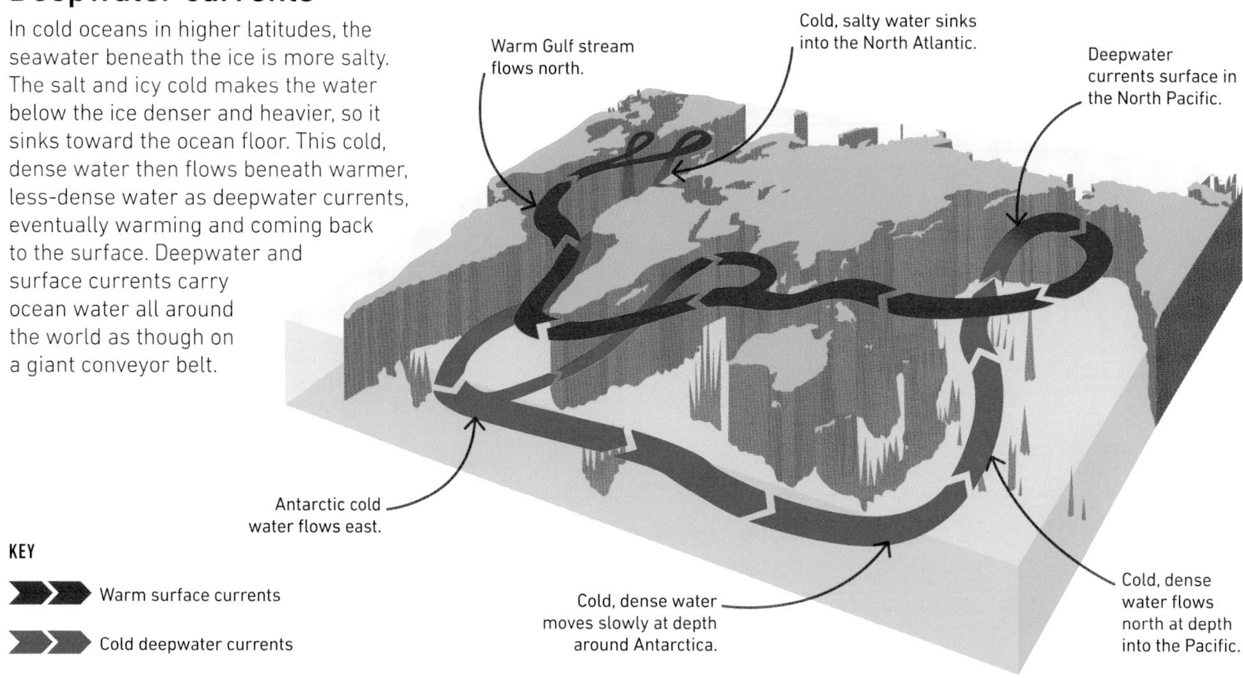
Warm Gulf stream flows north.
Cold, salty water sinks into the North Atlantic.
Deepwater currents surface in the North Pacific.
Antarctic cold water flows east.
Cold, dense water moves slowly at depth around Antarctica.
Cold, dense water flows north at depth into the Pacific.
KEY
Warm surface currents
Cold deepwater currents

Coastal upwelling and downwelling

When winds push surface water away from coasts, deeper water rises up to replace it in a process called "upwelling." A reverse process, called "downwelling," also occurs when winds cause surface water to build up against a coastline so that the surface water is forced to sink and replace the deeper water.

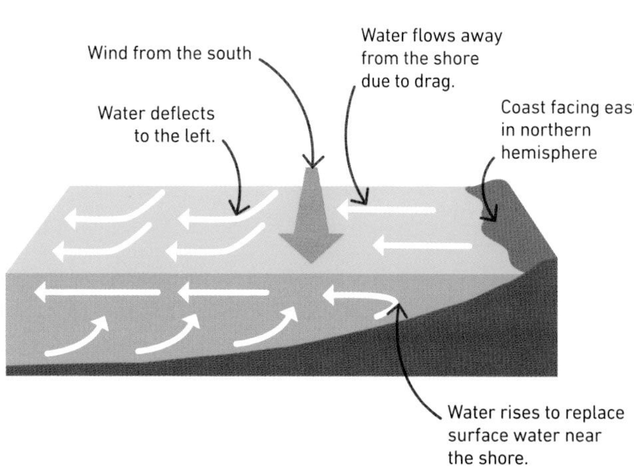
Wind from the south
Water flows away from the shore due to drag.
Water deflects to the left.
Coast facing east in northern hemisphere
Water rises to replace surface water near the shore.

△ **Coastal upwelling**
This process brings up deeper, cool water that is rich in nutrients, making the area where this happens very good for fishing.

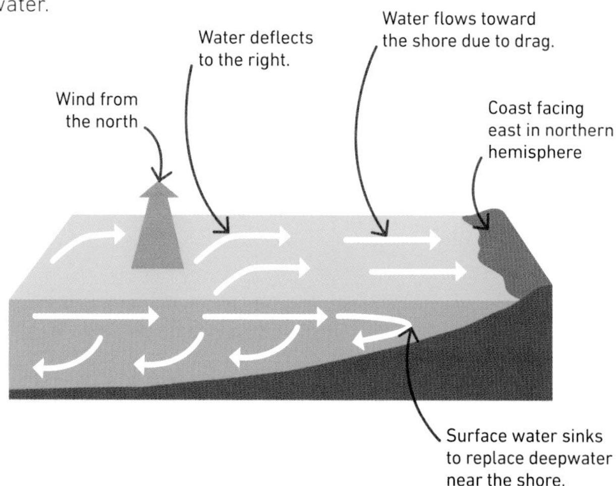
Water deflects to the right.
Water flows toward the shore due to drag.
Wind from the north
Coast facing east in northern hemisphere
Surface water sinks to replace deepwater near the shore.

△ **Coastal downwelling**
Salty water, rich in oxygen, is taken down into the ocean's depths, which helps maintain its chemical balance, in a process called downwelling.

Weather systems

LOCAL WEATHER CONDITIONS ARE DETERMINED BY GLOBAL PATTERNS OF AIR CIRCULATION CALLED WEATHER SYSTEMS.

SEE ALSO
❮ 78–79 Climate zones
❮ 82–83 Global winds
❮ 84–86 Ocean currents
Weather forecasting 90–91 ❯

Weather is the day-to-day change in the atmosphere, and includes wind strength and direction, precipitation, humidity and cloud cover, sunshine hours, and temperature. It is driven by air masses and giant patterns of air circulation called weather systems.

Air masses

Air masses are vast volumes of the atmosphere where the air is uniformly wet or dry, cold or warm. They form when air stays long enough over one surface, such as the sea, to take on its humidity and temperature. Far inland, an air mass can sit for many weeks, becoming very dry (and in winter, very cold). Unsettled, stormy weather occurs in the zones where two air masses meet, called fronts.

▽ **Source regions of air masses**
The weather an air mass brings is determined mostly by the conditions where it developed. This point of origin is known as the source region.

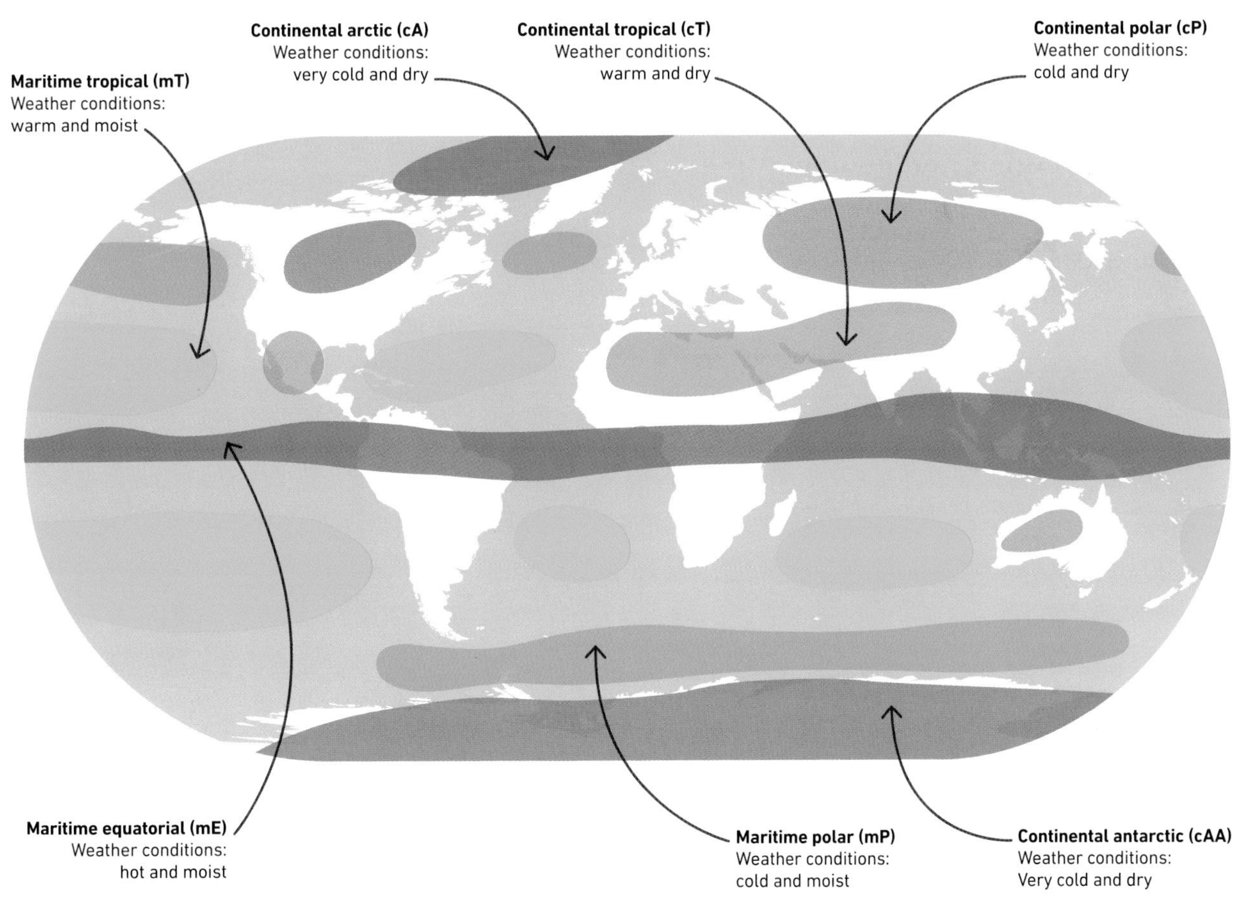
Maritime tropical (mT)
Weather conditions: warm and moist
Continental arctic (cA)
Weather conditions: very cold and dry
Continental tropical (cT)
Weather conditions: warm and dry
Continental polar (cP)
Weather conditions: cold and dry
Maritime equatorial (mE)
Weather conditions: hot and moist
Maritime polar (mP)
Weather conditions: cold and moist
Continental antarctic (cAA)
Weather conditions: Very cold and dry

Highs and lows

An anticyclone, or high, is a huge, high-pressure zone of stable air, which gently rotates and sinks. As it descends, the air begins to compress and warms, reducing humidity and preventing clouds from forming. A cyclone, or low or depression, is a low-pressure zone that forms in the unsettled zone where two air masses meet and rise. Air is pushed upward from the ground, causing it to expand and cool. As it rises, it becomes more humid, causing clouds to form.

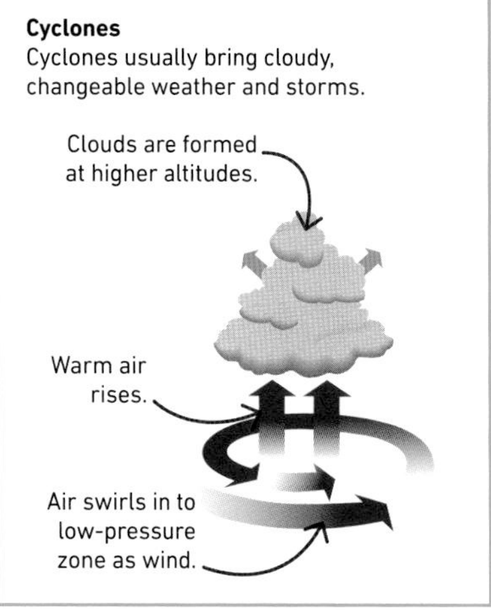
Cyclones
Cyclones usually bring cloudy, changeable weather and storms.
Clouds are formed at higher altitudes.
Warm air rises.
Air swirls in to low-pressure zone as wind.

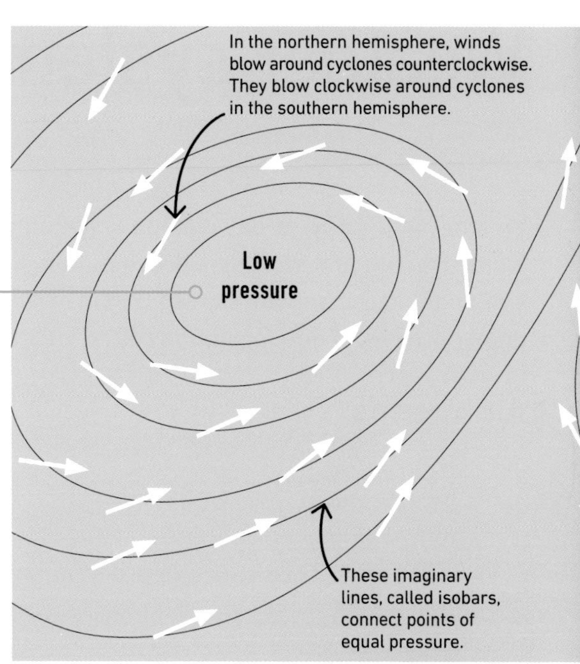
In the northern hemisphere, winds blow around cyclones counterclockwise. They blow clockwise around cyclones in the southern hemisphere.
Low pressure
These imaginary lines, called isobars, connect points of equal pressure.

Spiraling winds ▷
Winds blow from areas of high pressure to areas of low pressure, though the Earth's rotation causes them to blow in a spiral rather than straight from high to low.

Fronts

A front is the boundary between two air masses. In temperate latitudes, the worst weather is caused by lows that form along the polar front, where warm, wet air meets cold, polar air. Groups of lows are blown eastward by the prevailing westerly winds, bringing a distinct sequence of stormy weather as they pass over.

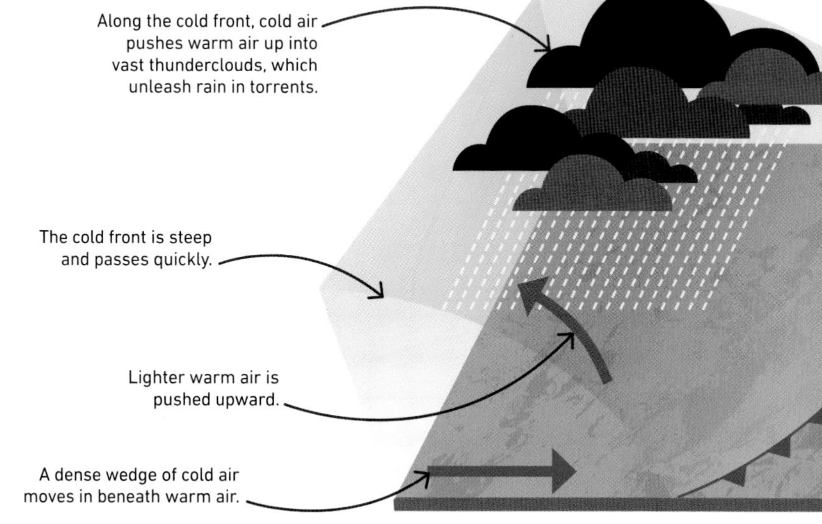
Along the cold front, cold air pushes warm air up into vast thunderclouds, which unleash rain in torrents.
The cold front is steep and passes quickly.
Lighter warm air is pushed upward.
A dense wedge of cold air moves in beneath warm air.

KEY

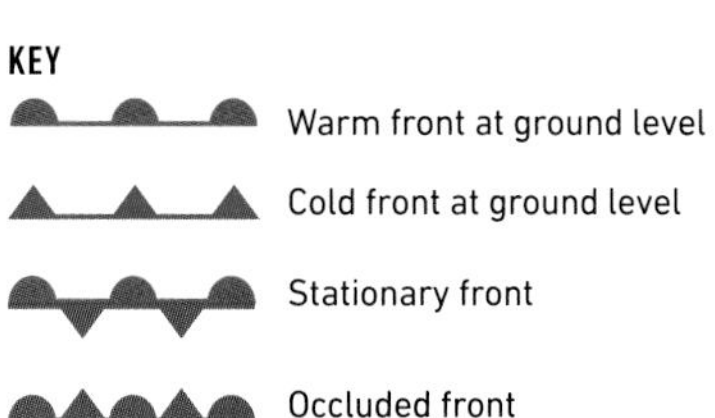
Warm front at ground level
Cold front at ground level
Stationary front
Occluded front

Developing frontal weather ▷
As a frontal storm comes in from the west, you experience first the cloudy, often rainy warm front, heralded by high cirrus clouds. Then comes a lull, and then the stormier cold front. So imagine you're on the ground at the right of this sequence as it passes over you, moving left to right.

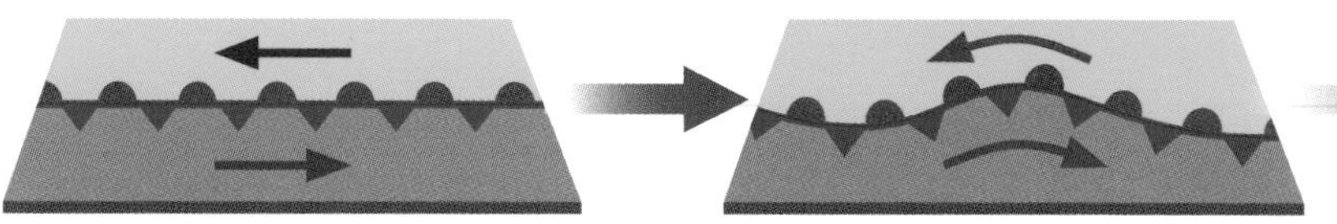

1 Air masses collide
The polar front develops where cold polar and warm tropical air masses collide. On one side, the cold air flows eastward and on the other, the warm air flows west.

2 The depression begins
The polar front develops a small wave, as warm air slides up over the cold air, creating the beginnings of a low-pressure zone or depression.

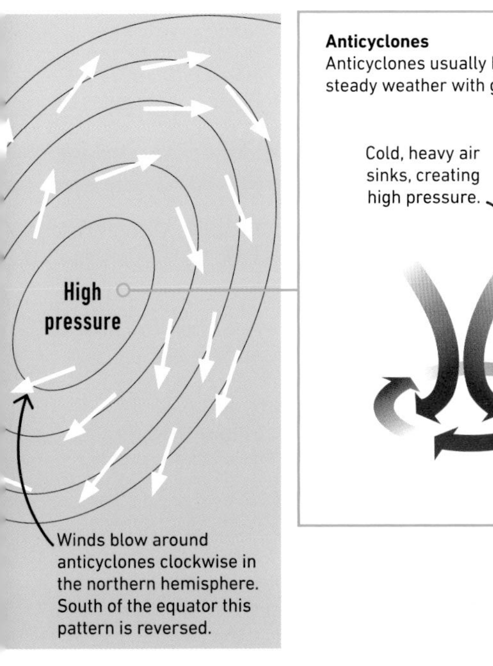
High pressure
Winds blow around anticyclones clockwise in the northern hemisphere. South of the equator this pattern is reversed.

Anticyclones
Anticyclones usually bring clear, steady weather with gentle winds.

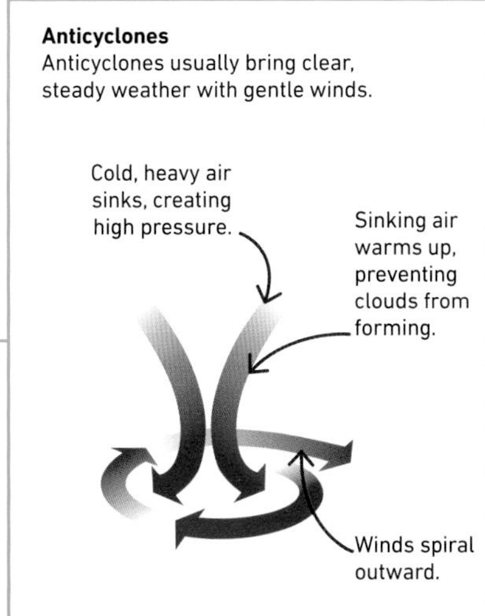
Cold, heavy air sinks, creating high pressure.
Sinking air warms up, preventing clouds from forming.
Winds spiral outward.

Jet streams

A jet stream is a narrow belt of high-altitude winds that blow at up to 230 mph (370 km/h). There are four main jets, blowing from the west at different latitudes. The most northerly, the polar jet, flows along the polar front, and plays a key role in the movement of frontal storms. Planes flying east across the Atlantic can take advantage of the polar jet, which can cut flight times considerably.

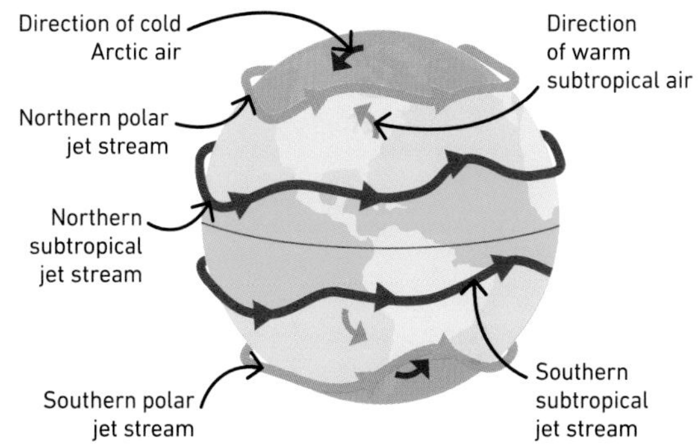
Direction of cold Arctic air
Direction of warm subtropical air
Northern polar jet stream
Northern subtropical jet stream
Southern polar jet stream
Southern subtropical jet stream

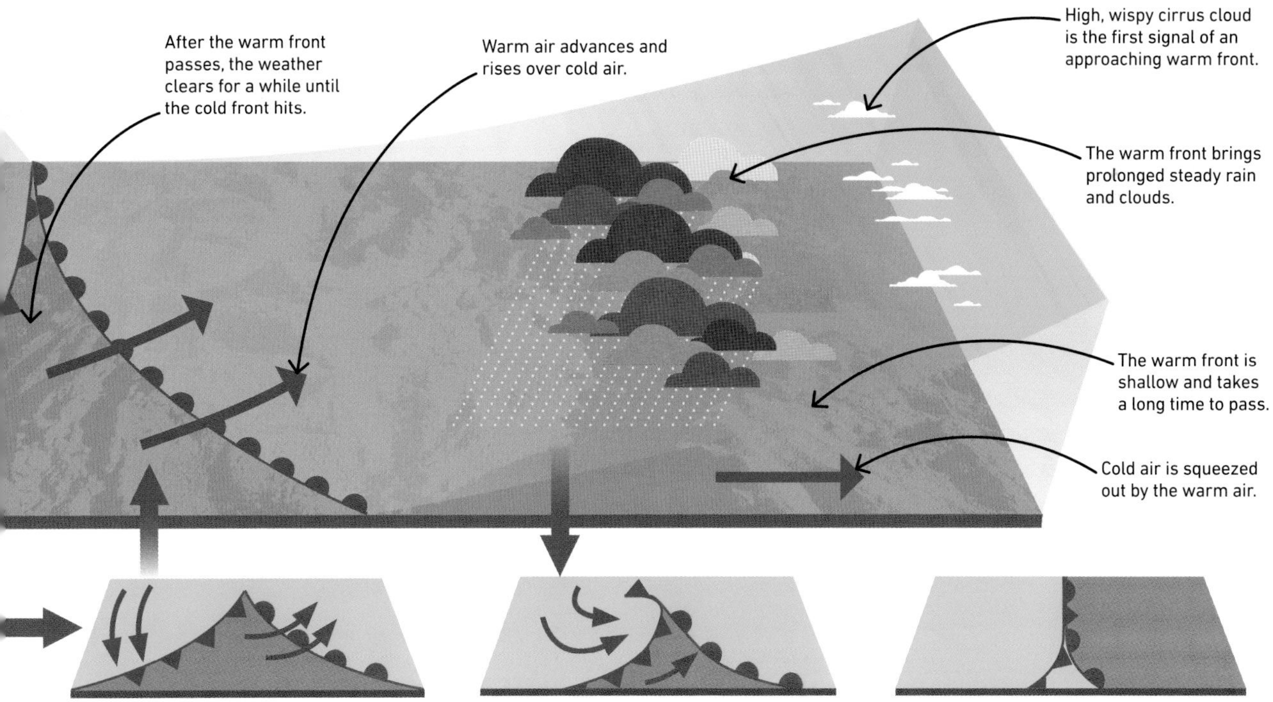

After the warm front passes, the weather clears for a while until the cold front hits.

Warm air advances and rises over cold air.

High, wispy cirrus cloud is the first signal of an approaching warm front.

The warm front brings prolonged steady rain and clouds.

The warm front is shallow and takes a long time to pass.

Cold air is squeezed out by the warm air.

3 Two fronts develop
The wave sharpens into a v-shaped notch, with two fronts—a warm front on the leading edge, and a cold front on the trailing edge. This brings stormy weather.

4 Fronts closing
The cold front moves faster than the warm front, and begins to catch up. The storm begins to lose some of its power and the depression weakens.

5 Occluded front
Eventually, the cold front can catch up entirely with the warm front, merging with it and lifting it right off the ground to form a single front called an occlusion.

Weather forecasting

PREDICTING THE WEATHER REQUIRES THE USE OF SCIENTIFIC MODELS AND THE INTERPRETATION OF DATA.

SEE ALSO
❮ 78–79 Climate zones
❮ 82–83 Global winds
❮ 87–89 Weather systems
Hurricanes and tornadoes 96–97 ❯

Weather forecasts don't just let us know whether we need sunscreen or an umbrella. They also help businesses such as farms and airlines plan, and they provide vital early warnings of oncoming storms.

Weather watching

Meteorologists (weather scientists) forecast the weather by making continuous observations of atmospheric conditions. By comparing this data with past patterns, they can forecast weather conditions in the near future. They are helped by powerful computers, which can handle huge amounts of data and make weather forecasting more accurate.

Temperature
Thermometers monitor temperature. They need to be kept in the shade and out of the wind.

Humidity
The air's moisture content is measured with a thermometer-like instrument called a hygrometer.

Pressure
Barometers are used to measure changes in air pressure caused by the weather.

Wind
Wind strength is measured with instruments called anemometers. Locally, weather vanes show wind direction.

Rainfall
Rainfall is measured by the depth of water collected in a container called a rain gauge.

Sharing data

Accurate forecasts rely on weather data being shared instantaneously around the world. Every minute of the day, weather observations from more than 10,000 data sources are swapped via the Global Telecommunications System (GTS) and analyzed at major weather centers worldwide.

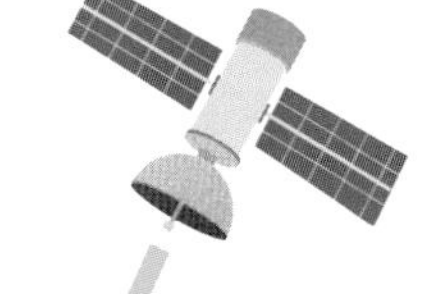

◁ **Weather satellites**
As well as monitoring cloud cover, satellites can measure temperatures, wind speeds, and water vapor in the atmosphere with great accuracy.

Global Telecommunications System (GTS)

◁ **Ships**
Out at sea, ships provide valuable data before weather hits land.

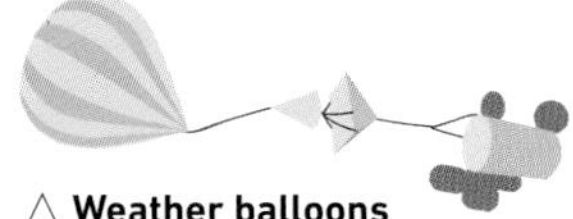

△ **Weather balloons**
Helium-filled balloons (radiosondes) are launched into the atmosphere and transmit the weather data they collect by radio.

◁ **Weather buoys**
Tethered buoys at sea gather data and transmit it by radio or satellite.

△ **Aircraft**
Specially modified research aircraft provide a close-up view of weather conditions. They can monitor storms at close range.

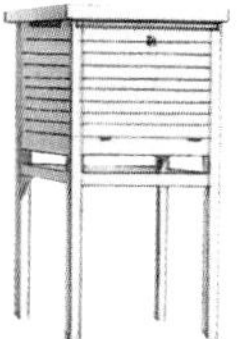

◁ **Stevenson screens**
These slatted boxes protect the weather instruments inside, such as thermometers, from wind, sun, and rain.

◁ **Automated stations**
In remote or extreme places, automated weather stations are set up to record conditions. They send the data via radio links.

What is a synoptic chart?

Weather maps show weather data using a system of lines and symbols. The most detailed and useful maps are called synoptic charts. "Synoptic" means "seen together," and these charts are based on observations of weather conditions all made at the same time. In practice, there will be variations in the times that data is observed, so it is computer-adjusted to become synoptic.

Isobars
These long, curving lines indicate places where the air pressure is the same. Measurements are given in millibars (mb). The closer the isobars are spaced together, the stronger the winds.

High-pressure zone
Rings of isobars with the highest pressure near the center mark a high-pressure zone or anticyclone. The weather is usually clear and sunny here.

Low-pressure zone
Rings of isobars with the lowest pressure near the center mark a low-pressure zone, cyclone, or depression. The weather here is usually dull and rainy, with strong winds.

Occluded front
A cold front joins a warm front and lifts warm air off the ground. This can produce heavy rain.

Warm front
Red semicircles on the isobar indicate a warm front. As the front approaches, the weather gets cloudy and rainy and stays dull as it passes.

Cold front
Blue triangles on the isobar indicate a cold front. The weather is cloudy along the front but clears when it passes.

Synoptic symbols

Symbols give local details about types of precipitation, the amount of cloud in the sky, and the wind strength and direction. Cloud cover is shown by how many eighths of the sky (oktas) are covered by cloud. Wind arrows show the wind direction by the way the arrow is pointing, and the wind speed is shown by the "tails" on the arrow.

PRECIPITATION	CLOUD COVER	WIND SPEED
Drizzle	Clear sky	Calm
Rain	One okta	1–2 knots
Heavy rain	Two oktas	5 knots
Snow	Three oktas	10 knots
Mist	Four oktas	15 knots
Fog	Five oktas	20 knots
Thunderstorm	Six oktas	28–32 knots
	Seven oktas	50 knots or more
	Eight oktas	
	Sky obscured	

Cloud and fog

AS PART OF THE WATER CYCLE, WATER VAPOR CAN CONDENSE IN THE AIR TO FORM CLOUDS. THERE ARE MANY DIFFERENT TYPES OF CLOUD.

SEE ALSO
❮ 80–81 The hydrological cycle
❮ 87–89 Weather systems
❮ 90–91 Weather forecasting
Precipitation 94–95 ❯

Clouds form when warm air rises and cools, turning into drops of water or tiny ice crystals that float in the air. Rain falls from clouds when the drops become too big and heavy to float.

Types of cloud

Different types of cloud form at different altitudes. There are three main kinds—cirrus, cumulus, and stratus. Cirrus are wispy clouds made of ice crystals. Cumulus are fluffy, heaped clouds that pile up as warm air rises. Stratus are layered clouds that form when layers of air cool to condensation point.

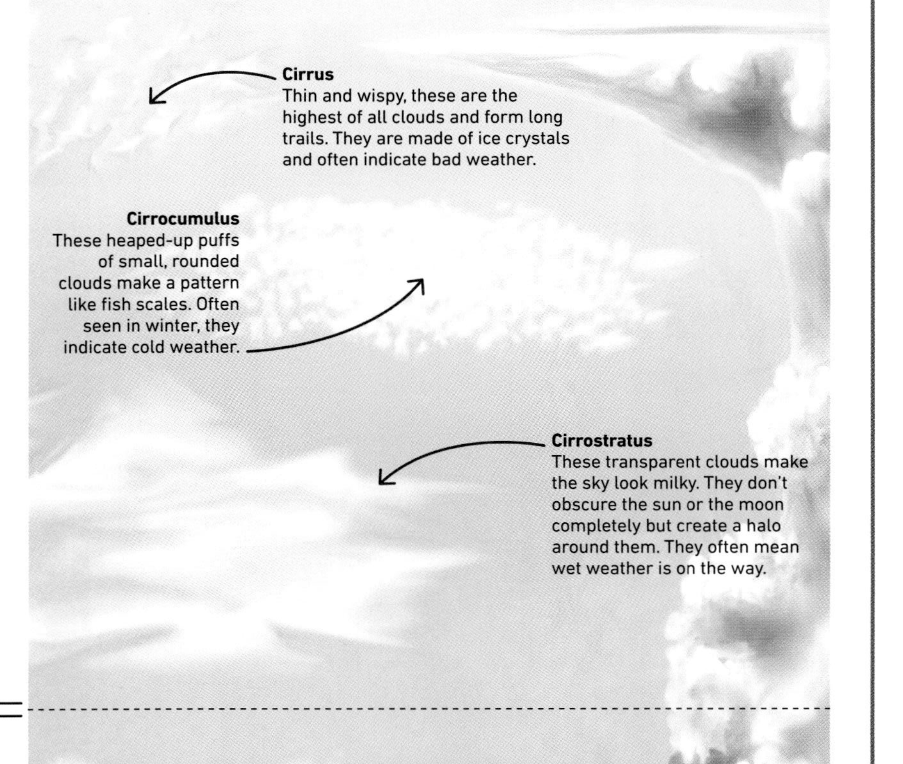
High level—above 20,000 ft (6,000 m)
Cirrus
Thin and wispy, these are the highest of all clouds and form long trails. They are made of ice crystals and often indicate bad weather.
Cirrocumulus
These heaped-up puffs of small, rounded clouds make a pattern like fish scales. Often seen in winter, they indicate cold weather.
Cirrostratus
These transparent clouds make the sky look milky. They don't obscure the sun or the moon completely but create a halo around them. They often mean wet weather is on the way.

How a convection cloud forms

Most clouds form when moist air rises high enough to cool and condense to form water drops. This uplift of air can be either convectional, when warm air rises; orographic, when winds are forced up by mountains; or frontal, when one air front meets another and air is forced up as a result.

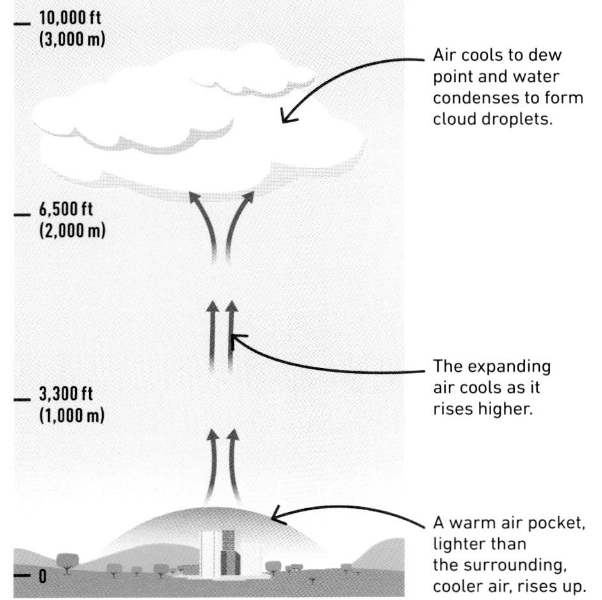
10,000 ft (3,000 m)
6,500 ft (2,000 m)
3,300 ft (1,000 m)
0
Air cools to dew point and water condenses to form cloud droplets.
The expanding air cools as it rises higher.
A warm air pocket, lighter than the surrounding, cooler air, rises up.

△ **Convection clouds**
When warm ground creates a bubble of warm air that rises, cools, and condenses, convection clouds form.

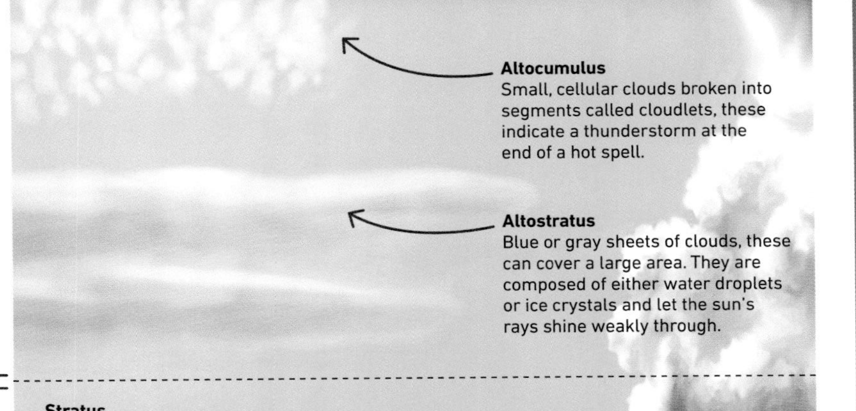
Middle level—6,500–20,000 ft (2,000–6,000 m)
Low level—0–6,500 ft (0–2,000 m)
Altocumulus
Small, cellular clouds broken into segments called cloudlets, these indicate a thunderstorm at the end of a hot spell.
Altostratus
Blue or gray sheets of clouds, these can cover a large area. They are composed of either water droplets or ice crystals and let the sun's rays shine weakly through.
Stratus
Sheets of layered clouds, these can cover the whole sky and make it look gray. Sometimes they sit right on the ground and form fog.
Stratocumulus
Lumpy layers of gray or white rounded clouds, these are the most common clouds. They can produce drizzle or light rain.
Cumulus
Fluffy heaps of cottony clouds, these are seen on sunny days. They form when rising warm-air bubbles, called thermals, cool down.
Nimbostratus
Dark and shapeless, these layered clouds blot out the sun completely, resulting in dull days. They cause prolonged heavy rain.
Cumulonimbus
Towering over 20,000 ft (6,000 m), these giant versions of cumulus clouds develop in deeply unstable conditions and cause heavy rain and thunderstorms.

Fog

Fog and mist are clouds that form at ground level as the moisture in the air condenses into tiny drops of water. Fog reduces visibility, making it harder to see clearly into the distance. For airlines, fog is declared when visibility is less than 3,300 ft (1,000 m). On the ground, fog becomes problematic when it reduces visibility to 650 ft (200 m).

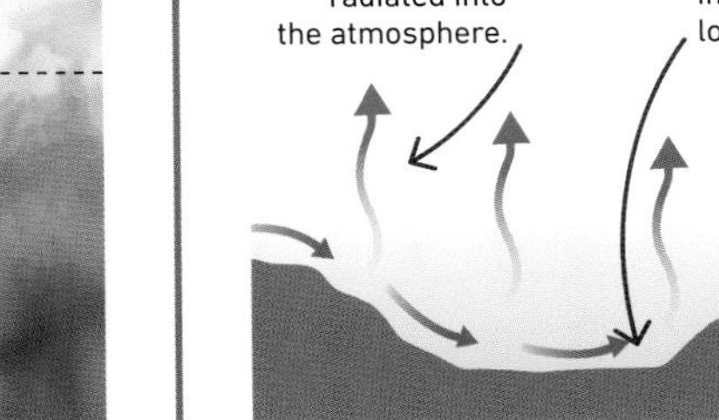
Warmth is radiated into the atmosphere.
Cool air sinks and forms fog in valleys and lowland areas.
Warm, moist air flows over cold water.
Water in the air condenses into fog.

◁ **Radiation fog**
On clear, cold nights when the ground cools the air above, moisture condenses, forming radiation fog. This is common in the fall. Cold air drains to form fog pockets in valleys.

◁ **Advection fog**
When warm, moist air flows over a cool surface (land or water), the air cools and the moisture in it condenses, forming advection fog.

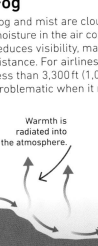

REAL WORLD

Cloud seeding experiments

When there isn't enough rain, authorities may try cloud seeding. This technique involves a plane releasing silver iodide powder into the air. Moisture from the air condenses on the powder, forming clouds and, eventually, rain. Cloud seeding experiments have had mixed success.

Precipitation

SEE ALSO
❮ 80–81 The hydrological cycle
❮ 87–89 Weather systems
❮ 90–91 Weather forecasting
❮ 92–93 Cloud and fog

WATER VAPOR IN THE AIR FALLS AS PRECIPITATION. COMMON FORMS INCLUDE RAIN, SNOW, AND DEW.

The processes of evaporation and condensation in the water cycle give rise to precipitation. This is in the form of water droplets or ice crystals, which fall to the ground or condense on surfaces.

Types of precipitation

Precipitation can be of different types, depending on the air conditions. Rain, drizzle, snow, sleet, hail, frost, and dew are all types of precipitation.

Drizzle
Fine rain with drops under 0.02 in (0.5 mm) fall from stratus clouds as drizzle.

Rain
Rain consists of spherical drops of water bigger than 0.02 in (0.5 mm) across.

Sleet
This is a mixture of wet snow and rain. It forms when snow partially melts.

Snow
Snow falls as snowflakes, which are formed from clusters of ice crystals.

Hail
Ice pellets that form inside cold thunderclouds and fall to the ground are called hail.

Dew
Drops of water that condense on surfaces when the air cools at night are called dew.

Frost
Frost is formed when moisture condenses on surfaces and freezes.

Rain, snow, and hail

Rain, snow, and hail fall when water droplets, ice droplets, or crystals in clouds become too big and heavy to float in the air. In warm clouds, water droplets grow by joining together, then fall as rain. Drizzle is light precipitation falling from shallow clouds. In cold clouds, ice crystals grow bigger, then either fall as snow if conditions are cold, or melt into rain as they fall. Hailstones are pellets of ice that are formed from ice crystals inside clouds.

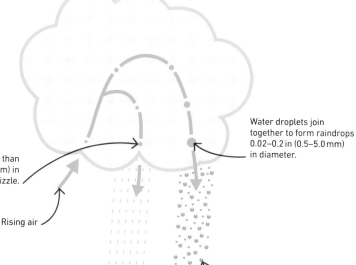

Water droplets join together to form raindrops 0.02–0.2 in (0.5–5.0 mm) in diameter.

Water droplets less than 0.02 in (0.5 mm) in diameter fall as drizzle.

Rising air

Drizzle

Rain

How rain forms ▷
In many parts of the world, the air is too warm for ice crystals to form. Here, raindrops are made when cloud droplets collide with each other, join, and grow bigger.

Types of rainfall

Rain forms when warm, moist air rises through the atmosphere and begins to cool. The cooling process causes the water vapor that is in the air to condense into droplets, forming clouds. The droplets collide in the cloud, combining to get larger. If they get heavy enough, they fall as rain. Every type of rain begins with an uplift of air. There are three types of rainfall: convectional, frontal, and relief.

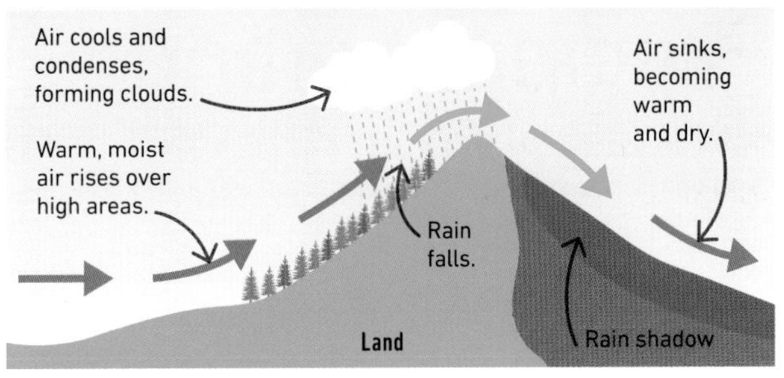

Air cools and condenses, forming clouds.
Air sinks, becoming warm and dry.
Warm, moist air rises over high areas.
Rain falls.
Land
Rain shadow

△ **Relief, or orographic rainfall**
When wind blows moist air over high areas, it cools, causing water vapor to condense into rain clouds. Most rainfall occurs on the windward side and little rain falls on the shadow side.

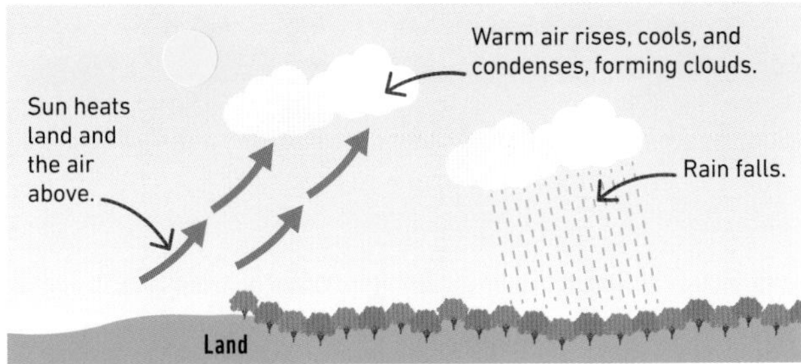

Warm air rises, cools, and condenses, forming clouds.
Sun heats land and the air above.
Rain falls.
Land

△ **Convectional rainfall**
The hot sun heats the land, warming the air above it. This air rises and cools, and water vapor in the air condenses into clouds that can produce thunderstorms.

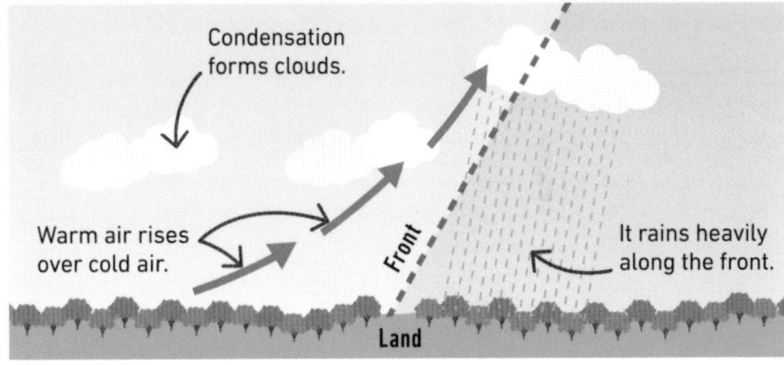

Condensation forms clouds.
Warm air rises over cold air.
Front
It rains heavily along the front.
Land

△ **Frontal rainfall**
A front is an area where two air masses of different temperatures meet. When warm air rises above cold air at a front, it cools and condenses to form clouds that produce frontal rainfall.

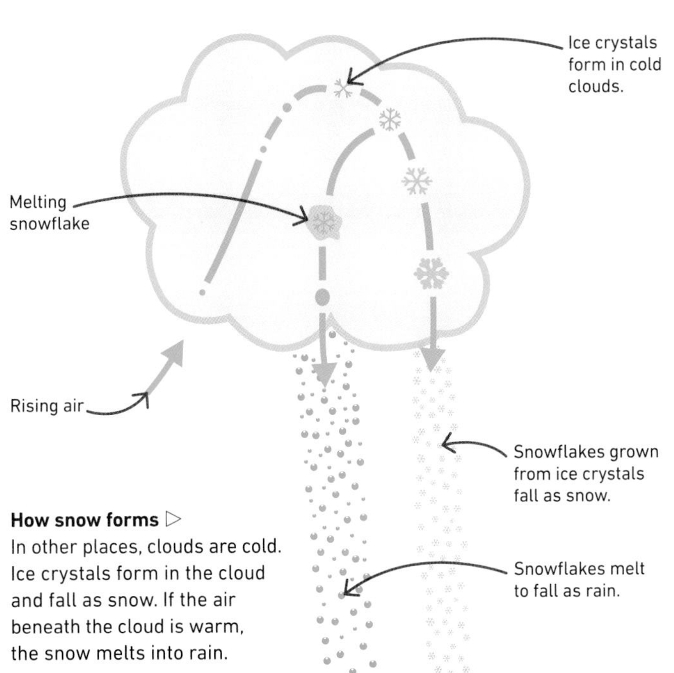

Ice crystals form in cold clouds.
Melting snowflake
Rising air
Snowflakes grown from ice crystals fall as snow.
Snowflakes melt to fall as rain.

How snow forms ▷
In other places, clouds are cold. Ice crystals form in the cloud and fall as snow. If the air beneath the cloud is warm, the snow melts into rain.

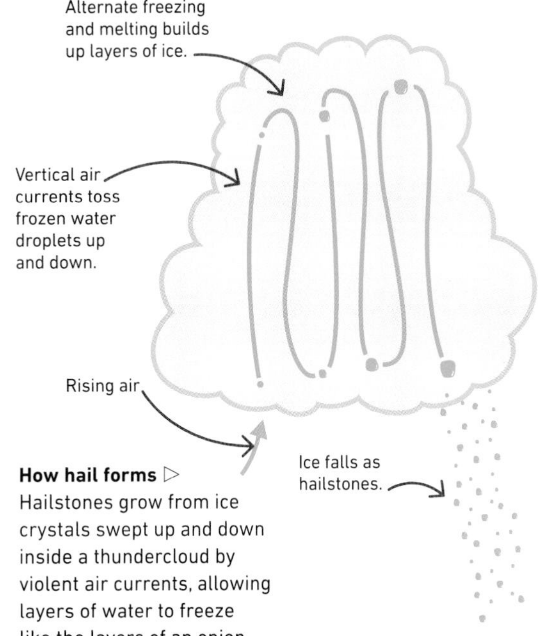

Alternate freezing and melting builds up layers of ice.
Vertical air currents toss frozen water droplets up and down.
Rising air
Ice falls as hailstones.

How hail forms ▷
Hailstones grow from ice crystals swept up and down inside a thundercloud by violent air currents, allowing layers of water to freeze like the layers of an onion.

Hurricanes and tornadoes

SEE ALSO

❮ 80–81 The hydrological cycle
❮ 87–89 Weather systems
❮ 92–93 Cloud and fog

AREAS OF LOW PRESSURE CAN SOMETIMES DEVELOP INTO DESTRUCTIVE HURRICANES AND TORNADOES.

Hurricanes are giant tropical storms, which circulate across hundreds of miles. Tornadoes are fiercely spiraling columns of air that drop from thunderclouds and spread over a few hundred meters.

Where do tropical storms develop?

Hurricanes start just north or south of the equator on the eastern edge of the Atlantic, Indian, and Pacific oceans. They usually sweep westward, then swing away from the equator before finally petering out. Some loop back northwest as they hit the western edge of the oceans.

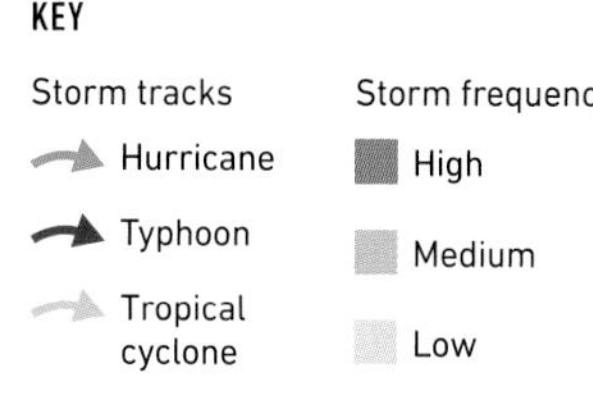

KEY
Storm tracks
Hurricane
Typhoon
Tropical cyclone
Storm frequency
High
Medium
Low

▽ **Storm tracks**
Hurricanes are so called only in the Atlantic Ocean. They are known as typhoons off the coast of East Asia and tropical cyclones elsewhere.

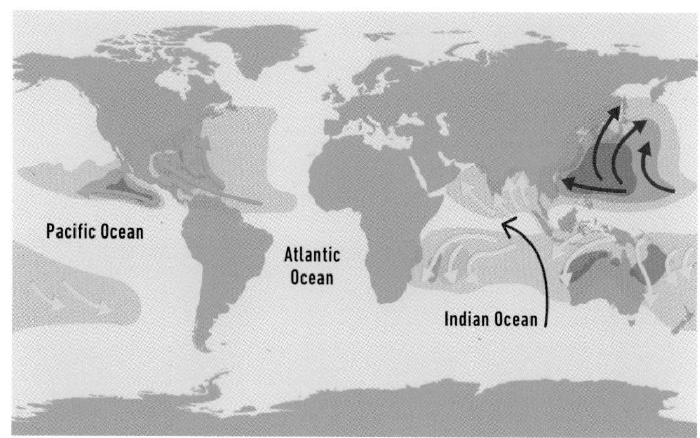

Pacific Ocean
Atlantic Ocean
Indian Ocean

How do tropical storms form?

Hurricanes are stirred up as the late summer sun heats the ocean water. Water vapor from the ocean builds giant thunderclouds. High above, strong winds blow from the east, swirling the clouds together into a big spiral storm. The storm swirls westward, gathering in more clouds.

→N
Strong offshore winds at high and low levels.

1 Tropical disturbance
In the Atlantic, hurricanes typically begin near the Cape Verde islands in Africa, with clouds forming as the sun turns the water of the ocean into water vapor.

Clouds join together to form a comma shape, with its tail pointing toward the east.

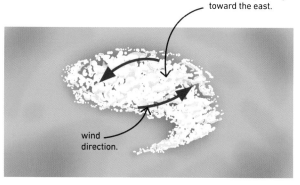

wind direction.

2 Building storm
Winds twist the isolated thunderclouds together into one single storm. Air rises sharply in the center drawing in a spiral of winds.

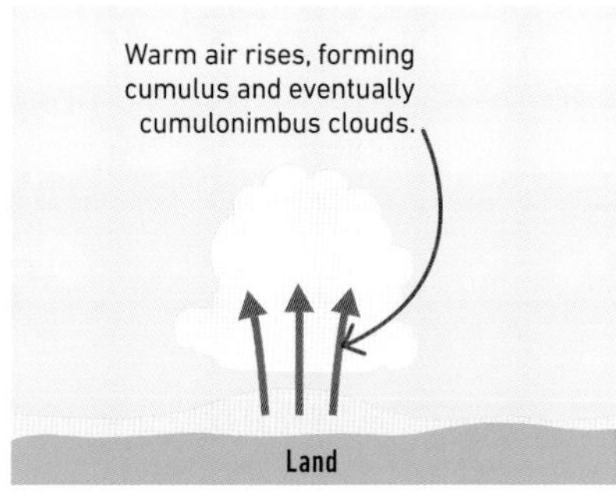
Warm air rises, forming cumulus and eventually cumulonimbus clouds.
Land

1 Stage one
Tornadoes begin when the summer sun warms the air. The heated air rises to build up a massive thundercloud, which is called a supercell.

Tornadoes

Tornadoes, also known as twisters and whirlwinds, can spin at speeds of up to 300 mph (480 km/h). They are whirling funnels of air that form over land during summer thunderstorms. At the edges, the winds spiral at ferocious speeds; in the centre, the air pressure is very low and can suck objects up like a giant vacuum cleaner.

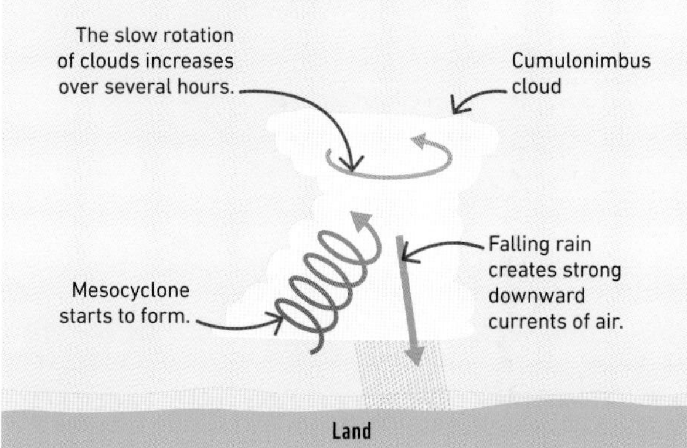
The slow rotation of clouds increases over several hours.
Cumulonimbus cloud
Falling rain creates strong downward currents of air.
Mesocyclone starts to form.
Land

2 Stage two
The updrafts gather momentum and come up against cold winds blowing over the cloud. The clash makes the updraft spin, creating a twisting column of air, or mesocyclone.

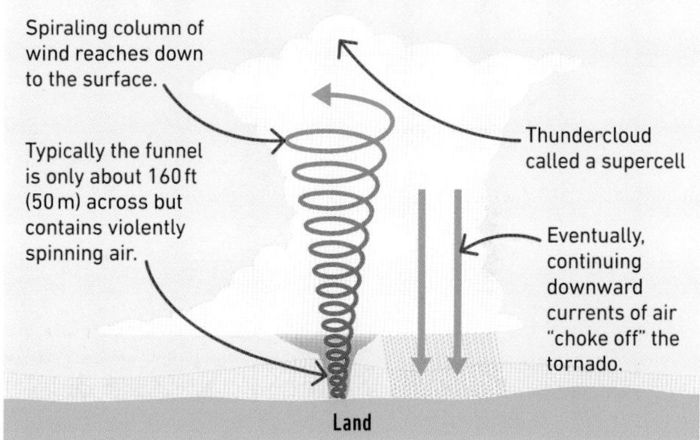
Spiraling column of wind reaches down to the surface.
Thundercloud called a supercell
Typically the funnel is only about 160 ft (50 m) across but contains violently spinning air.
Eventually, continuing downward currents of air "choke off" the tornado.
Land

3 Stage three
As the storm intensifies and rain falls, the mesocyclone column drops from the bottom of the cloud in a violently spiraling funnel and the tornado starts.

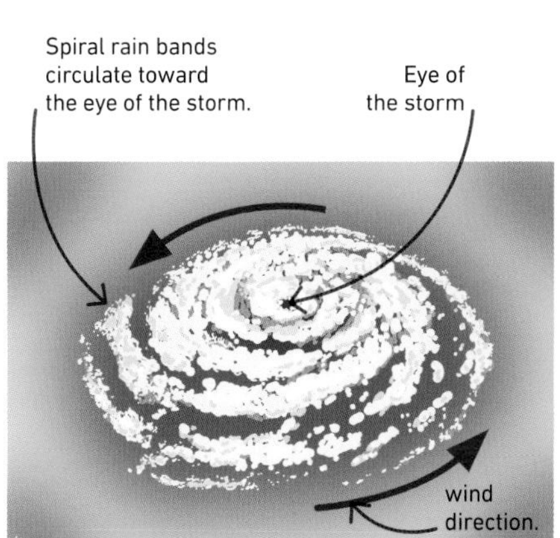
Spiral rain bands circulate toward the eye of the storm.
Eye of the storm
wind direction.

3 Tropical storm
The storm spirals westward slowly, gaining power as it gathers in more clouds and spinning with increasing momentum.

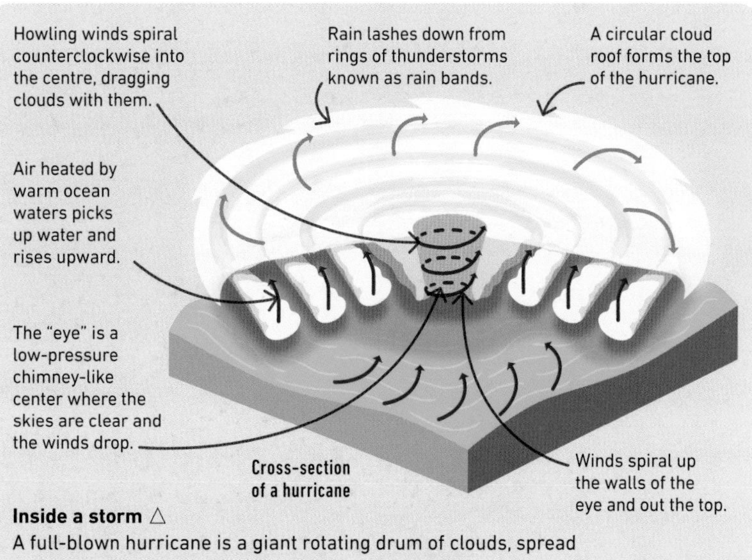
Howling winds spiral counterclockwise into the centre, dragging clouds with them.
Rain lashes down from rings of thunderstorms known as rain bands.
A circular cloud roof forms the top of the hurricane.
Air heated by warm ocean waters picks up water and rises upward.
The "eye" is a low-pressure chimney-like center where the skies are clear and the winds drop.
Cross-section of a hurricane
Winds spiral up the walls of the eye and out the top.

Inside a storm △
A full-blown hurricane is a giant rotating drum of clouds, spread across hundreds of miles, with a hole in the middle.

Biomes

A BIOME IS A LARGE AREA OF THE EARTH THAT HAS A SPECIFIC CLIMATE AND IS HOME TO SEVERAL DISTINCT SPECIES OF PLANTS AND ANIMALS.

SEE ALSO
Distribution of species 100–101 ❯
Ecosystems 102–103 ❯
Deserts 106–107 ❯
Boreal forests and tundra 110–111 ❯

The regions of the Earth can be divided into biomes according to the climate, plants, and animals that are dominant there. The same biome can be found on different continents.

Biomes around the world

Most biogeographers (geographers who study biomes) identify around 10 main terrestrial biomes on our planet. These vary from arid deserts to lush rain forests teeming with life. Each biome has its own climate, soils, landscape, and communities of plants and animals that live there. These plants and animals are adapted to their own environment.

Boreal forest is the largest biome, covering almost **one-fifth** of the Earth's surface.

Polar ▷ The ice-covered polar regions are very cold, with temperatures well below freezing. They are as dry as deserts, and only algae, fungi, and bacteria live here.

Mountain ▷ While the peaks can be cold and icy with little or no vegetation, valleys are often warm and fertile and home to forests and meadows.

◁ Mediterranean This shrubland has a summer dry season and a winter rainy season. Bush fires are common in the summer.

KEY
- Polar
- Mountain
- Mediterranean
- Desert
- Tropical grasslands
- Tropical forest
- Temperate rain forest
- Temperate grassland
- Boreal forest
- Tundra

How latitude affects biomes

Plants need warmth, water, and light to grow, so they are adapted to the temperature, rainfall, and amount of daylight where they live. A warm, wet climate supports broadleaf evergreen trees. Drier climates with distinct seasons support broadleaf deciduous trees or evergreen trees. North and south of the equator, trees give way to shrublands, grasslands, and deserts.

Arctic Circle
Tropic of Cancer
Equator
Tropic of Capricorn
Antarctic Circle

The regions within the Arctic and Antarctic circles receive the least intense sunlight. Summers are cool, while winters are long, dark, and freezing.

The temperate zones have a moderate climate where extremes of hot and cold or wet and dry are unusual. But there are distinct seasonal changes.

Centered on the equator, between the two tropics, the tropical zone receives the most intense sunlight and is warmer than other latitudes.

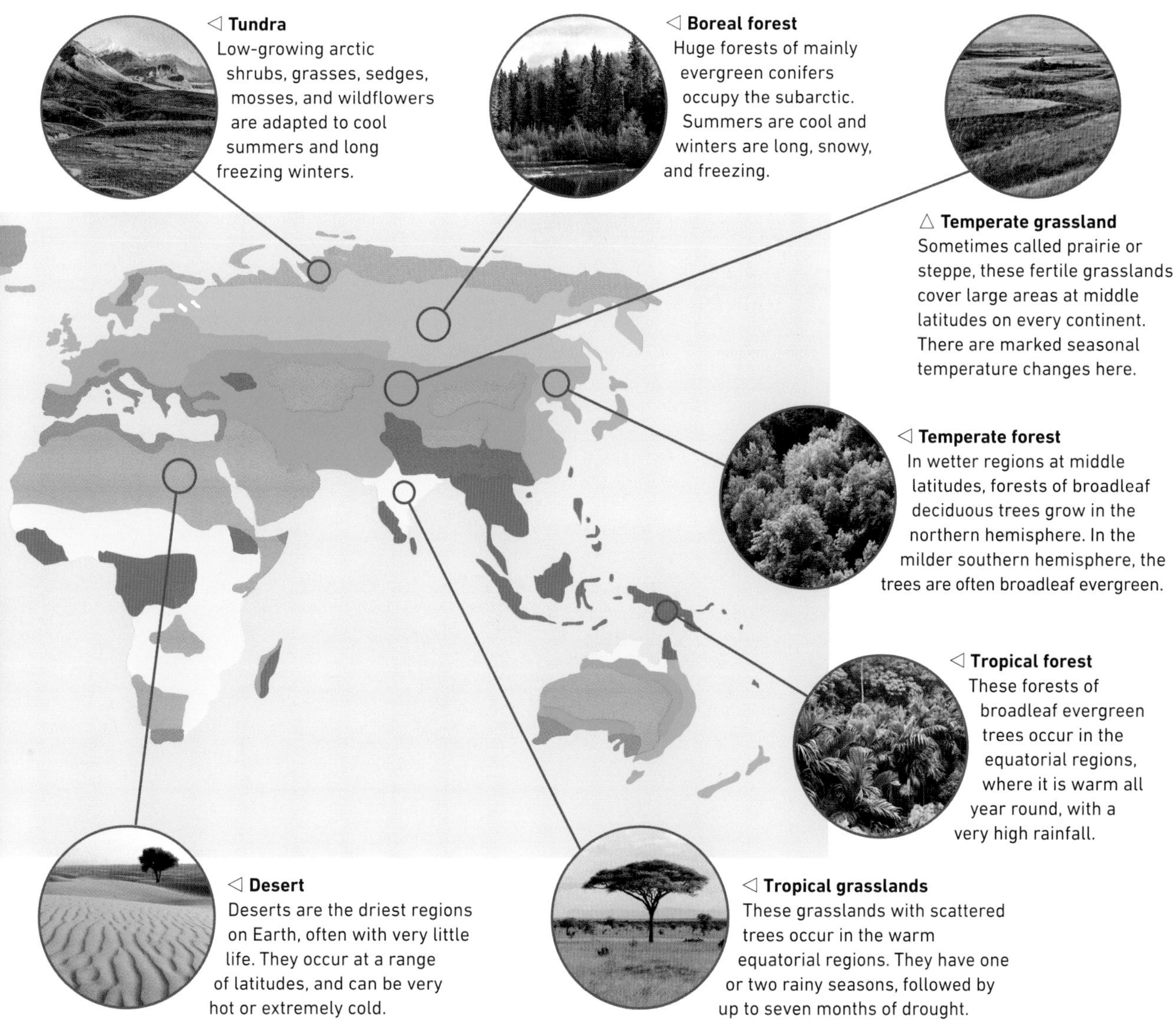

◁ **Tundra**
Low-growing arctic shrubs, grasses, sedges, mosses, and wildflowers are adapted to cool summers and long freezing winters.

◁ **Boreal forest**
Huge forests of mainly evergreen conifers occupy the subarctic. Summers are cool and winters are long, snowy, and freezing.

△ **Temperate grassland**
Sometimes called prairie or steppe, these fertile grasslands cover large areas at middle latitudes on every continent. There are marked seasonal temperature changes here.

◁ **Temperate forest**
In wetter regions at middle latitudes, forests of broadleaf deciduous trees grow in the northern hemisphere. In the milder southern hemisphere, the trees are often broadleaf evergreen.

◁ **Tropical forest**
These forests of broadleaf evergreen trees occur in the equatorial regions, where it is warm all year round, with a very high rainfall.

◁ **Desert**
Deserts are the driest regions on Earth, often with very little life. They occur at a range of latitudes, and can be very hot or extremely cold.

◁ **Tropical grasslands**
These grasslands with scattered trees occur in the warm equatorial regions. They have one or two rainy seasons, followed by up to seven months of drought.

How altitude affects biomes

Plant and animal life changes with altitude (height) because it becomes increasingly cold the higher you are. Animals and plants that live among the trees at the base of a mountain are different from those that make the colder, steeper slopes near the top their home.

Life in the mountains ▷
The foothills are covered in deciduous broadleaf forest, while the higher altitude vegetation is a variation of the arctic tundra and polar biomes.

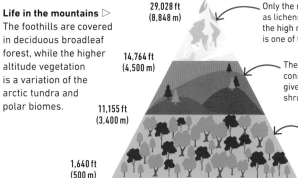

29,028 ft (8,848 m)

Only the most hardy tundra plants, such as lichens and mosses, can survive in the high mountains. The snow leopard is one of the few predators found here.

14,764 ft (4,500 m)

The climate is cold here. There are coniferous forests of cedar, which give way to alpine birch and juniper shrublands. Wild goats live here.

11,155 ft (3,400 m)

The lower slopes have a subtropical to temperate climate with deciduous forests. This region is home to the bears, monkeys, and other animals.

1,640 ft (500 m)

Distribution of species

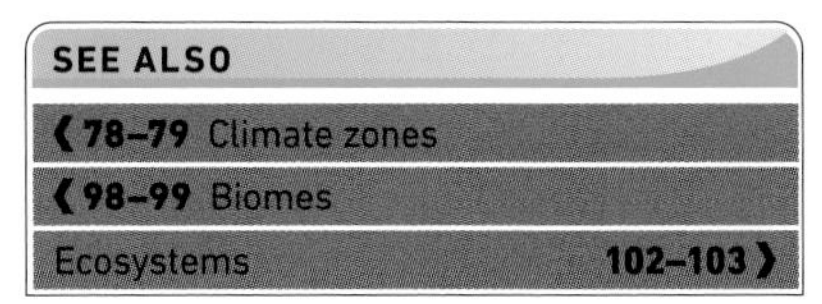
SEE ALSO
❮ 78–79 Climate zones
❮ 98–99 Biomes
Ecosystems 102–103 ❯

WHERE A SPECIES LIVES DEPENDS ON HOW WELL IT HAS ADAPTED TO ITS ENVIRONMENT OVER MANY GENERATIONS.

An animal or plant's distribution depends on local conditions, such as climate or geology. It also depends on the species' ability to spread into new regions and adapt to different conditions over many generations.

Evolution

A species is a group of closely related animals or plants that can breed with each other. Over time, generations may evolve certain characteristics, or evolve into new species, better suited to the environmental conditions. English naturalist Charles Darwin explained this process of "natural selection" in 1859, showing that those animals and plants that were best suited to their environment were most likely to survive and pass on their genes.

▽ **Darwin's finches**
Each species of finch in the Galápagos Islands has a slightly different bill depending on what it eats. They are all descended from a common ancestor but have adapted to their new environments.

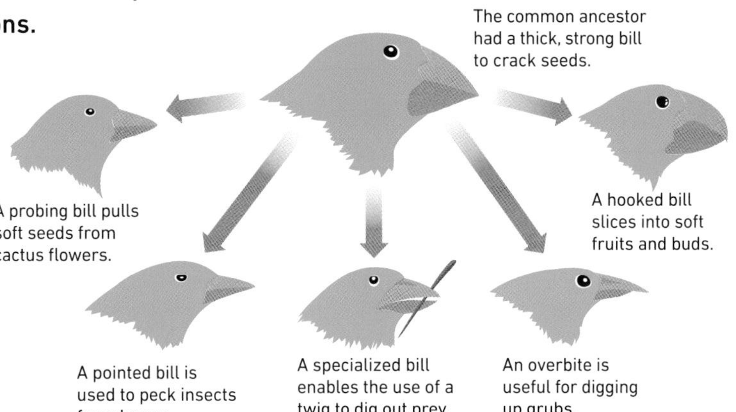
The common ancestor had a thick, strong bill to crack seeds.
A probing bill pulls soft seeds from cactus flowers.
A hooked bill slices into soft fruits and buds.
A pointed bill is used to peck insects from leaves.
A specialized bill enables the use of a twig to dig out prey.
An overbite is useful for digging up grubs.

Dispersal of species

The range of a species depends on how easily it can disperse (spread). The species' survival will depend on its ability to live successfully in its new local environment and how well it can spread its seeds or find a mate.

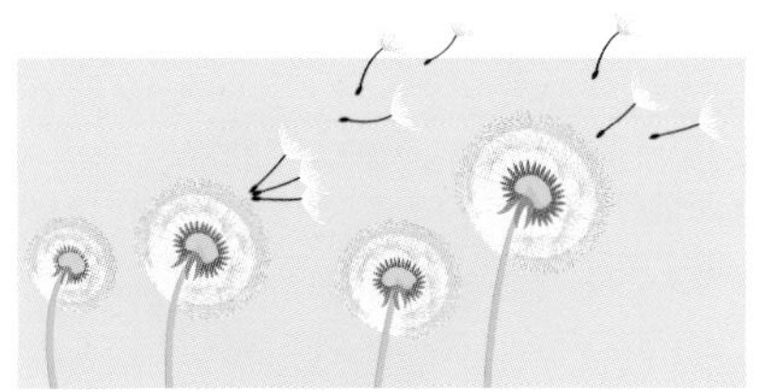

△ **Seeds carried by wind**
Some seeds are small and light with downy attachments. These catch the wind and are blown great distances.

△ **Seeds carried by water**
The seeds of some land plants, such as coconuts, can float and may be carried by ocean currents to distant shores.

△ **Dispersal of animals**
Most animals have the ability to move by themselves, walking, swimming, or flying away from their home range.

△ **Seeds carried by animals**
Some seeds are dispersed by animals. They have specially adapted hooks or teeth that get caught in the animal's fur or wool.

△ **Seeds carried by birds**
The seeds of some fleshy fruits pass through a bird's digestive tract and are deposited at a new location in their droppings.

Patterns of distribution

Some species are found on almost all continents or throughout most of the world's oceans, while other species may be found only in the areas where they evolved. This may be because of physical barriers such as mountains or the oceans, or because they have only recently evolved and have not had time to spread from their centers of origin.

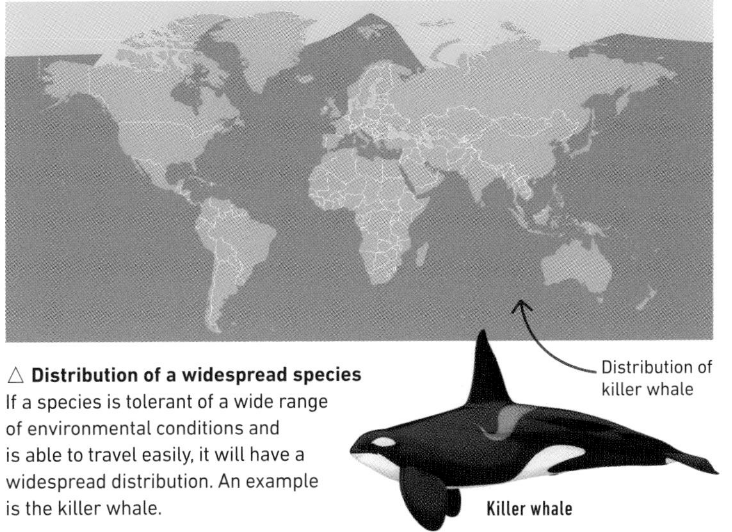
Distribution of killer whale
Killer whale

△ **Distribution of a widespread species**
If a species is tolerant of a wide range of environmental conditions and is able to travel easily, it will have a widespread distribution. An example is the killer whale.

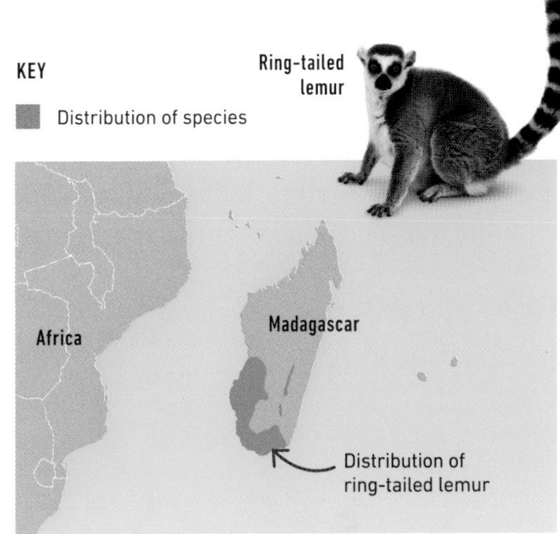
KEY
Distribution of species
Ring-tailed lemur
Africa
Madagascar
Distribution of ring-tailed lemur

△ **Distribution of a native species**
Most animal species that evolve on islands cannot disperse far because of the ocean. The ring-tailed lemur, for example, lives only on Madagascar.

Biogeographical regions

In the mid-19th century, naturalists exploring the world divided the planet into separate regions depending on the plants and animals that lived there. English naturalists Philip Lutley Sclater and Alfred Russel Wallace identified six regions of animal life, each of which has distinct wildlife. These regions are less well known today, and most biogeographers divide the planet into "biomes."

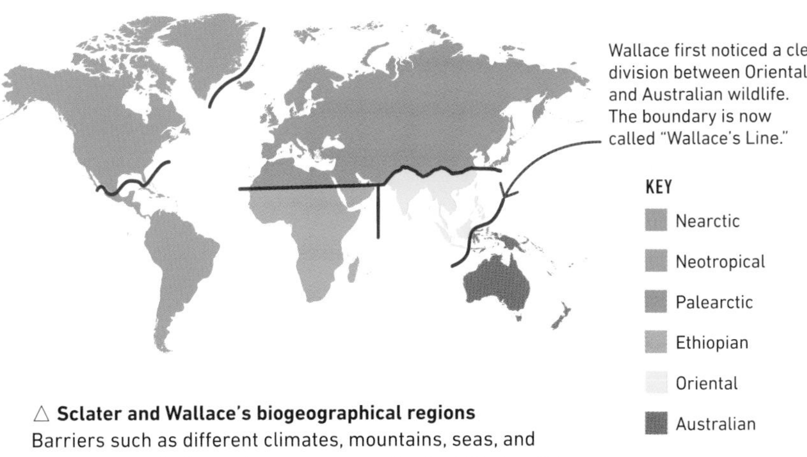
Wallace first noticed a clear division between Oriental and Australian wildlife. The boundary is now called "Wallace's Line."
KEY
Nearctic
Neotropical
Palearctic
Ethiopian
Oriental
Australian

△ **Sclater and Wallace's biogeographical regions**
Barriers such as different climates, mountains, seas, and oceans separate the regions, preventing the spread of some species of animal from one biogeographical region into another.

REAL WORLD

Invasive species

Some species of plants and animals flourish as new arrivals in an area, threatening previously healthy native ecosystems. One of the most invasive plants in the United States is the Asian vine kudzu. Climbing trees and shrubs, kudzu forms a dense blanket through which very little light can enter.

Ecosystems

SEE ALSO
❮ 78–79 Climate zones
❮ 98–99 Biomes
❮ 100–101 Distribution of species

THE COMMUNITY OF PLANTS AND ANIMALS THAT LIVE AND INTERACT IN A PARTICULAR PLACE IS CALLED AN ECOSYSTEM.

An ecosystem can be as small as a puddle or as large as the entire globe. In an ecosystem, living things interact with each other and with the non-living parts of the environment that surrounds them.

How organisms fit in an ecosystem

Each organism occupies a particular place in an ecosystem, and interacts with the other lifeforms that share its ecosystem. An organism is adapted to its place so that it can compete for various resources, such as space, water, food, and mates. These adaptations emerge by chance, over many generations, and include size, shape, color, and behavior.

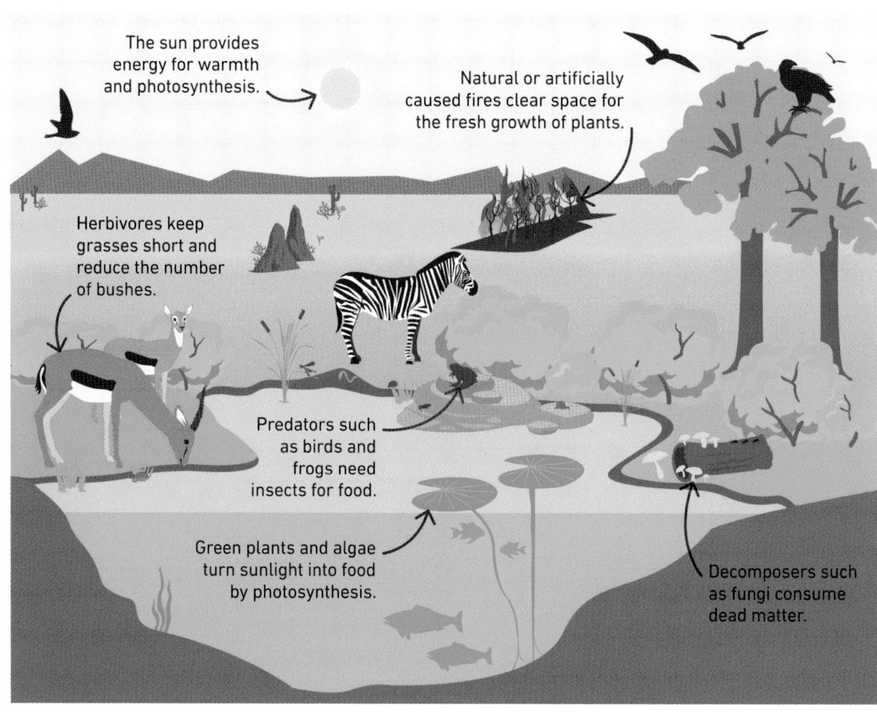
The sun provides energy for warmth and photosynthesis.
Natural or artificially caused fires clear space for the fresh growth of plants.
Herbivores keep grasses short and reduce the number of bushes.
Predators such as birds and frogs need insects for food.
Green plants and algae turn sunlight into food by photosynthesis.
Decomposers such as fungi consume dead matter.

A stable ecosystem ▷
In a healthy ecosystem, there is a balanced relationship between the different organisms living there and the environment in which they live.

Fuel for an ecosystem

Organisms need energy and nutrients to grow and function. Sunlight provides solar energy, which is used by green plants to produce food for themselves in a process called photosynthesis. Decomposing materials release nutrients into the soil, which are converted into food by organisms in a process known as the nutrient cycle.

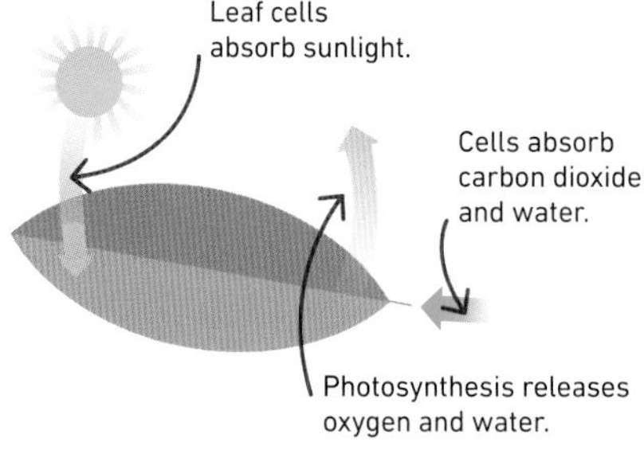
Leaf cells absorb sunlight.
Cells absorb carbon dioxide and water.
Photosynthesis releases oxygen and water.

△ Photosynthesis
The energy absorbed from sunlight is used to combine carbon dioxide, minerals, and water to make food in the form of glucose.

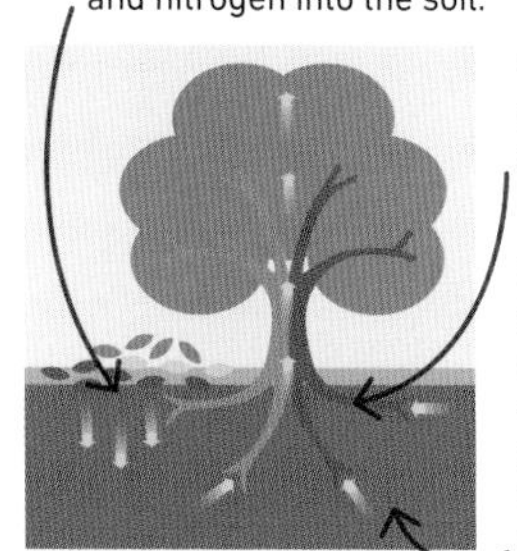
Plant decay releases carbon and nitrogen into the soil.
The plant absorbs nutrients through its system of roots.
Worms and fungi decompose organic matter to release carbon dioxide. Bacteria convert nitrogen into plant food.

△ Nutrient cycle
In an ecosystem, nutrients flow from the non-living to the living and back to the non-living components.

Food chains

Living things depend on one another for food. A food chain, such as this one for a prairie habitat, shows how plants and animals are linked together by who eats what. It begins with producers, such as plants that make their own food. Primary consumers eat the producers and in turn are eaten by secondary consumers. A food chain always ends with a top predator.

Eaten by

Eaten by

Sun's energy

Cacti

Grass

Shrubs

Mule deer

Mouse

Beetle

Black-footed ferret

Rattlesnake

Meadow lark

Coyote

Golden eagle

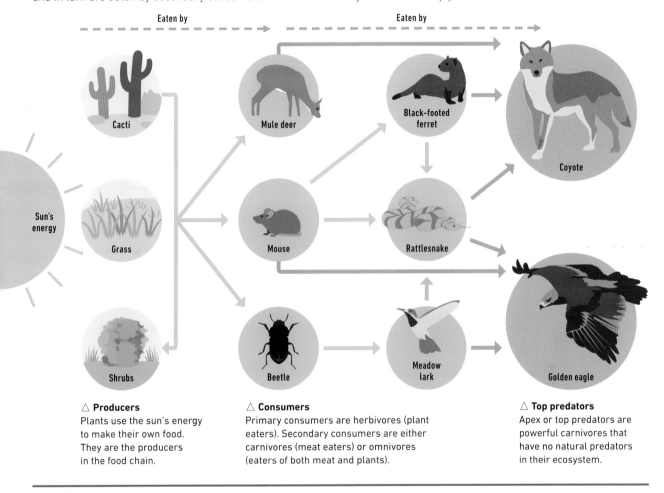

△ **Producers**
Plants use the sun's energy to make their own food. They are the producers in the food chain.

△ **Consumers**
Primary consumers are herbivores (plant eaters). Secondary consumers are either carnivores (meat eaters) or omnivores (eaters of both meat and plants).

△ **Top predators**
Apex or top predators are powerful carnivores that have no natural predators in their ecosystem.

Sensitivity to change

The living and the non-living parts of an ecosystem are in a complex balance, and change in one part can impact the other parts. Natural disasters, such as earthquakes, flood, disease, or invasion by a new species can impact the ecosystem. Today, human activities are disturbing many ecosystems.

△ **Drought**
A period of prolonged drought may create an imbalance in the ecosystem and even cause some species to go extinct.

△ **Spraying fertilizers**
Fertilizers sprayed on fields run off into streams and lakes causing algae to bloom, which damages marine ecosystems.

Tropical grasslands and rainforests

SEE ALSO
❮ 98–99 Biomes
❮ 100–101 Distribution of species
❮ 102–103 Ecosystems
Deforestation 172–174 ❯

TROPICAL BIOMES ARE FOUND IN THE REGIONS ON EITHER SIDE OF THE EQUATOR, WHERE THE SUN IS MOST INTENSE.

The tropical regions are warm all year, but rainfall patterns vary. There are one or two rainy seasons in the grasslands, followed by months of drought. In the rainforest, it rains almost every day.

KEY
Tropical grassland
Tropical rainforest

△ Distribution of tropical biomes Tropical grasslands and rainforests span the equator, between about 25°N and 25°S. Over half of all tropical grasslands are in Africa. The largest rainforest is in South America.

Tropical grasslands

This biome typically has a ground cover of grasses, with a scattering of shrubs or trees, such as on the African savanna. Grasses grow quickly in the rainy season, sometimes to heights of many feet. Plants and animals have adapted to survive the seasonal droughts and frequent wildfires that occur here.

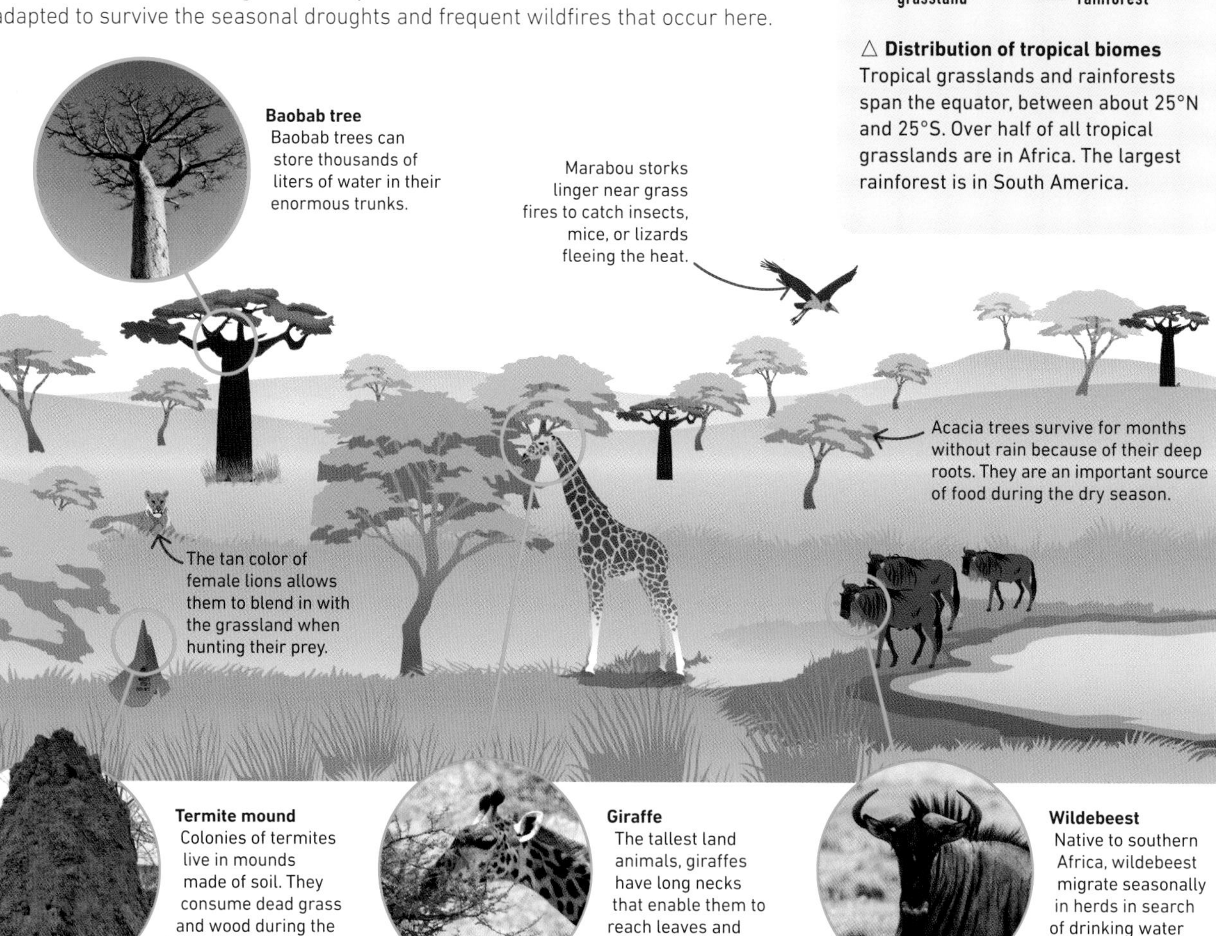
Baobab tree
Baobab trees can store thousands of liters of water in their enormous trunks.
Marabou storks linger near grass fires to catch insects, mice, or lizards fleeing the heat.
Acacia trees survive for months without rain because of their deep roots. They are an important source of food during the dry season.
The tan color of female lions allows them to blend in with the grassland when hunting their prey.
Termite mound
Colonies of termites live in mounds made of soil. They consume dead grass and wood during the harsh dry season.
Giraffe
The tallest land animals, giraffes have long necks that enable them to reach leaves and twigs in the treetops.
Wildebeest
Native to southern Africa, wildebeest migrate seasonally in herds in search of drinking water and fresh grasses.

Tropical rainforests

Rainforests are home to about half of the known plant and animal species that live on land. The constant warm and wet conditions enable year-round growth and activity, with a continual supply of nectar, fruit, nuts, seeds, and young leaves for plant-eating animals. So much growth creates a struggle for light, which influences the structure of the forest and the variety of plant forms.

▽ **Tropical rainforest wildlife**
The animals and plants shown on this illustration live on separate continents.

Orangutan
Using their long arms and large hands to clamber through the rainforests of Southeast Asia, orangutans forage for fruits and leaves.

Red-eyed tree frog
These Central American frogs hide in the forest canopy at night, ambushing passing insects with their long, sticky tongues.

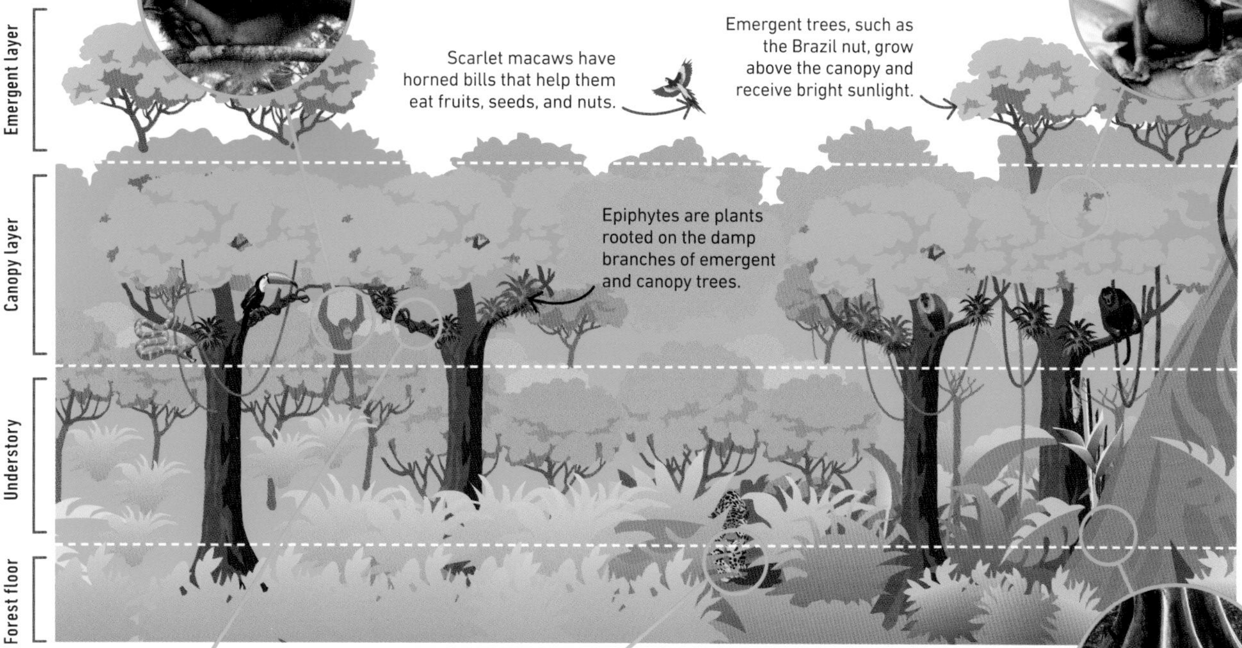

Emergent trees, such as the Brazil nut, grow above the canopy and receive bright sunlight.

Scarlet macaws have horned bills that help them eat fruits, seeds, and nuts.

Epiphytes are plants rooted on the damp branches of emergent and canopy trees.

Emergent layer

Canopy layer

Understory

Forest floor

Buttress roots
The biggest trees in the rainforest develop buttress roots that provide support and anchor them in the shallow soil.

Lianas
These woody vines have roots in the ground but climb up trees in order to reach the sunlight above, where their leaves and flowers grow.

Jaguar
The largest cats in South America, jaguars are agile carnivores. They can run, leap, swim, and climb trees when hunting their prey.

REAL WORLD

People of the rainforest

The rainforest is home to groups of hunter-gatherers, people who obtain food through hunting animals or gathering fruit, plants, and nuts. Many rainforest peoples (such as these Baka people from central Africa) depend on the complex forest ecosystem for almost all their daily needs. Forest clearance means this indigenous knowledge is being lost.

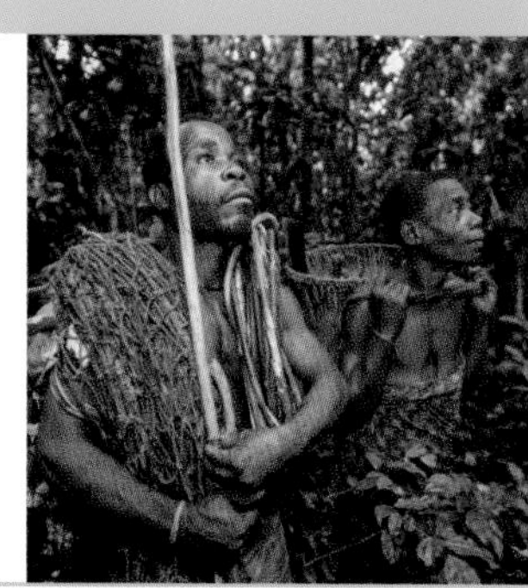

Deserts

DESPITE THEIR HARSH, DRY CLIMATE, DESERTS ARE HOME TO SOME SPECIALLY ADAPTED PLANTS AND ANIMALS.

SEE ALSO
❮ 78–79 Climate zones
❮ 84–86 Ocean currents
❮ 98–99 Biomes
Deforestation 172–174 ❯

Deserts are the driest regions on Earth, receiving less than 10 in (250 mm) of precipitation a year. They can be hot or cold and are often windy, with extreme differences in temperature between day and night.

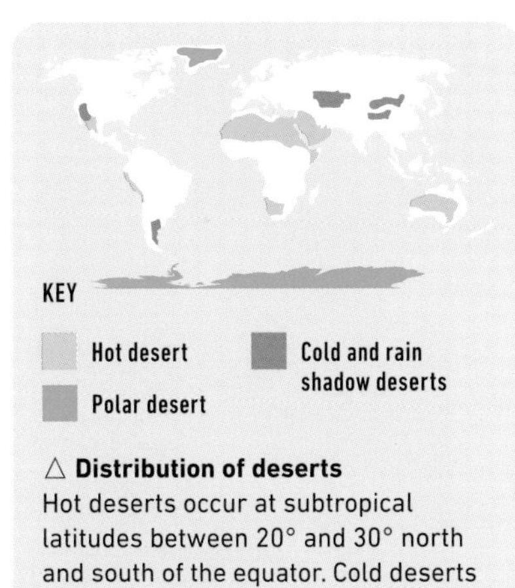
KEY
Hot desert
Polar desert
Cold and rain shadow deserts

△ **Distribution of deserts**
Hot deserts occur at subtropical latitudes between 20° and 30° north and south of the equator. Cold deserts are found at middle and polar latitudes.

Hot deserts

Found around the subtropics, north and south of the equator, hot deserts typically have thin, stony soils. Daytime temperatures rise to over 104°F (40°C) in the summer, while temperatures at night can approach freezing. There is little water, and only a few drought-resistant plants grow here. Desert animals have special adaptations to help them survive the harsh conditions.

▽ **Hot-desert wildlife**
The animals and plants shown on this illustration live on different continents.

Saguaro cactus
These cacti from North America have thick green stems that store water for use in dry periods.

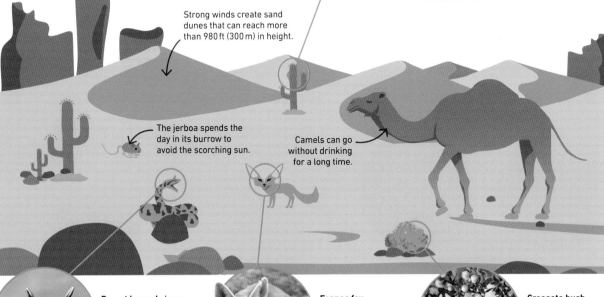

Strong winds create sand dunes that can reach more than 980 ft (300 m) in height.
The jerboa spends the day in its burrow to avoid the scorching sun.
Camels can go without drinking for a long time.

Desert horned viper
Also known as sand vipers, these snakes from northern Africa can move with great agility on loose sand.

Fennec fox
The large ears of these foxes allow heat to be radiated from their bodies, keeping them cool.

Creosote bush
This medium-size shrub has small, thick leaves with waxy skin to reduce water loss.

Cold deserts

Cold deserts often occur far inland, where the air is dry, but can also be found in coastal regions alongside cool ocean currents. Mountain ranges can cause rain shadow deserts. Summer days may be warm, but temperatures far below freezing are common throughout the winter.

Polar desert

Antarctica is a landscape made up of snow and ice. It is also a desert because it hardly ever rains and there is almost no liquid water. A few hardy algae, bacteria, and fungi manage to survive in these conditions, and two species of flowering plant live in the more moist conditions near the coast.

Golden eagle
Among the largest hunting birds in the world, golden eagles are swift and have sharp talons to snatch their prey.

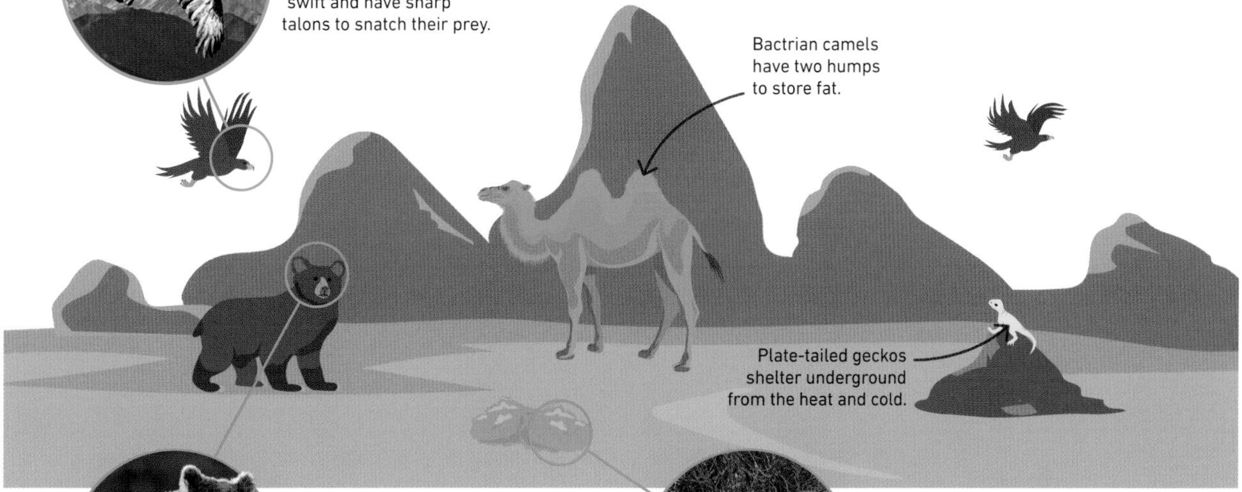

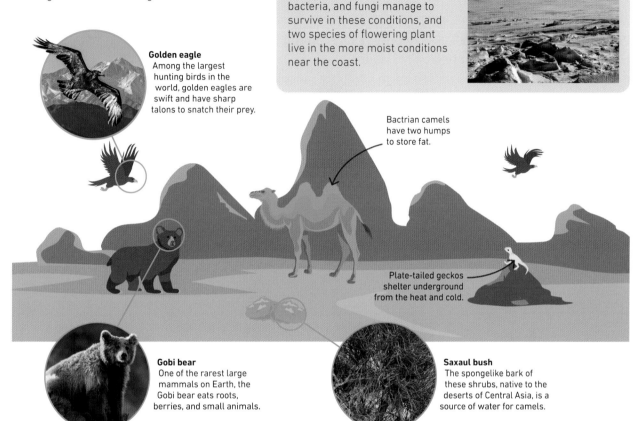

Bactrian camels have two humps to store fat.

Plate-tailed geckos shelter underground from the heat and cold.

Gobi bear
One of the rarest large mammals on Earth, the Gobi bear eats roots, berries, and small animals.

Saxaul bush
The spongelike bark of these shrubs, native to the deserts of Central Asia, is a source of water for camels.

Rain shadow deserts

The downwind side of a mountain range is sometimes home to a rain shadow desert. For example, the Himalayan mountains and Tibetan plateau block precipitation from the Indian Ocean, creating the Gobi Desert in the rain shadow to the north.

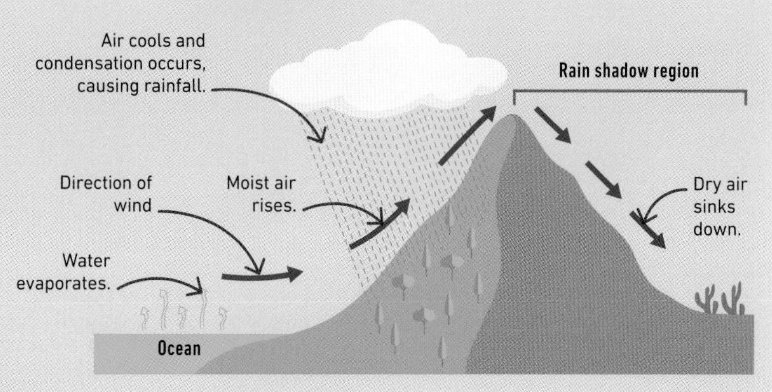
Air cools and condensation occurs, causing rainfall.
Rain shadow region
Direction of wind
Moist air rises.
Dry air sinks down.
Water evaporates.
Ocean

Creating arid conditions ▷
Moist air is forced upward and falls as rainfall on the near side of the mountain. Only dry air reaches the far side, forming a rain shadow.

Temperate forests and grasslands

SEE ALSO
❮ 98–99 Biomes
❮ 104–105 Tropical grassland and rain forests
Deforestation 172–173 ❯

LYING BETWEEN THE TROPICS AND THE POLAR LATITUDES, THE TEMPERATE ZONE IS HOME TO TWO MAJOR BIOMES.

The seasons are very different in temperate regions. Spring is a time of growth, while grasslands are prone to wildfire in summer. Trees lose their leaves in the fall, ready for a cold or snowy winter.

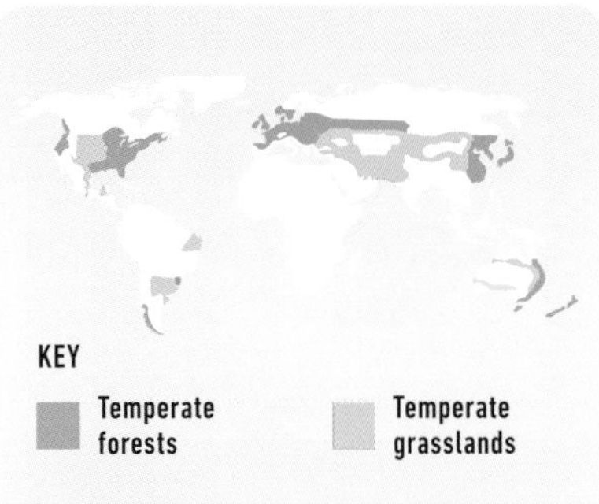
KEY
Temperate forests
Temperate grasslands

△ Distribution of temperate biomes
Temperate forests and grasslands lie between latitudes 25°N and 50°N in the north, as well as in parts of South America, Australia, and New Zealand.

Temperate forests

The forests of the northern hemisphere are home to broadleaf trees that lose their leaves in the fall. In early spring, bare branches let sunlight through, allowing new growth below. Wildflowers bloom first, and taller trees come into leaf last. In the milder southern hemisphere, broadleaf forests are evergreen.

▽ Fall
Northern hemisphere trees have thin, wide leaves that are shed in fall. This means the tree does not need to protect its leaves from winter frosts.

Beech tree
Forests of beech extend across central Europe. Fungi and wild garlic grow in the shade beneath the dense canopy.

Fall

The subcanopy layer includes smaller trees, as well as younger trees that have not yet reached the canopy.

The shrub layer is closer to the ground and consists of bushes and young trees.

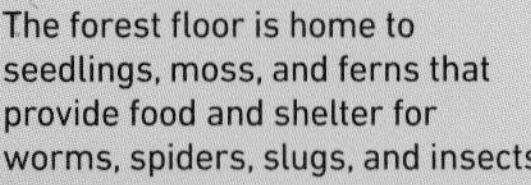

The forest floor is home to seedlings, moss, and ferns that provide food and shelter for worms, spiders, slugs, and insects.

Spring

American black bear
These omnivores hibernate through the winter, living on the body fat they have built up by eating all summer and fall.

Red squirrel
Also found in conifer forests, red squirrels do not hibernate but rather stay inside nests in the winter, relying on their stores.

Bluebells
In spring, when the forest floor gets plenty of light, wildflowers such as bluebells bloom.

Temperate grasslands

Each spring, temperate grasslands come alive. Grasses send out shoots, colorful wildflowers bloom into the summer, and insects abound. As winter approaches, plants die back, but because the underground root systems remain alive, they sprout again next spring. Large herbivores were once very common, but hunting has reduced their populations.

▽ **Temperate grassland wildlife**
The animals and plants shown on this illustration live on separate continents.

Red kangaroo
Native to Australia, red kangaroos avoid predators by using their powerful hind legs to escape at speed. Their body fat helps them survive droughts.

Przewalski's horse
The last surviving wild horse subspecies lives in Mongolia. Once driven to extinction in the wild, they have been reintroduced to the steppe.

Large-headed grasshopper
These plant-eating insects thrive among the tall grasses during the summer months.

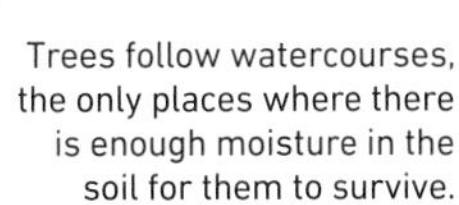

Trees follow watercourses, the only places where there is enough moisture in the soil for them to survive.

Animals drink from water sources that are derived from melted snow or limited rainfall.

Prairie dog
These burrow-dwelling rodents from North America have sharp claws for digging and sharp teeth for eating tough prairie plants.

Prairie grass
The roots of prairie grass can reach groundwater far below the surface, helping the plant survive summer droughts.

Guanaco
Found in the pampas of South America, this deerlike animal has soft sensitive lips to help it find its preferred plants.

Bison
These heavy herbivores from North America and eastern Europe trample the ground, forming areas of soil where new plants can grow.

Boreal forests and tundra

SEE ALSO
❮ 78–79 Climate zone
❮ 98–99 Biomes
❮ 102–103 Ecosystems
Deforestation 172–174 ❯

VAST, COLD BOREAL FORESTS AND THE TREELESS PLAINS OF THE TUNDRA DOMINATE THE FAR NORTH OF OUR PLANET.

The boreal forest extends across the northern part of Eurasia and North America, where summers are cool and winters long. Further north, trees give way to tundra plants adapted to the extreme cold.

KEY
Boreal forest
Tundra

△ **Distribution of the boreal forest and tundra biomes**
Boreal forest stretches across Eurasia and North America, between 50°N and the Arctic Circle. The tundra lies farther to the north.

Boreal forests

Evergreen conifers, such as spruce, pine, or fir, are the most common type of tree in the boreal forest. However, broad-leaved deciduous trees, including birch and alder, may be quick to colonize clearings opened up by forest fires. Beneath the trees, a carpet of moss covers much of the forest floor.

Black spruce
Like most northern conifers, spruce have waxy, needlelike leaves that retain moisture and help them survive the winter.

Moose
The largest species in the deer family, the moose has a varied diet that includes pine and spruce needles in the winter.

Grizzly bears can sleep through the winter for six months without eating or drinking.

The conical shape of fir trees means that the snow falls off easily.

Balsam fir
The cones of this North American conifer contain seeds eaten by several species of birds and rodents.

Snowshoe hare
Named after their large feet, which help them hop on snow, these hares have coats that turn white in winter.

Lichens
Composed of algae and fungi, lichens grow on rocks or trees. They survive the harshest winters and are a precious food source for animals.

Tundra

The tundra is home to short, shallow-rooted plants, such as dwarf shrubs, grasses, sedges, wildflowers, mosses, and lichens. For most of the year, these plants are covered in a blanket of snow. In spring, the top few inches of permafrost thaw and snow melts. The meltwater fills streams, lakes, and bogs. Plants make use of the 24-hour summer sunlight to quickly grow and reproduce.

Arctic wolf
Working in packs, these wolves are excellent hunters because of their superb hearing and white coats, which hide them in the snow.

Musk ox
Protected by their thick coats and insulating body fat, these hardy mammals are able to withstand the bitter tundra winter.

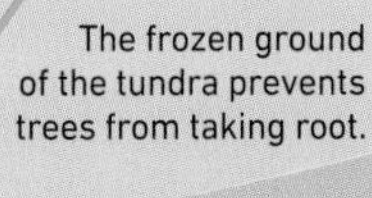

The frozen ground of the tundra prevents trees from taking root.

Summer

Winter

Caribou
Also known as reindeer, caribou use their hooves and snouts to uncover grass and lichen under the snow.

Arctic poppy
One of the many flowering plants found on the tundra, these poppies have tiny flowers that turn their heads to follow the sun.

Snow goose
These migratory birds breed on the Arctic tundra each spring but spend the winter much farther south.

Permafrost

Soil or gravel that remains frozen for two or more years is called permafrost. It may be hundreds of feet thick and has a much thinner "active" top layer that freezes and thaws each winter and spring. Farther south, the permafrost becomes more patchy, while the active layer gets thicker.

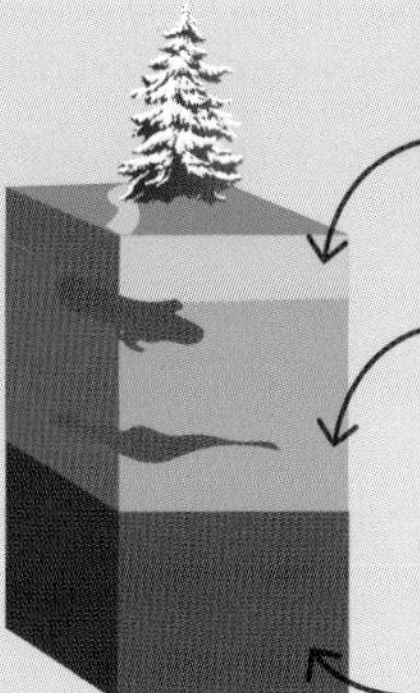

The active layer freezes and thaws each year.

The permafrost layer remains frozen all year.

Unfrozen ground called talik can lie below the permafrost or between the active layer and the permafrost.

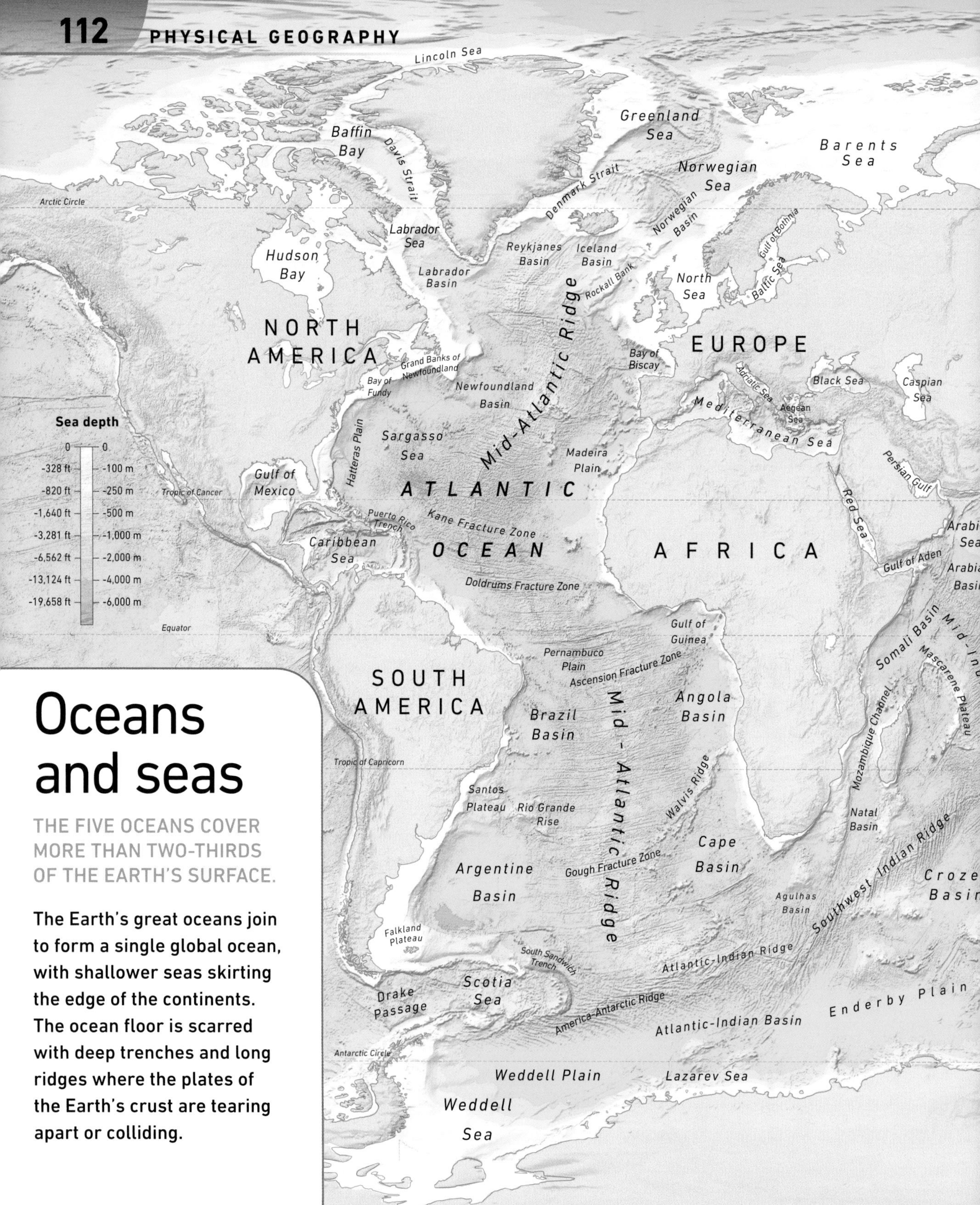
Lincoln Sea
Greenland Sea
Baffin Bay
Davis Strait
Barents Sea
Norwegian Sea
Denmark Strait
Arctic Circle
Norwegian Basin
Gulf of Bothnia
Labrador Sea
Reykjanes Basin
Iceland Basin
Hudson Bay
Labrador Basin
Rockall Bank
North Sea
Baltic Sea
NORTH AMERICA
EUROPE
Grand Banks of Newfoundland
Bay of Biscay
Bay of Fundy
Newfoundland Basin
Mid-Atlantic Ridge
Adriatic Sea
Black Sea
Caspian Sea
Aegean Sea
Mediterranean Sea
Sea depth
0 / 0
-328 ft / -100 m
-820 ft / -250 m
-1,640 ft / -500 m
-3,281 ft / -1,000 m
-6,562 ft / -2,000 m
-13,124 ft / -4,000 m
-19,658 ft / -6,000 m
Hatteras Plain
Sargasso Sea
Madeira Plain
Persian Gulf
Gulf of Mexico
Tropic of Cancer
ATLANTIC OCEAN
Red Sea
Puerto Rico Trench
Kane Fracture Zone
Caribbean Sea
AFRICA
Gulf of Aden
Doldrums Fracture Zone
Equator
Gulf of Guinea
Somali Basin
Mascarene Plateau
Pernambuco Plain
Ascension Fracture Zone
SOUTH AMERICA
Angola Basin
Brazil Basin
Mid-Atlantic Ridge
Mozambique Channel
Tropic of Capricorn
Walvis Ridge
Santos Plateau
Rio Grande Rise
Natal Basin
Cape Basin
Southwest Indian Ridge
Argentine Basin
Gough Fracture Zone
Agulhas Basin
Falkland Plateau
South Sandwich Trench
Atlantic-Indian Ridge
Scotia Sea
Drake Passage
America-Antarctic Ridge
Enderby Plain
Atlantic-Indian Basin
Antarctic Circle
Weddell Plain
Lazarev Sea
Weddell Sea

Oceans and seas

THE FIVE OCEANS COVER MORE THAN TWO-THIRDS OF THE EARTH'S SURFACE.

The Earth's great oceans join to form a single global ocean, with shallower seas skirting the edge of the continents. The ocean floor is scarred with deep trenches and long ridges where the plates of the Earth's crust are tearing apart or colliding.

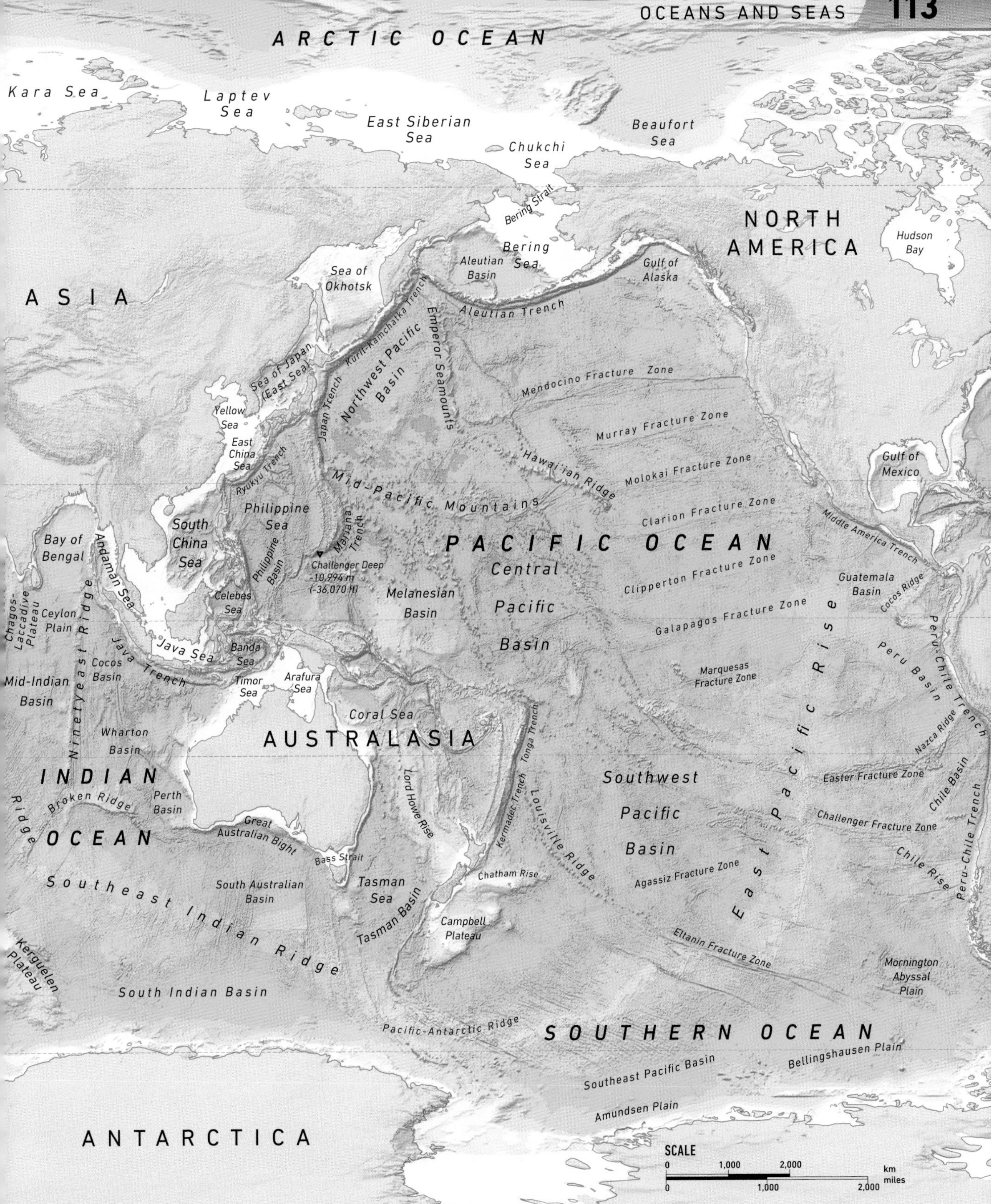
ARCTIC OCEAN
Kara Sea
Laptev Sea
East Siberian Sea
Chukchi Sea
Beaufort Sea
Bering Strait
NORTH AMERICA
Hudson Bay
Bering Sea
Aleutian Basin
Gulf of Alaska
Sea of Okhotsk
ASIA
Aleutian Trench
Kuril-Kamchatka Trench
Emperor Seamounts
Northwest Pacific Basin
Sea of Japan (East Sea)
Japan Trench
Mendocino Fracture Zone
Yellow Sea
East China Sea
Murray Fracture Zone
Ryukyu Trench
Hawai'ian Ridge
Molokai Fracture Zone
Gulf of Mexico
Mid-Pacific Mountains
Philippine Sea
Clarion Fracture Zone
South China Sea
Mariana Trench
Middle America Trench
Bay of Bengal
Andaman Sea
Philippine Basin
PACIFIC OCEAN
Challenger Deep -10,994 m (-36,070 ft)
Central Pacific Basin
Clipperton Fracture Zone
Guatemala Basin
Cocos Ridge
Melanesian Basin
Chagos-Laccadive Plateau
Ceylon Plain
Celebes Sea
Galapagos Fracture Zone
Peru-Chile Trench
Ninetyeast Ridge
Java Sea
Banda Sea
Peru Basin
Cocos Basin
Java Trench
Timor Sea
Arafura Sea
Marquesas Fracture Zone
Mid-Indian Basin
Coral Sea
East Pacific Rise
Wharton Basin
AUSTRALASIA
Tonga Trench
Nazca Ridge
Lord Howe Rise
INDIAN OCEAN
Southwest Pacific Basin
Easter Fracture Zone
Chile Basin
Broken Ridge
Perth Basin
Kermadec Trench
Louisville Ridge
Great Australian Bight
Challenger Fracture Zone
Peru-Chile Trench
Ridge
Chile Rise
Bass Strait
Chatham Rise
Agassiz Fracture Zone
Southeast Indian Ridge
South Australian Basin
Tasman Sea
Tasman Basin
Campbell Plateau
Kerguelen Plateau
Eltanin Fracture Zone
Mornington Abyssal Plain
South Indian Basin
Pacific-Antarctic Ridge
SOUTHERN OCEAN
Bellingshausen Plain
Southeast Pacific Basin
Amundsen Plain
ANTARCTICA
SCALE
0 1,000 2,000 km
0 1,000 2,000 miles

Oceans, seas, and lakes

WATER COVERS MORE THAN 70 PERCENT OF THE EARTH. IT IS HELD IN THE WORLD'S MANY OCEANS, SEAS, AND LAKES.

SEE ALSO
❮ 58–61 Rivers
❮ 112–113 Oceans and seas map
Continents and oceans 202–203 ❯
Hemispheres and latitude 210–211 ❯

Two main types of ecosystems are based in water. Marine ecosystems contain salt water and can be found in seas and oceans; nonmarine ecosystems usually contain fresh water and are found in lakes and rivers.

Life in the oceans and seas

The greatest variety of plants and animals can be found in the Earth's shallow seas, which are enriched with nutrients carried by rivers. Shallow tropical waters are hot spots for marine diversity. In the depths of the oceans, fish and other animals must adapt in order to survive the harsh conditions.

Oceanic zones
As the depth of the ocean increases, light and temperature decrease, and pressure increases. Only specially adapted organisms can live at great depths.

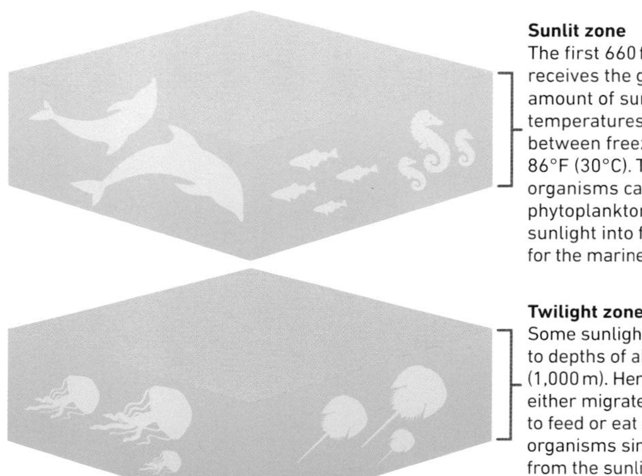

Sunlit zone
The first 660 ft (200 m) receives the greatest amount of sunlight, and temperatures vary between freezing and 86°F (30°C). Tiny organisms called phytoplankton convert sunlight into food energy for the marine ecosystem.

Twilight zone
Some sunlight penetrates to depths of about 3,300 ft (1,000 m). Here, animals either migrate upward to feed or eat dead organisms sinking down from the sunlit zone.

Types of lakes

Lakes are open bodies of standing water surrounded by land. They range in size from ponds to huge lakes with depths of more than 5,000 ft (1,500 m). Many are freshwater, though some are saline. Altogether, they cover about 2 percent of the Earth's surface, providing important habitats for plants and animals.

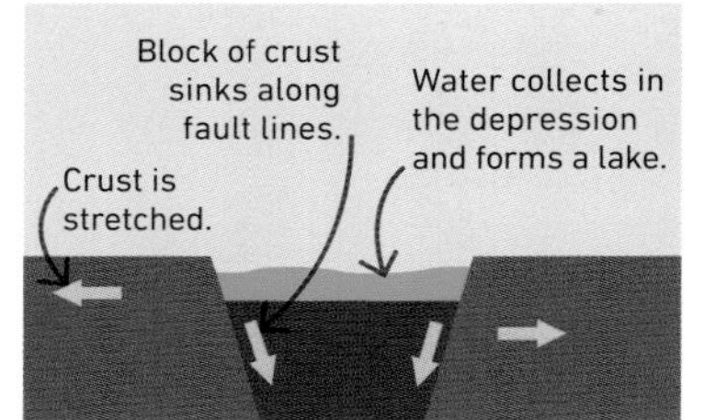
Block of crust sinks along fault lines.
Water collects in the depression and forms a lake.
Crust is stretched.

△ **Tectonic lakes**
Folding and faulting sometimes create depressions. When rivers drain into them, freshwater lakes are formed.

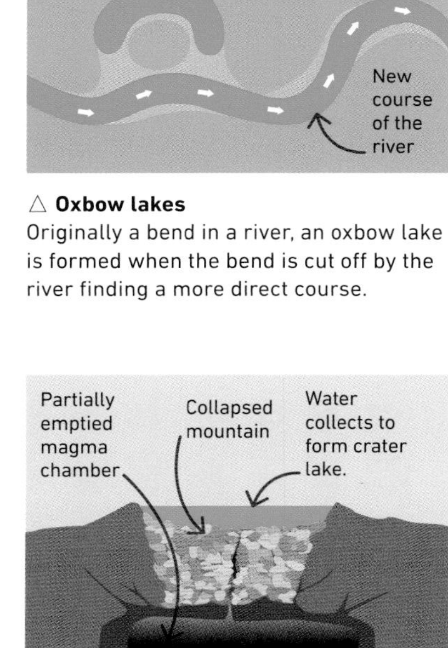
Oxbow lake forms in an abandoned river channel.
New course of the river
Partially emptied magma chamber
Collapsed mountain
Water collects to form crater lake.

△ **Oxbow lakes**
Originally a bend in a river, an oxbow lake is formed when the bend is cut off by the river finding a more direct course.

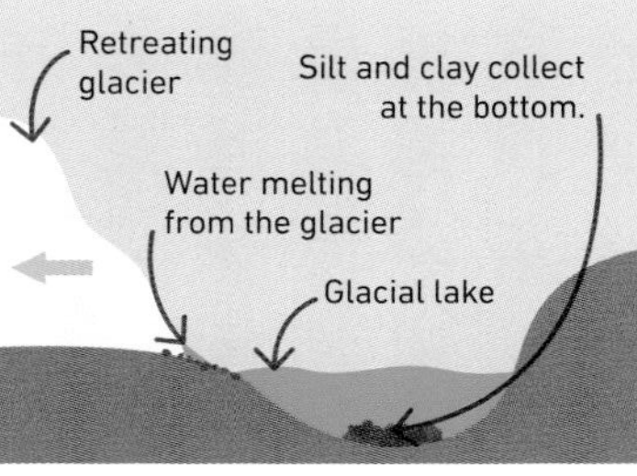
Retreating glacier
Silt and clay collect at the bottom.
Water melting from the glacier
Glacial lake

△ **Glacial lakes**
The majority of the Earth's lakes were originally formed when meltwater from a glacier filled a depression or valley.

△ **Crater lakes**
A few lakes are formed in the craters of extinct volcanoes. Acidic lakes may cover the vents of active volcanoes.

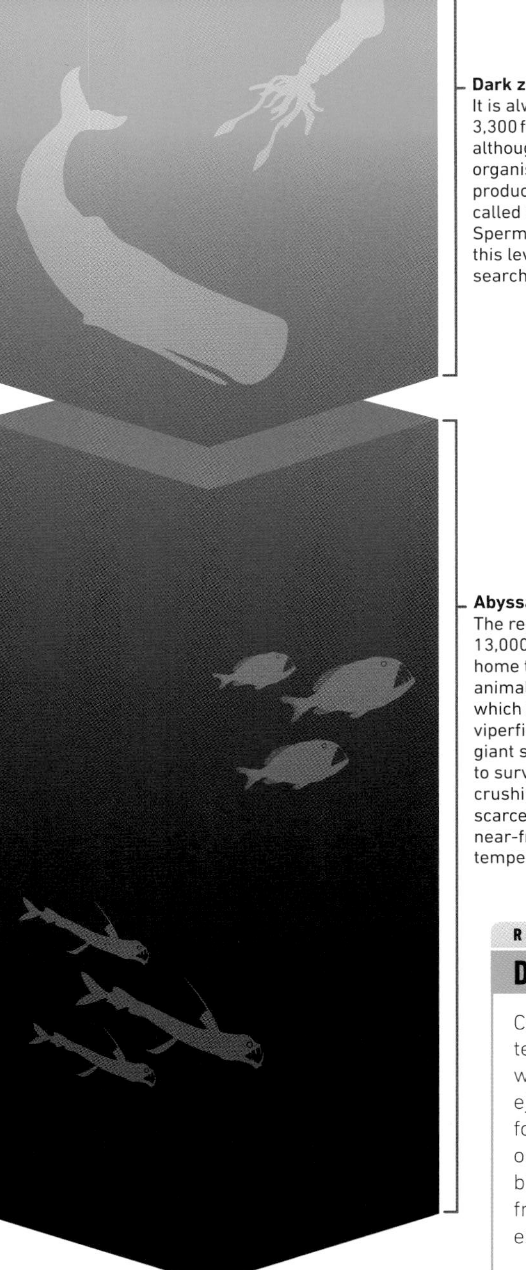

Dark zone
It is always dark below 3,300 ft (1,000 m), although some organisms are able to produce their own light, called bioluminescence. Sperm whales dive to this level when searching for food.

Abyssal zone
The region below 13,000 ft (4,000 m) is home to only a few animals. These animals, which include anglerfish, viperfish, jellyfish, and giant squid, have adapted to survive total darkness, crushing water pressure, scarce food, and near-freezing temperatures.

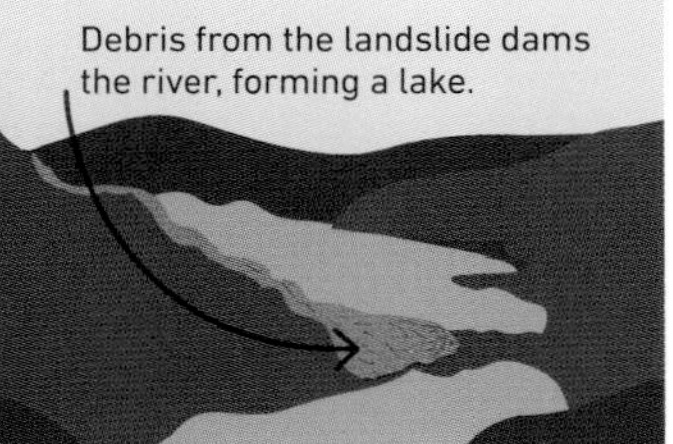
Debris from the landslide dams the river, forming a lake.

△ **Landslide lakes**
Landslides sometimes block or dam streams and rivers, creating a lake that may only be short-lived.

Water collects behind a dam and fills the artificially created basin.

△ **Artificial lakes**
People make artificial lakes for a variety of reasons, including electricity generation, water storage, and recreation.

How latitude affects lakes

Lakes at high latitudes freeze in winter, while lakes at lower latitudes may dry out. Seasonal changes at different latitudes affect how water mixes, redistributing nutrients and oxygen that are important for the lake's ecology.

△ **Temperate lakes**
In summer, a warm layer of water forms over a cooler layer, while in winter the top layer may freeze.

△ **Polar lakes**
Ice covers lakes for most or all of the year. Mixing of water layers occurs during the short summer if the ice melts.

REAL WORLD

Dying reefs

Coral reefs need stable ocean temperatures. If the water gets too warm, corals become stressed and eject the colorful algae they rely on for energy, causing bleaching. The oceans are heating up, and this bleaching is happening more frequently, endangering the very existence of coral reefs.

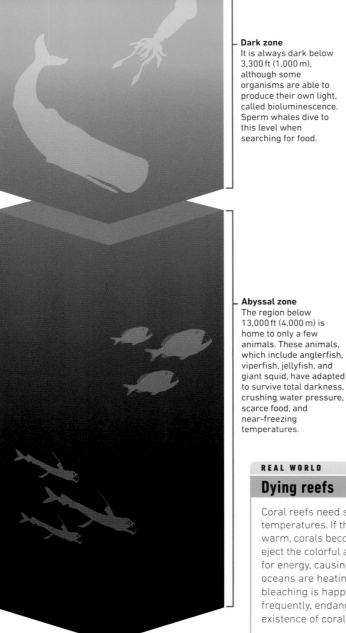

2

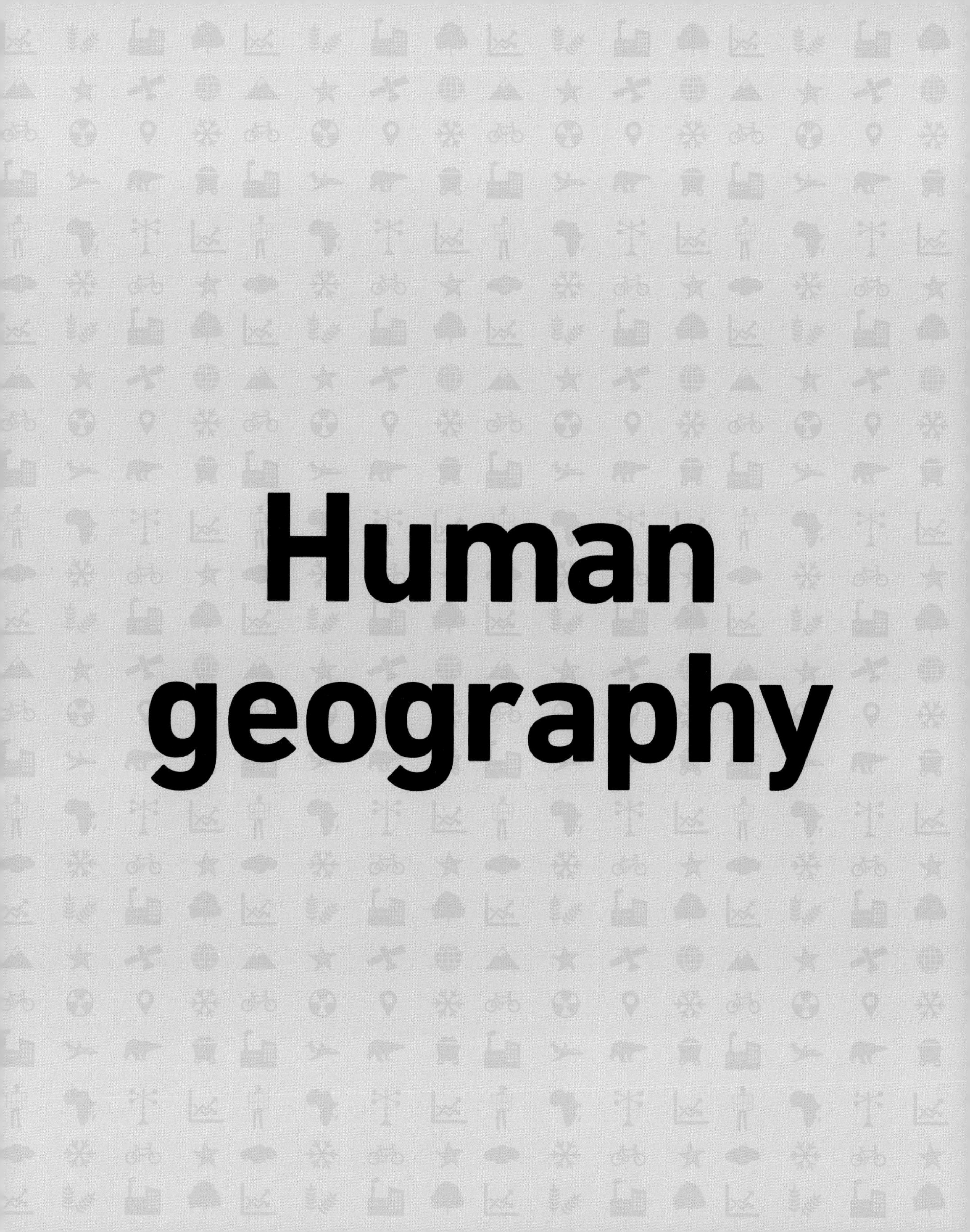

Human geography

What is human geography?

HUMAN GEOGRAPHY IS THE STUDY OF THE RELATIONSHIP BETWEEN PEOPLE AND THE PLACES AND ENVIRONMENTS AROUND THEM.

Geographers are interested in how people change the places around them as well as how places influence the people who live there. This may be in natural environments, such as the rainforest, or in built environments, such as the city.

Population and settlement

Geographers record where people live and how many people live there. They look at how a population changes over time and investigate how different types of settlements (villages, towns, and cities) are affected as the population changes.

▽ **Towns and cities**
Most people in the world live in towns and cities. Urban areas may change rapidly as the local population changes.

Economic and social geography

The Earth's natural resources provide people and societies with the means to make a living, providing us with food, water, and energy. Geographers study how economies have grown, how societies organize work, the distribution of goods and services, and how all this affects environments around us.

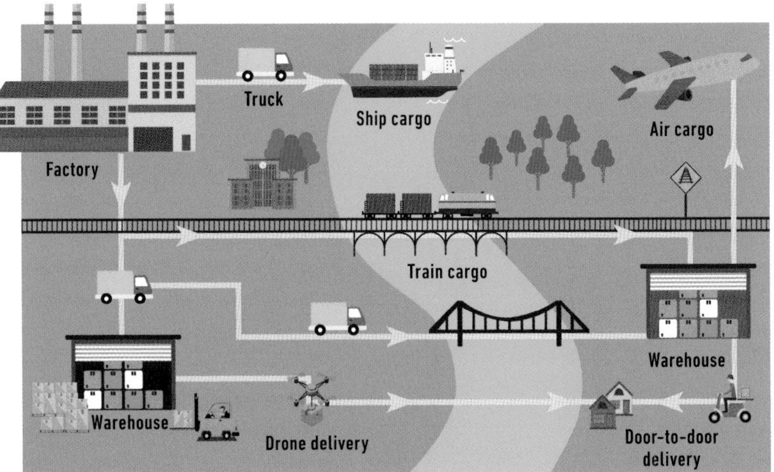
Truck
Ship cargo
Air cargo
Factory
Train cargo
Warehouse
Warehouse
Drone delivery
Door-to-door delivery

Transportation and distribution ▷
Products need to be transported, by different methods, from the factories where they are made to where their customers buy them.

People and the environment

Humans are part of the environments in which they live, so their actions can have serious environmental consequences. Geographers study the impact of humans on the environment, such as the destruction of forests, overfishing, or the emission of greenhouse gases into the atmosphere. They also look at the ways people or governments can reduce the harm caused by these activities.

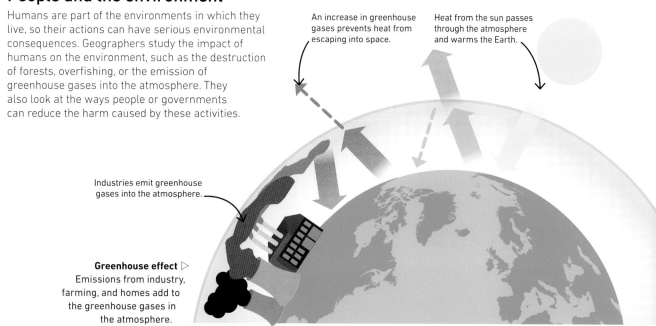

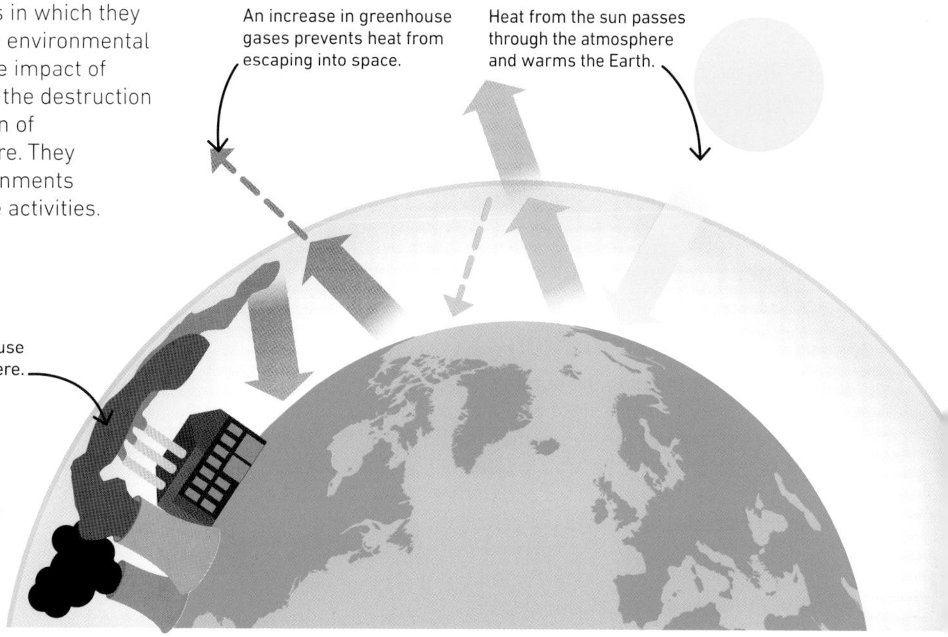

An increase in greenhouse gases prevents heat from escaping into space.
Heat from the sun passes through the atmosphere and warms the Earth.
Industries emit greenhouse gases into the atmosphere.

Greenhouse effect ▷ Emissions from industry, farming, and homes add to the greenhouse gases in the atmosphere.

Big geographical issues

Our planet and population are facing some serious issues. These include the effects of uneven development around the world, population growth projections and poverty, and problems with accessing sufficient food and water supplies. Some parts of the world are also in conflict, despite the fact we are all linked globally and locally by trade.

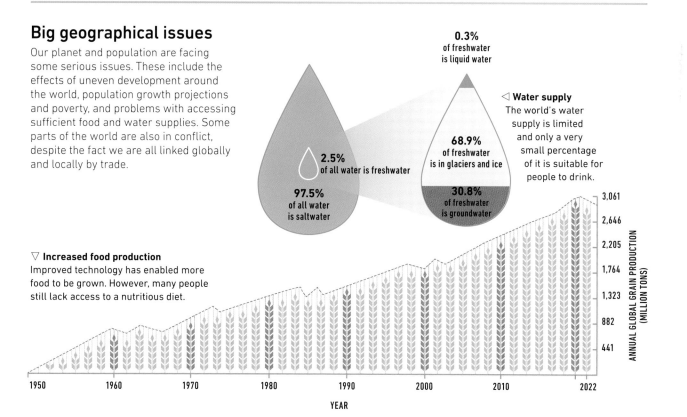

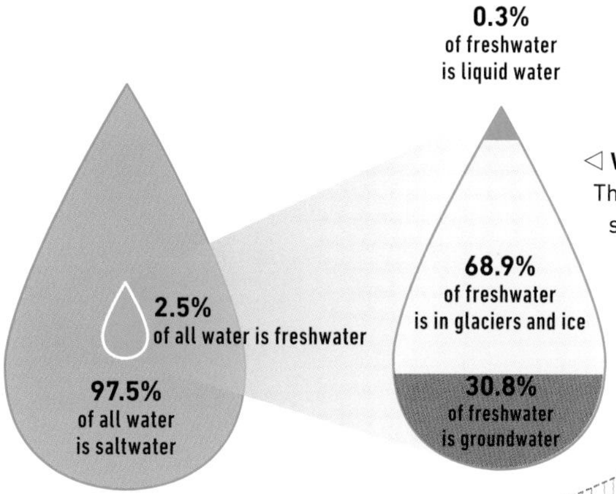

0.3%
of freshwater is liquid water
2.5%
of all water is freshwater
68.9%
of freshwater is in glaciers and ice
97.5%
of all water is saltwater
30.8%
of freshwater is groundwater

◁ **Water supply** The world's water supply is limited and only a very small percentage of it is suitable for people to drink.

▽ **Increased food production** Improved technology has enabled more food to be grown. However, many people still lack access to a nutritious diet.

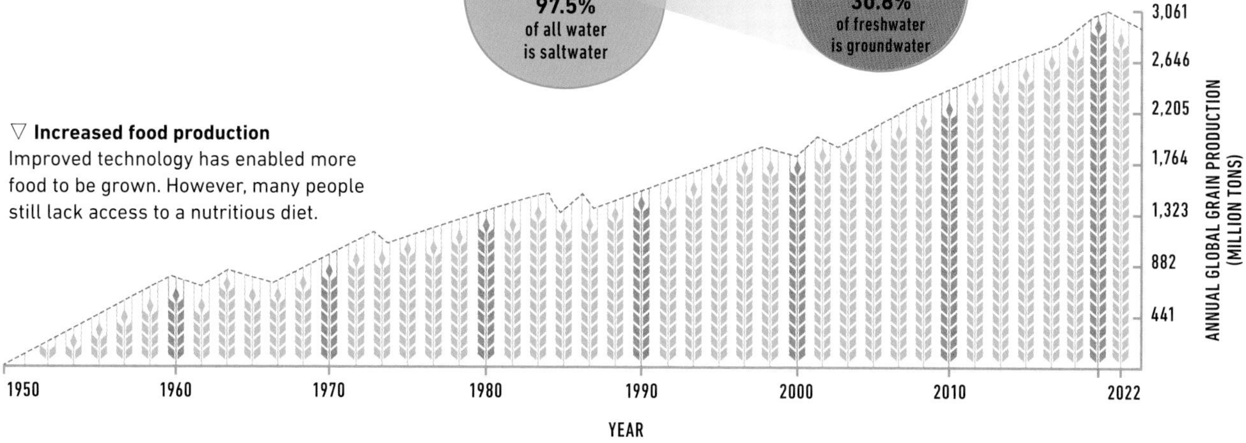

3,061
2,646
2,205
1,764
1,323
882
441
ANNUAL GLOBAL GRAIN PRODUCTION (MILLION TONS)
1950
1960
1970
1980
1990
2000
2010
2022
YEAR

Where people live

THE WORLD'S POPULATION IS ESTIMATED TO BE 8 BILLION. PEOPLE ARE UNEVENLY DISTRIBUTED AROUND THE WORLD.

SEE ALSO
Migration 124–125 ›
Human settlements 128–129 ›
Megacities 132–133 ›
Urbanization 136–137 ›

The world's population is not evenly distributed due to a combination of physical and human factors, which affect the number of people living in an area.

Density and distribution

Population density refers to the number of people living in a given area, usually 1 sq mile. It is calculated by dividing the total population of an area by the size of the area (in sq mile) and is a measure of how crowded the area is. Density figures make it easier to compare the populations of different areas. Distribution refers to how people are spread out across the world.

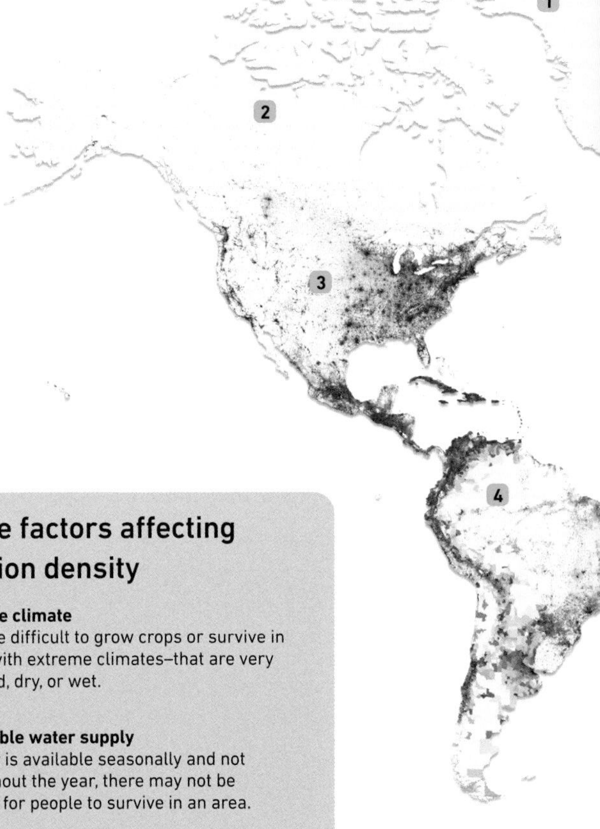
1
2
3
4

Factors affecting population density

Positive factors encourage settlement and lead to high population density, while negative factors discourage settlement and lead to low population density.

✔ Positive factors affecting population density

Reliable water supply
Areas with a regular supply of freshwater not only enable people to survive but also allow them to farm and produce crops.

Accessible location
Places located on a coast, along a river, or in a gap between mountains are more attractive to live in, as they allow trade and tourism.

Stable government
Political stability encourages investment in industry and infrastructure, creating jobs that attract people to live in an area.

Availability of natural resources
A plentiful supply of fuel, water, and minerals to use or to trade encourages people to live in an area.

Easy-to-clear vegetation and fertile soil
An area where it is easy to clear the vegetation for construction, or that has fertile soil to grow crops, is attractive.

✖ Negative factors affecting population density

Extreme climate
It can be difficult to grow crops or survive in areas with extreme climates–that are very hot, cold, dry, or wet.

Unreliable water supply
If water is available seasonally and not throughout the year, there may not be enough for people to survive in an area.

Inaccessible location
People are less likely to live in areas that are remote and difficult to access, such as high mountain regions or dense forests.

Lack of natural resources
Areas with insufficient mineral resources and energy supplies do not attract people because fewer jobs are available.

Hard-to-clear vegetation and infertile soil
Areas with hard-to-clear vegetation, or with infertile soil where growing crops is difficult, are sparsely populated.

The **population of China** is almost **60 times** more than that of **Australia**.

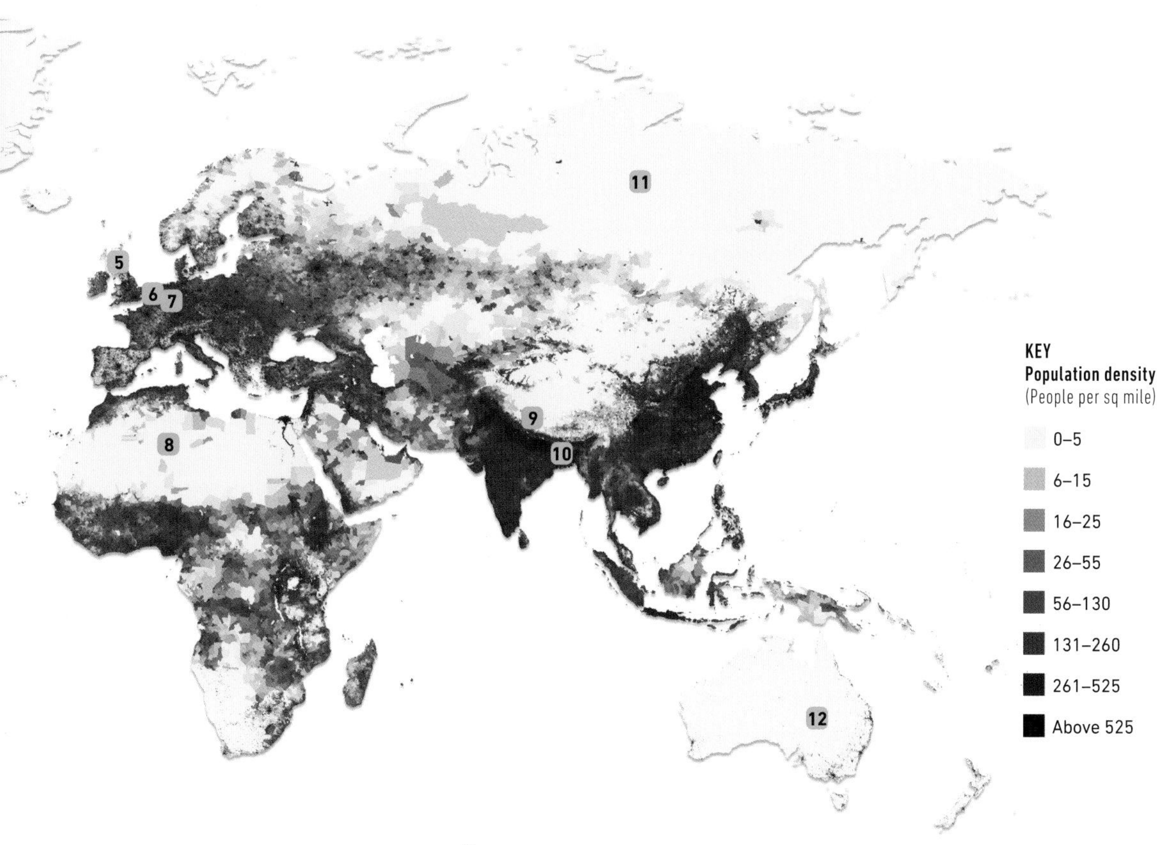
11
5
6
7
9
8
10
12
KEY
Population density
(People per sq mile)
0–5
6–15
16–25
26–55
56–130
131–260
261–525
Above 525

1 Greenland
Mostly covered in ice, Greenland experiences an extreme climate and has a population density of 0.07 people per sq mile.

2 Northern Canada
With some areas that are very inaccessible and with an extreme climate, Northern Canada has a very low population density.

3 The United States
With a stable government, good infrastructure, sufficient fuel, and plentiful food and water supplies, the US supports a large population.

4 The Amazon Rainforest
A dense tropical rainforest, the Amazon has a low population density due to its climate and inaccessibility.

5 The United Kingdom
With a moderate climate, fertile soil, and few inaccessible areas, the UK has a high population density of 719 people per sq mile.

6 The Netherlands
A stable country with flat, fertile soil and an accessible coastline, this is the most densely populated country in the European Union.

7 The Ruhr Valley, Germany
This area is rich in coal and iron ore. This enabled the development of heavy industry, and the population has grown around it.

8 Sahara
With an extreme climate and little water to grow crops, the Sahara desert has a low population density of 0.3 people per sq mile.

9 Himalayas
The highest mountain range in the world, it is difficult to grow crops here, and the Himalayas are largely inaccessible, with a low population density.

10 Lower Ganges Valley
Low-lying and flat with rich, fertile soil and a warm, wet climate ideal for farming, the Lower Ganges Valley has a high population density.

11 Siberia, Northern Russia
While Siberia has an extreme climate, it is rich in natural resources. This has encouraged a small population to live there.

12 Australia
While large parts of the country are too dry to support farming, coastal areas are more densely populated. Australia has a population density of 9 people per sq mile.

Demography

DEMOGRAPHY IS THE STUDY OF POPULATION AND CHANGES IN POPULATION STATISTICS OVER TIME AND SPACE.

SEE ALSO
❮ 120–121 Where people live
Migration 124–125 ❯
Health 140–141 ❯
Uneven development 158–161 ❯

Overall, the global population is increasing quickly. Most of this growth is in low-income countries (LICs). In contrast, the population of some high-income countries (HICs) is declining.

Changes in population

Changes in a country's population size are due mainly to changes in its birth and death rates. If the birth rate is higher than the death rate, then the population will increase, and if the death rate outweighs the birth rate, the population will decrease. These natural causes of population change are called the natural increase or decrease. Infant mortality rates, life expectancy, and migration (the movement of people in and out of a country) also affect population size.

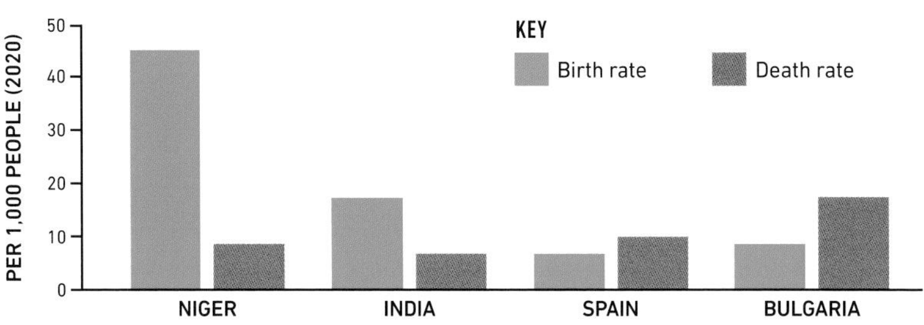
PER 1,000 PEOPLE (2020)
50
40
30
20
10
0
NIGER
INDIA
SPAIN
BULGARIA
KEY
Birth rate
Death rate

△ **Birth and death rates**
With birth rates rising almost five times faster than death rates, Niger has one of the fastest growing populations in the world. Bulgaria has one of the fastest declining populations, with its birth rate exactly half its death rate.

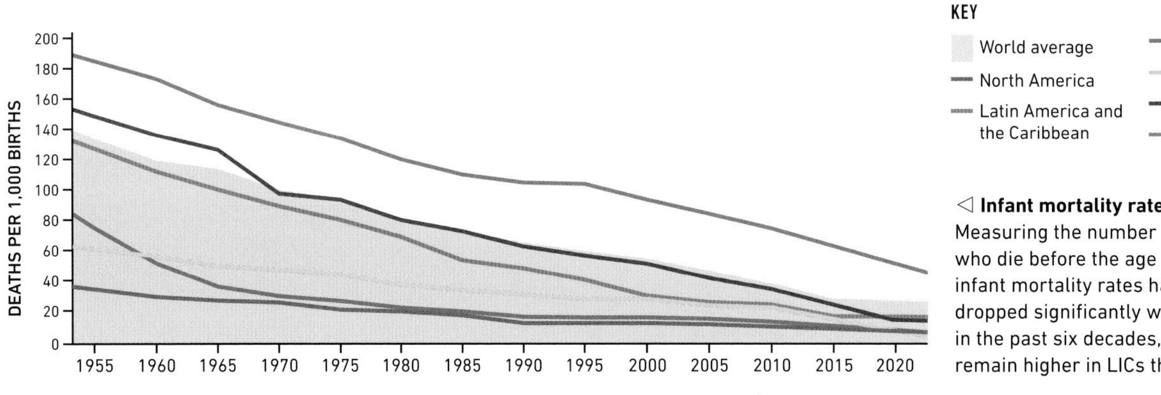
DEATHS PER 1,000 BIRTHS
200
180
160
140
120
100
80
60
40
20
0
1955 1960 1965 1970 1975 1980 1985 1990 1995 2000 2005 2010 2015 2020
KEY
World average
North America
Latin America and the Caribbean
Europe
Oceania
Asia
Africa

◁ **Infant mortality rate**
Measuring the number of babies who die before the age of 1, infant mortality rates have dropped significantly worldwide in the past six decades, but remain higher in LICs than HICs.

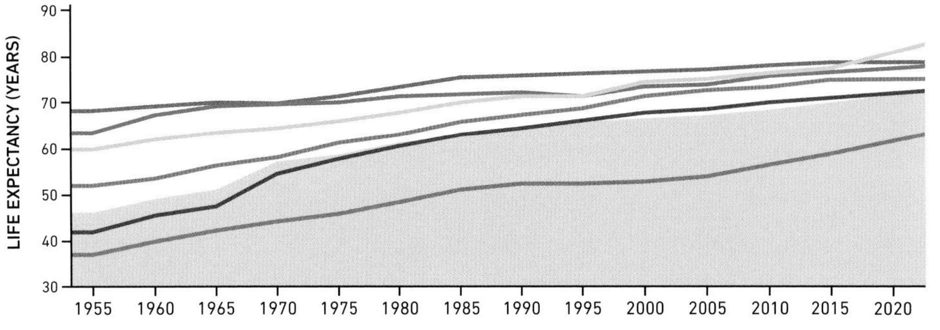
LIFE EXPECTANCY (YEARS)
90
80
70
60
50
40
30
1955 1960 1965 1970 1975 1980 1985 1990 1995 2000 2005 2010 2015 2020

◁ **Life expectancy**
As life expectancy—the average number of years from birth a person is expected to live—increases, death rates fall. Life expectancy has improved around the world over the last 60 years.

Migration

International migration figures show the number of people who move to a country (immigration) and away from a country (emigration). There are many reasons for people to migrate, including natural disasters, war, and economic factors.

Many people from LICs, such as parts of Latin America, seek to emigrate to HICs such as the US and Canada.

Many people have been forced to leave Syria because of war. Türkiye and Germany have taken in many Syrian refugees.

Many skilled workers leave China to find better-paid jobs abroad.

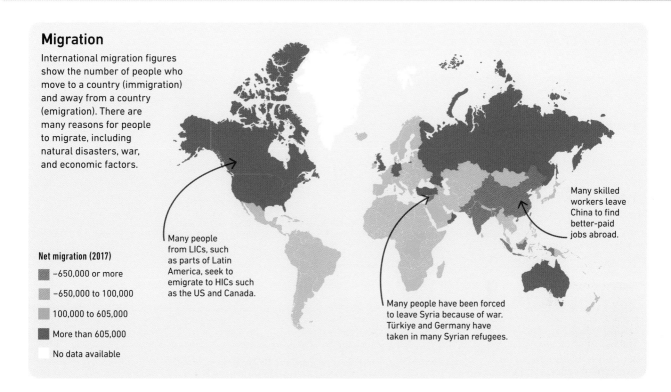

Net migration (2017)

- –650,000 or more
- –650,000 to 100,000
- 100,000 to 605,000
- More than 605,000
- No data available

Population structure

Population structure refers to the number of people in each age group in a country and how they are divided by gender. These statistics are usually plotted using population pyramids. The shapes of these pyramids tend to vary depending on the level of development of a country.

In Africa, **41 percent** of people are **under the age of 15**. In Europe, the figure is **16 percent**.

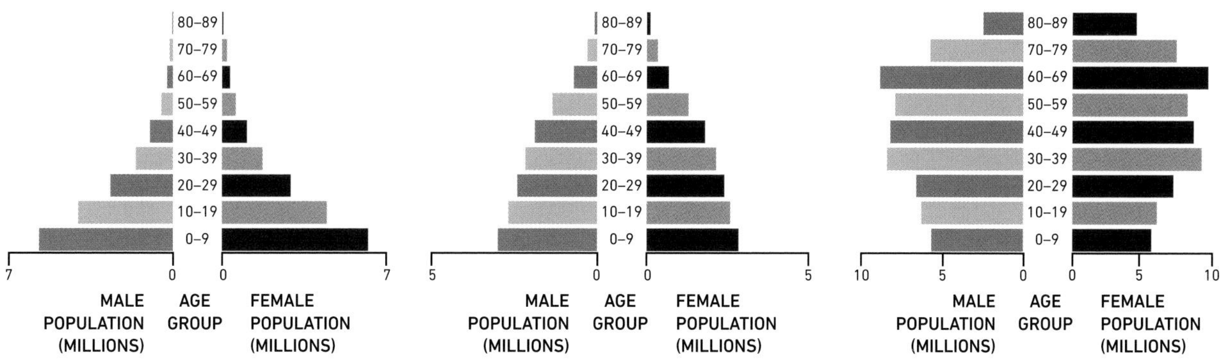
80–89
70–79
60–69
50–59
40–49
30–39
20–29
10–19
0–9
7
0
0
7
MALE POPULATION (MILLIONS)
AGE GROUP
FEMALE POPULATION (MILLIONS)
5
0
0
5
10
5
0
0
5
10

△ **Young population (Uganda: LIC)**
A pyramid with a broad base but a narrow tip indicates a fast-growing population with a high birth rate and low life expectancy. This is typical of a LIC such as Uganda.

△ **Stable population (Malaysia: NEE)**
A tapering pyramid indicates slower population growth with a constant birth rate and relatively low death rate, typical of Newly Emerging Economies (NEEs).

△ **Aging population (Japan: HIC)**
A narrow-based, bulging pyramid shows a low proportion of young people. Although life expectancy is long, the birth rate is low and Japan's population is falling.

Migration

THE MOVEMENT OF PEOPLE FROM ONE AREA TO ANOTHER IS KNOWN AS MIGRATION.

SEE ALSO
❮ 122–123 Demography
Urbanization 136–137 ❯
The spread of cultures 138–139 ❯
Uneven development 158–161 ❯
Globalization 162–163 ❯

Throughout human history, people have migrated, often seeking water, food, or space. Migration may involve a long journey to a new country or a relocation in the same country.

Types of migration

Migration can be of different types depending on where people are moving from, where they are moving to, and the reasons for their movement. People who move into a place, or country, are known as immigrants, and people who move out of a place, or country, are called emigrants.

Type of migration	Movement
External (international) migration	**Moving from one country to another**
Voluntary migration	People may choose to move from one country to another. It may be for a new job, to join family members, or to retire in a new country.
Forced migration	People may be forced to move from one country to another. They may be escaping war or famine.
Internal (domestic) migration	**Moving from one part of a country to another**
Regional migration	In many countries, people migrate from regions that are isolated or suffering from decline and unemployment, to others where work is plentiful.
Rural–urban migration	People may move from the countryside to a town or city. In low-income countries, unskilled workers often move to cities in search of jobs, a better education, and other opportunities.
Counterurbanization	People may move from the city, or town, to the countryside. This happens especially in high-income countries, where people move to the countryside for a better quality of life.
Semi-permanent migration	**Moving from one place to another for months or years, but not permanently**
Seasonal migration	People may move for a fixed period at a specific time of the year. For example, people may move to work at a ski resort for the ski season or to a farming area to pick crops at harvest time.
Daily, or weekly migration	This involves people commuting to work in a different town or city from where they live, but returning home in the evening, or at the end of the week.
Economic migrants	People may migrate to a different country to find better jobs and quality of life. Economic migrants are not always poor, nor less skilled or undereducated. They may return home or settle permanently.
Refugees	Some people may have been forced by conflict, or political or religious persecution to leave their country, often at great risk. Refugees have a right to asylum under international law.

Reasons for migration

People move from one place to another due to push and pull factors. Push factors drive people to leave, while pull factors attract people into an area. Sometimes, one kind of factor may be stronger than the other, but more often it is a combination of factors that helps people make the decision to move.

▽ **Push out, pull in**
Different factors can either drive people away or draw them to other locations.

Better educational opportunities

Better jobs with higher rates of pay

Religious or political freedom

Better services, such as health care

PULL
FACTORS

PUSH
FACTORS

Natural disasters

Slavery or conflict

Poverty

Overcrowding

Impacts of migration

Immigrants moving to another country, or another part of a country, can have an impact on both the area they are leaving and the area they are moving to. These impacts can be financial—affecting the economy—and social, affecting the lives of people and communities.

Positive impacts on destination country

- ✔ Immigrants pay taxes, which helps the economy.
- ✔ Immigrants introduce new cultures, food, and languages.
- ✔ Immigrants bring skills, trades, and experiences, such as nursing, plumbing, and caring.
- ✔ Immigrants add to the workforce.

Negative impacts on country of origin

- ✖ There is a loss of a young and fit workforce.
- ✖ Skilled workers move out.
- ✖ The birth rate falls due to young people migrating.
- ✖ Family units break up.

Why migration is controversial

Immigration can be controversial. New people arriving in a country can make the people already living there feel anxious about competition for jobs, living space, and local services. Sometimes immigrants work hard for low pay, which may be seen as a threat. Immigrants bring with them their own religion and ways of living and these can seem strange to the settled population. All this can put pressure on governments to reduce the number of people they allow into a country.

Slavery

One terrible form of forced migration is slavery. For 300 years until the transatlantic slave trade was finally abolished, millions of Africans were forced onto crowded ships and carried to the Americas to be slaves. Slavery is no longer legal, but there are still 27 million people enslaved around the world. The largest group are victims of human trafficking, the secret trade of people across borders for sexual exploitation and forced labor.

Population change

THE SIZE AND STRUCTURE OF THE WORLD'S HUMAN POPULATION CHANGES OVER TIME AND SPACE.

SEE ALSO
❮ 120–121 Where people live
❮ 122–123 Demography
Urbanization 136–137 ❯
Sustainability 184–185 ❯

The population of the world is increasing and is unevenly distributed across countries and regions. The age structure of the population and proportion of men to women also vary from one place to another.

Rising global population

Birth and death rates affect how the population changes across the globe and with the passage of time. The current global population is around 7.8 billion and is expected to reach 10 billion by 2056.

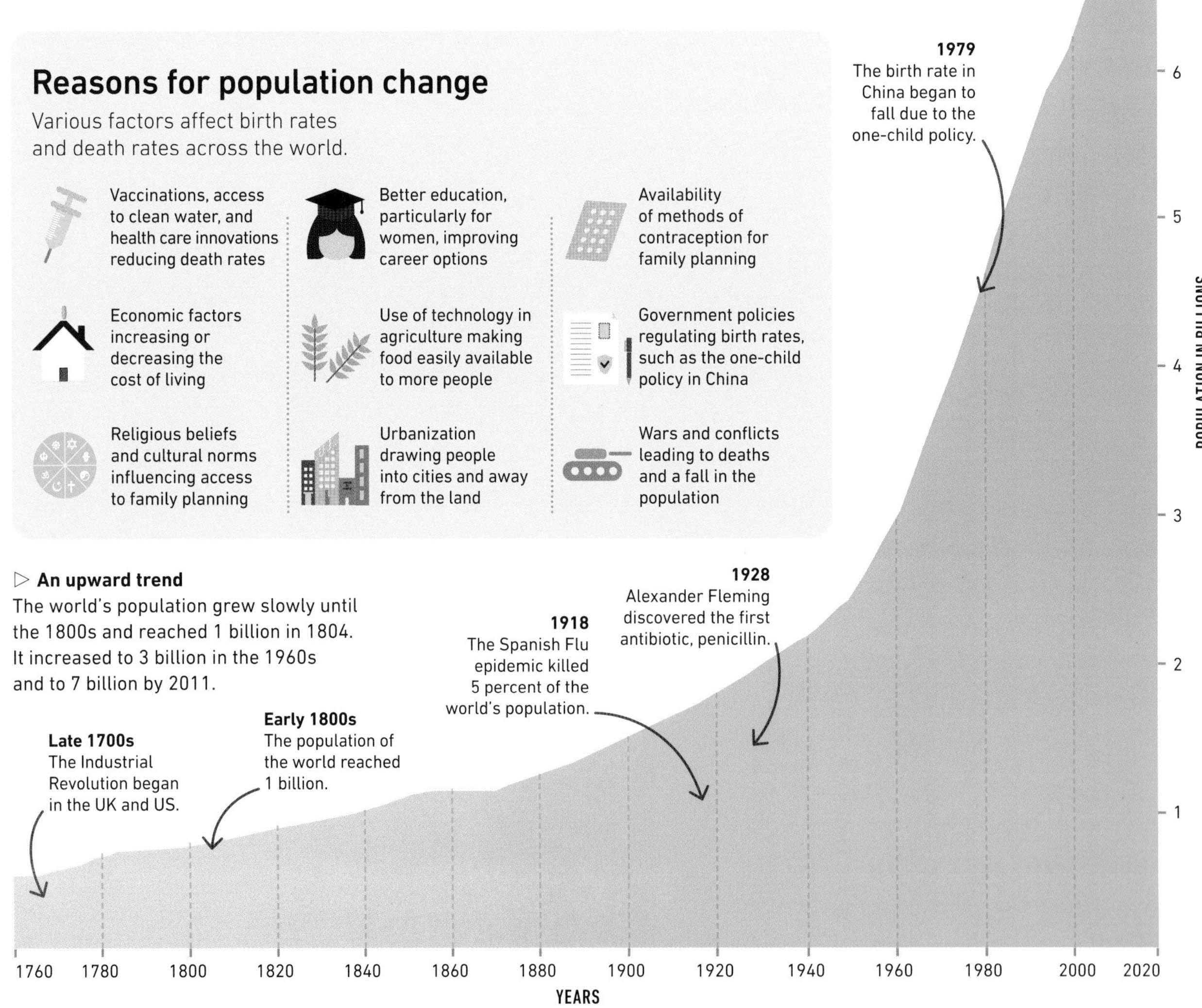
Reasons for population change
Various factors affect birth rates and death rates across the world.
Vaccinations, access to clean water, and health care innovations reducing death rates
Better education, particularly for women, improving career options
Availability of methods of contraception for family planning
Economic factors increasing or decreasing the cost of living
Use of technology in agriculture making food easily available to more people
Government policies regulating birth rates, such as the one-child policy in China
Religious beliefs and cultural norms influencing access to family planning
Urbanization drawing people into cities and away from the land
Wars and conflicts leading to deaths and a fall in the population
2020 The rate of population growth is showing signs of a slowdown.
2011 The population of the world reached 7 billion.
1979 The birth rate in China began to fall due to the one-child policy.
1928 Alexander Fleming discovered the first antibiotic, penicillin.
1918 The Spanish Flu epidemic killed 5 percent of the world's population.
Early 1800s The population of the world reached 1 billion.
Late 1700s The Industrial Revolution began in the UK and US.
1
2
3
4
5
6
7
8
POPULATION IN BILLIONS
1760
1780
1800
1820
1840
1860
1880
1900
1920
1940
1960
1980
2000
2020
YEARS

▷ **An upward trend**
The world's population grew slowly until the 1800s and reached 1 billion in 1804. It increased to 3 billion in the 1960s and to 7 billion by 2011.

Population change around the world

Population growth is fastest in low-income countries (LICs), where improvements in health care have reduced the death rate while the birth rate remains high. In high-income countries (HICs), both the birth rate and death rate have fallen due to better family planning, good health care, and universal education.

▽ **Shifting population**
The percentage of the global population on each continent has changed over time, and will keep changing in the years to come.

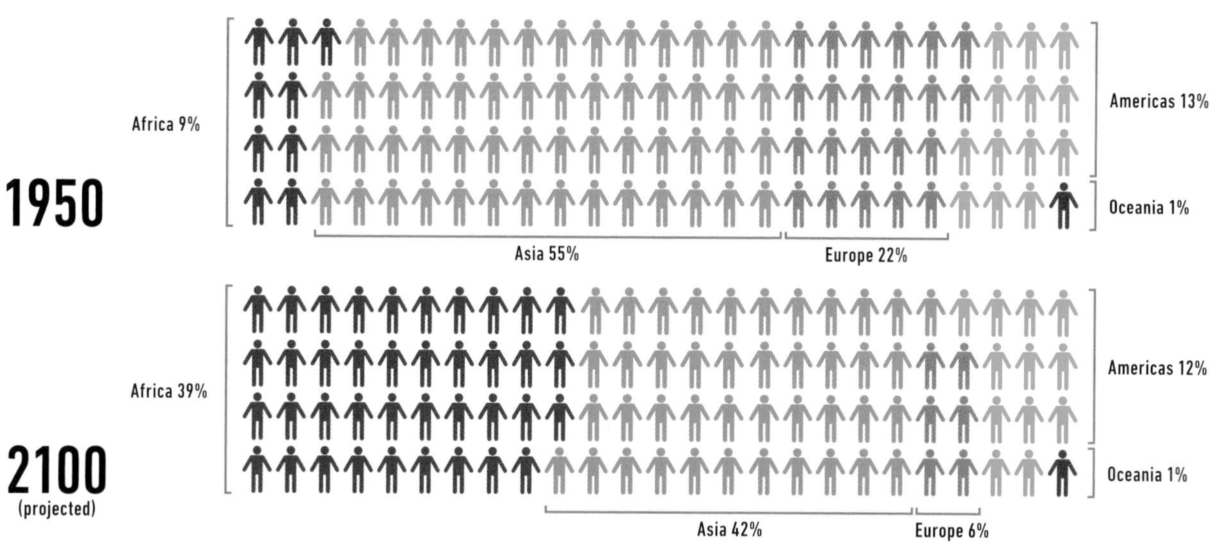
1950
Africa 9%
Americas 13%
Oceania 1%
Asia 55%
Europe 22%
2100
(projected)
Africa 39%
Americas 12%
Oceania 1%
Asia 42%
Europe 6%

The demographic transition model

A good way to illustrate how birth and death rates relate to a country's economic development is the demographic transition model. This graph shows how populations move through five stages depending on their birth and death rates. The model provides a very useful way to analyze population change from country to country.

▽ **Stages of population change**
The model shows five stages. The population of a country becomes sustainable at Stage 4, when the birth rate and death rate are low.

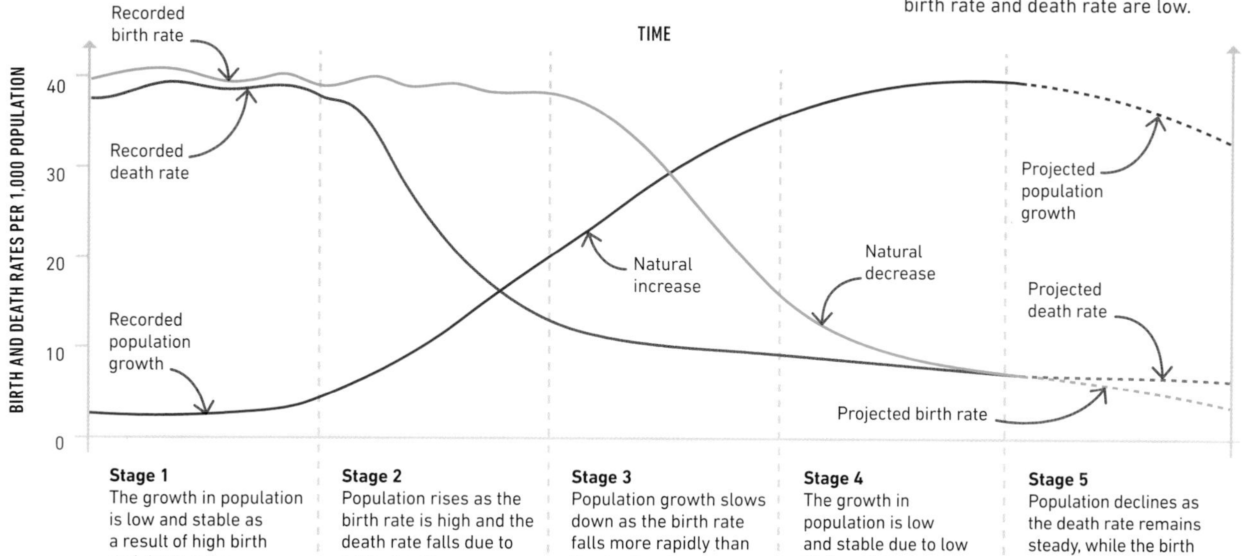
Recorded birth rate
TIME
BIRTH AND DEATH RATES PER 1,000 POPULATION
40
30
20
10
0
Recorded death rate
Recorded population growth
Natural increase
Natural decrease
Projected population growth
Projected death rate
Projected birth rate

Stage 1
The growth in population is low and stable as a result of high birth and death rates.

Stage 2
Population rises as the birth rate is high and the death rate falls due to better health care.

Stage 3
Population growth slows down as the birth rate falls more rapidly than the death rate.

Stage 4
The growth in population is low and stable due to low birth and death rates.

Stage 5
Population declines as the death rate remains steady, while the birth rate decreases.

Human settlements

SEE ALSO
❮ 120–121 Where people live
Rural settlements 134–135 ❯
Urbanization 136–137 ❯

A NUMBER OF PHYSICAL FEATURES DETERMINE THE LOCATION OF WHERE PEOPLE LIVE.

A settlement is a place where people live and can be large or small. Early settlements gradually expanded over time to become larger towns and cities.

Local rock can provide building materials. Other minerals may be used to make objects, such as pots and tools, and also for trade.

A water body, such as a river, usually guarantees a reliable, year-round supply of fresh water for people and agriculture. Rivers can also provide food and a means of transportation.

Nearby forests can provide fuel for cooking and heating, and building materials.

Site and situation

The site of a settlement describes the place where it is located, such as by a river or on a hill. The situation of a settlement is its location in relation to surrounding features, such as other towns and natural routes. Site and situation together determine the location of a settlement.

Settlement sizes

Settlements can be classified according to their size and the range and number of services, such as shops and schools, they provide. When the size of a settlement increases, so does the range and number of services offered. There are many smaller settlements and fewer large ones.

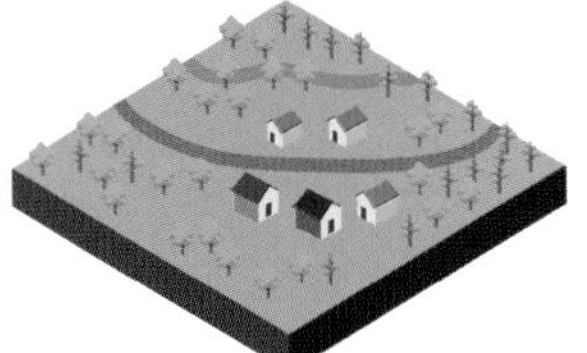

△ **Hamlet**
Consisting of only five or six farms or houses in rural areas, hamlets do not have shops or services.

△ **Village**
A small community with several hundred people, a village may have a small shop and a primary school.

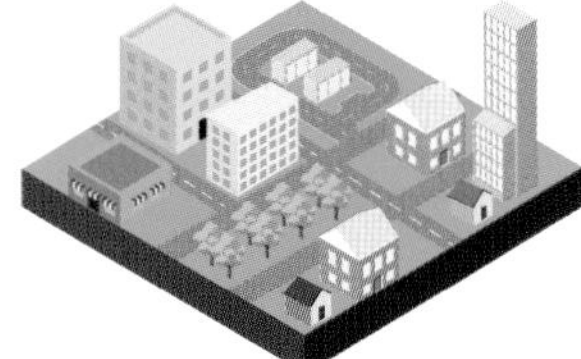

△ **Town**
With a wider range of services, transportation links, and jobs, towns serve not only their residents but also surrounding villages.

Settlement patterns

The pattern of settlements across the landscape is often determined by the physical terrain and the availability of natural resources such as water. Accessibility plays an important role, too, along with proximity to other settlements.

△ **Dispersed**
These are isolated buildings or a hamlet separated from the next by some distance and exist where natural resources are insufficient to support a large population.

△ **Nucleated**
Buildings may be clustered together, initially for defense and later for social or economic reasons. There may be a central marketplace or water supply.

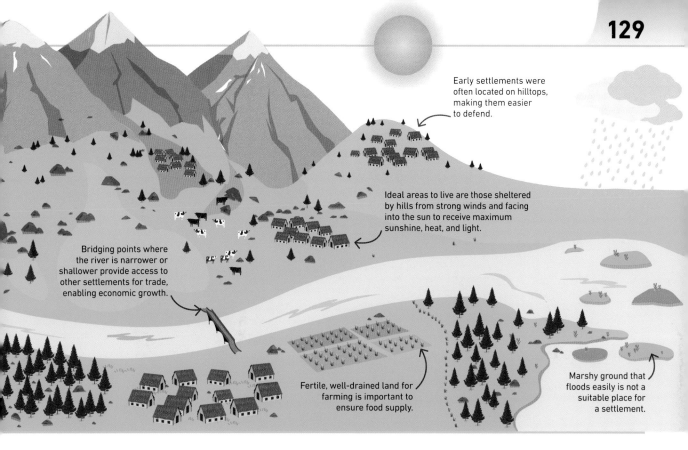

Early settlements were often located on hilltops, making them easier to defend.

Ideal areas to live are those sheltered by hills from strong winds and facing into the sun to receive maximum sunshine, heat, and light.

Bridging points where the river is narrower or shallower provide access to other settlements for trade, enabling economic growth.

Fertile, well-drained land for farming is important to ensure food supply.

Marshy ground that floods easily is not a suitable place for a settlement.

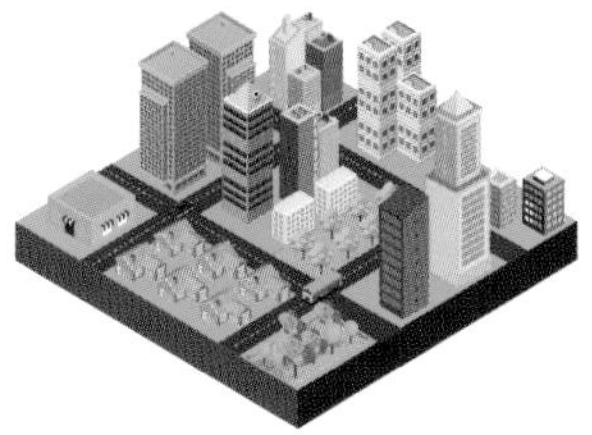

△ **City**
With specialized services such as universities and public and private offices, cities attract visitors as well as migrants who wish to live and work there.

△ **Linear**
These settlements have buildings stretched out in a line along a physical feature, such as at the foot of a hill, beside a river, or along a human feature, such as a road.

Functions of settlements

The function of a settlement is its main purpose or activity, such as trading. Settlements have arisen and expanded due to their original purpose or a change in their purpose. Today, most settlements are multifunctional.

Port
Located on the coast or on navigable rivers, ports enable goods to move from one place to another.

Industrial town
Areas close to the supply of raw materials for industries develop into industrial towns.

Religious center
Major religious sites of worship give rise to settlements that function as religious centers.

Educational towns and cities
Some places are famous for education, such as Oxford, in the UK, where the university dominates employment.

Administrative town
Often developed around law and order and local council offices, administrative towns run the local area.

Tourist areas
Linked to different types of tourism, tourist areas grow in locations such as coastal or forested areas.

Cities

SEE ALSO
❮ 120–121 Where people live
❮ 128–129 Human settlements
Megacities 132–133 ❯
Urbanization 136–137 ❯

MOST CITIES CAN BE DIVIDED INTO DISTINCT ZONES ACCORDING TO HOW THEIR LAND IS USED.

A city typically includes residential areas where people live, industrial areas for factories, and commercial areas for stores and offices. The city's layout may reflect its growth and history.

Land use patterns

Some cities can be divided into distinct rings that surround the city center. At the center is the central business district (CBD), where many stores, banks, or institutions can be found. Surrounding this are rings containing different types of housing and industry. Each city is different, however, and many cities follow patterns that differ from this one.

Rural-urban fringe
The edge of the city is home to larger houses, open spaces, and retail parks. It may have a number of transportation routes running through it.

Central business district
The center contains stores, businesses, offices, and entertainment. It is well connected to the other zones.

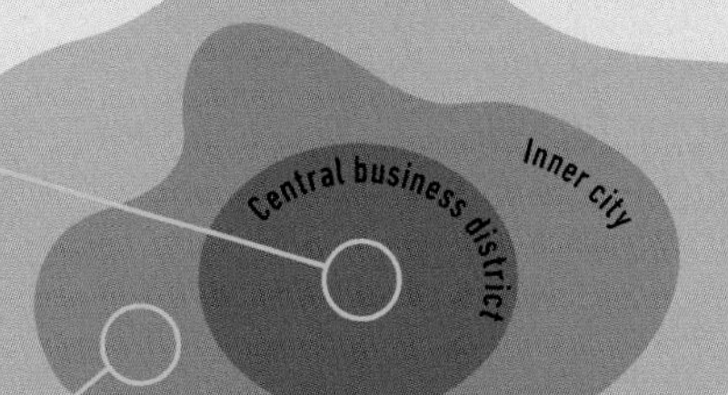
Suburbs
Rural-urban fringe
Inner city
Central business district

Inner city
This area once contained factories and workers' housing. In many cases, they have been redeveloped and may contain apartment complexes and converted warehouses.

Suburbs
These residential areas contain more modern housing with gardens and some services, such as local schools and shops.

In the inner city

Inner-city areas are located next to the central business district, which means property and land can be very expensive. They are densely packed areas with many small houses originally built for the working population and modern high-rise apartments. This part of the city is very accessible and in high demand.

1 The inner city contains older buildings such as museums and libraries. Abandoned factories and warehouses may have been put to new uses.

2 The area is crowded with little open space. There are few parks or gardens, and buildings are usually very close to each other.

3 Land and property is very expensive in this zone because of its proximity to workplaces and centers of entertainment.

4 Traffic jams are a problem because many people drive into the center. There is usually a good choice of public transportation here, too.

5 The proximity to jobs in the central business district attracts people, leading to a rich cultural mix of people, shops, and services.

Change in the city

The inner city was traditionally an area of industry and housing for workers. In many cities, industry has moved abroad or to more accessible sites out of town, bringing about major changes to inner-city areas. The use of land in this part of a city is now more diverse to reflect these changes.

PROBLEM	SOLUTION
As industry has moved out of this zone, unemployment has increased, especially for low-skilled workers.	New industries such as media, advertising, and finance have moved into the old factory buildings, creating jobs.
Many former factories have been left empty or become derelict, attracting vandals and social problems.	Abandoned buildings have been converted into housing, art galleries, and museums, attracting more people into the area.
Poor infrastructure means businesses leave the area. Traffic congestion makes it difficult for people to commute.	Governments have provided grants to develop modern, environmental public transportation systems, such as trams or light-rail.

Since 1950, the city of **Detroit** has lost ***more than*** 60 percent of its population.

Megacities

AS CITIES GROW LARGER, MANY HAVE BECOME HUGE MEGACITIES, WITH POPULATIONS BIGGER THAN SOME COUNTRIES.

SEE ALSO

❮ 120–121 Where people live
❮ 130–131 Cities
Urbanization 136–137 ❯
Capitals and major cities 208–209 ❯

Cities with a population of over 10 million are known as megacities. New York was the only megacity in 1950; Tokyo, Mexico City, and São Paulo were added to the list in 1975.

How megacities form

As populations move from the countryside to the cities for work and better services, cities grow larger and more sprawling, leading to the formation of megacities. There were 33 megacities in 2018, most of them in low-income countries (LICs) and newly emerging economies (NEEs), especially in Asia. It is estimated that there will be more than 50 megacities by 2050.

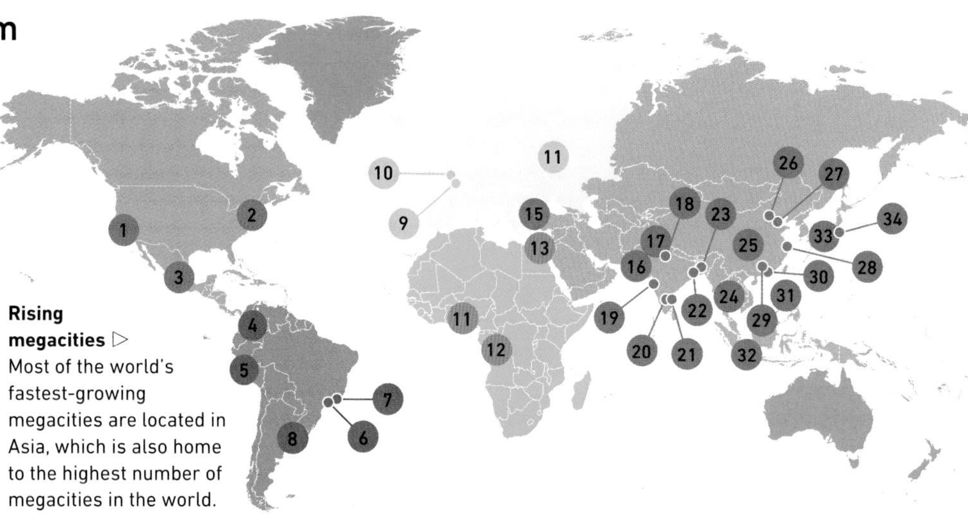

Rising megacities ▷
Most of the world's fastest-growing megacities are located in Asia, which is also home to the highest number of megacities in the world.

KEY

NORTH AMERICA	SOUTH AMERICA	EUROPE	AFRICA	ASIA				
1 Los Angeles	4 Bogota	9 Paris	12 Lagos	15 Istanbul	20 Bengaluru	25 Chongqing	30 Shenzhen
2 New York-Newark	5 Lima	10 London	13 Kinshasa	16 Karachi	21 Chennai	26 Beijing	31 Manila
3 Mexico City	6 São Paulo	11 Moscow	14 Cairo	17 Lahore	22 Kolkata	27 Tianjin	32 Jakarta
	7 Rio de Janeiro			18 Delhi	23 Dhaka	28 Shanghai	33 Osaka
	8 Buenos Aires			19 Mumbai	24 Bangkok	29 Guangzhou	34 Tokyo

Rapidly increasing population

With more people in LICs moving from rural areas to cities in search of jobs, the population of cities such as Delhi is expected to grow rapidly. However, in high-income countries (HICs), the population of cities such as Tokyo is expected to grow slowly.

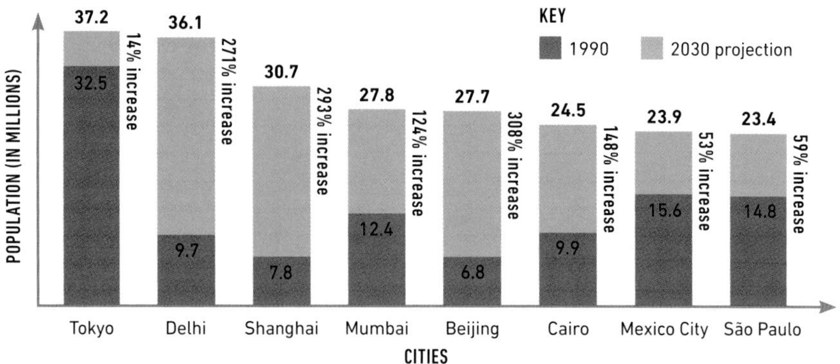
KEY
1990
2030 projection
POPULATION (IN MILLIONS)
CITIES
Tokyo 32.5 37.2 14% increase
Delhi 9.7 36.1 271% increase
Shanghai 7.8 30.7 293% increase
Mumbai 12.4 27.8 124% increase
Beijing 6.8 27.7 308% increase
Cairo 9.9 24.5 148% increase
Mexico City 15.6 23.9 53% increase
São Paulo 14.8 23.4 59% increase

Development of edge cities

Edge cities form on the edges of city suburbs as some people move out from city centers. In HICs, where the car is more important than public transportation, these edge cities have grown. Stores, services, jobs, and centers for leisure also move out to the edges of the city. The first edge cities were formed in the US in the 1980s.

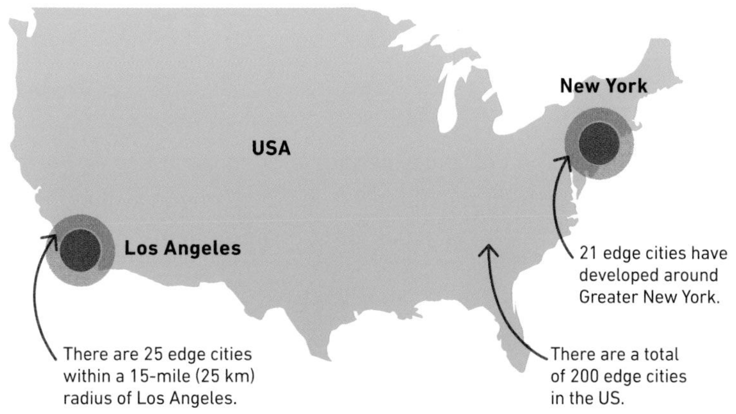
New York
USA
Los Angeles
21 edge cities have developed around Greater New York.
There are 25 edge cities within a 15-mile (25 km) radius of Los Angeles.
There are a total of 200 edge cities in the US.

Edge cities in the US ▷ With large shopping malls, concentrated areas for offices and housing, and high-tech industries, a number of edge cities have emerged in the US.

Issues facing shanty towns

In LICs, a different kind of edge city can develop—the shantytown. Also called squatter settlements, shantytowns developed in LICs due to rapid urbanization and are often the only space where migrants from rural areas can settle. As migrants have limited money, makeshift structures and poor living conditions are common in shantytowns.

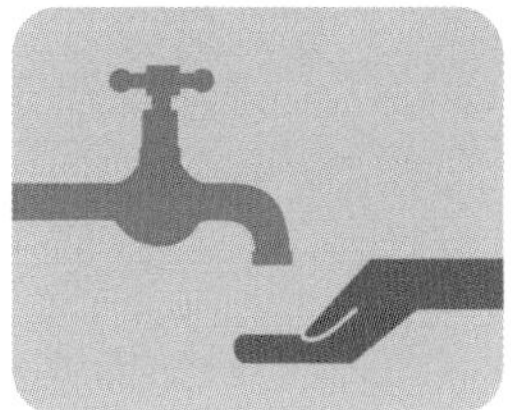

△ **Lack of basic services** Limited clean water and sanitation, and an unreliable power supply, lead to waterborne diseases and fire-related accidents.

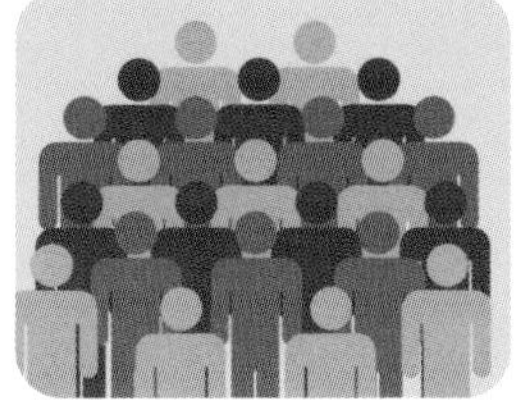

△ **Overcrowding** The population in shantytowns is high and is always increasing. Houses are often built very close to each other.

△ **Dangerous location** Some shantytowns are located in dangerous areas such as on steep slopes or next to garbage dumps, where land is cheap.

△ **Unemployment** Rural migrants often lack skills for the job market and either fail to find jobs or end up doing casual work that is poorly paid.

△ **Crime and gangs** Being unemployed can lead people to commit crime, an activity that is often controlled by gangs.

△ **Illegal settlements** Some shantytowns are illegal, built on land that does not belong to the residents, and that can be cleared without notice.

REAL WORLD

Cramped spaces

Kibera is a shantytown in Nairobi, Kenya. It is home to over 1.2 million people in an area of 1 sq mile (2.5 sq km). People live close together in homes built using mud, wood, and corrugated iron sheeting. Despite such living conditions, the homes are kept clean.

Rural settlements

SMALL TOWNS AND VILLAGES, OFTEN SURROUNDED BY FARMLAND, ARE KNOWN AS RURAL SETTLEMENTS.

SEE ALSO
❮ 128–129 Human settlements
Urbanization 136–137 ❯
Food and farming 144–145 ❯

For hundreds of years, most people lived in rural areas. With industrialization, people started to move to towns and cities in search of jobs, and rural areas went through many changes.

Characteristics of rural settlements

Rural settlements vary considerably in different parts of the world, from the people who live there to the services available. However, there are certain characteristics that are common to rural settlements in both high-income countries (HICs) and low-income countries (LICs).

Area and population
Rural settlements are smaller, both in terms of the number of people living there and the area they cover.

Isolated
Rural areas are often situated far from urban areas and so are self-sufficient in their services for the people living there.

Industry
Rural settlements are generally based on a primary industry, such as farming, forestry, mining, or fishing.

Rural and urban areas

Rural areas have historically been self sufficient, but this has become more difficult in recent times. Their development often depends on how close they are to urban areas. In HICs, while many people have started to move from cities to accessible rural areas, people are still moving out of remote rural areas.

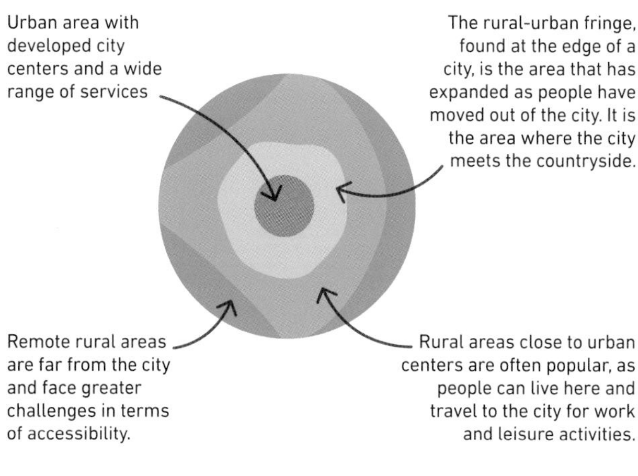

Urban area with developed city centers and a wide range of services

The rural-urban fringe, found at the edge of a city, is the area that has expanded as people have moved out of the city. It is the area where the city meets the countryside.

Remote rural areas are far from the city and face greater challenges in terms of accessibility.

Rural areas close to urban centers are often popular, as people can live here and travel to the city for work and leisure activities.

The changing world

In the 19th century, the Industrial Revolution encouraged people to move to towns and cities in Europe and North America. This rural-to-urban migration is a continuous process.

▽ **Rural-to-urban shift**
In 1950, 84 percent of people around the world lived in rural areas. By 2022, only 43 percent of people lived in the countryside, and these are mainly in LICs.

KEY
Most people living in rural areas
Most people living in urban areas

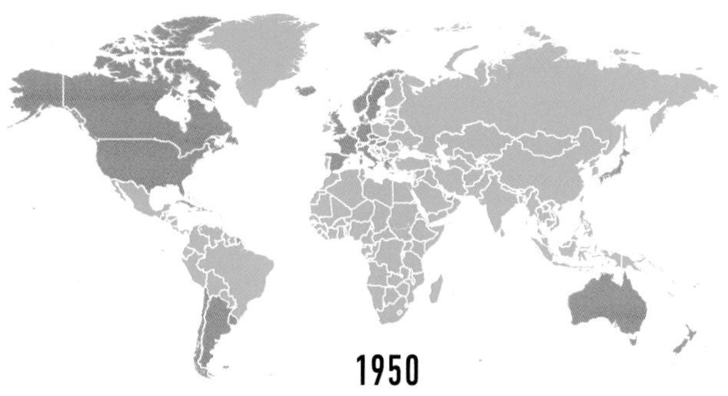
1950

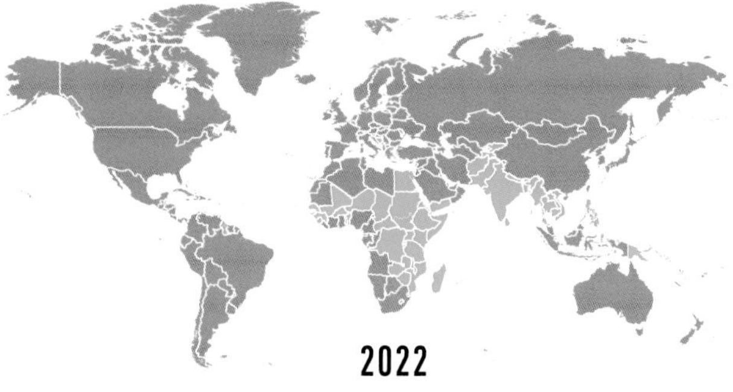
2022

Rural depopulation

Around the world rural areas tend to lose people, or depopulate. This is a result of migration, especially of younger people, to urban areas where opportunities for employment are greater and access to educational and health facilities is easier. This often leaves the rural areas increasingly impoverished.

The **Industrial Revolution** of the 19th century **caused massive** rural-to-urban **movements**.

IN HIGH-INCOME COUNTRIES

- Loss of jobs in the rural economy, including agriculture and quarrying, due to mechanization
- Loss of rural jobs due to a decline in agriculture and availability of cheaper, imported goods
- Decline in public transportation and other services, which may increase hardship in rural areas

IN LOW-INCOME COUNTRIES

- Not enough land for people to grow food to feed their families
- Poor basic services, such as health and education, in rural areas
- Low wages and limited job opportunities

Changes to rural areas

Many villages in HICs that are within a short distance of urban centers have been "suburbanized," which means they have become more like the suburbs of the towns. Rural areas can also be commuter settlements, where people live but travel to nearby towns and cities to work. Life in rural areas has changed a great deal.

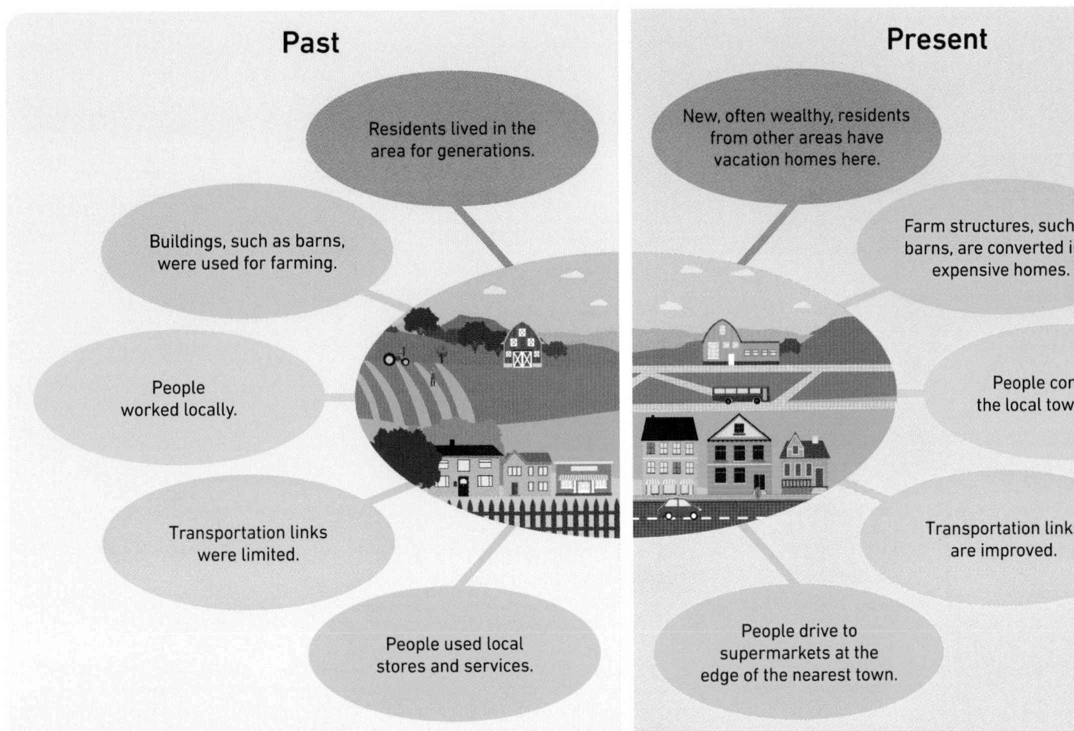
Past
Residents lived in the area for generations.
Buildings, such as barns, were used for farming.
People worked locally.
Transportation links were limited.
People used local stores and services.
Present
New, often wealthy, residents from other areas have vacation homes here.
Farm structures, such as barns, are converted into expensive homes.
People commute to the local town for work.
Transportation links are improved.
People drive to supermarkets at the edge of the nearest town.

Urbanization

MORE PEOPLE NOW LIVE IN URBAN CENTERS THAN IN THE COUNTRYSIDE.

SEE ALSO
❮ 124–125 Migration
❮ 126–127 Population change
❮ 128–129 Human settlements
Uneven development 158–161 ❯
Urban housing 164–165 ❯

As countries develop economically, more people move from the countryside to live in urban areas. Cities usually have more opportunities than rural areas, though they may be overcrowded.

Causes of urbanization

Organized urban settlements were a feature of early civilization, but most people still lived in the countryside where they engaged in farming. Over time, as agriculture became mechanized, people lost their farming jobs and moved to cities to look for work in industries.

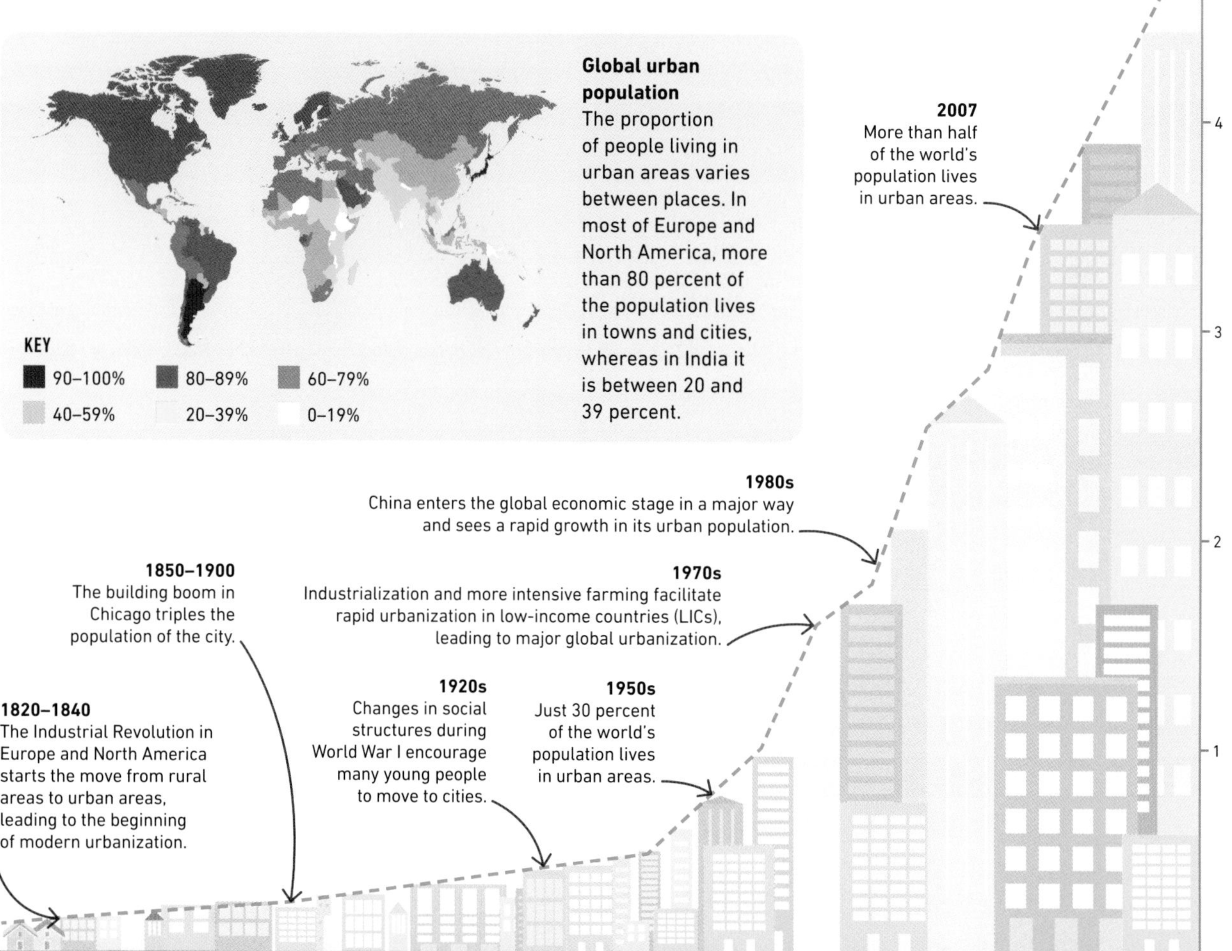
KEY
90–100%
80–89%
60–79%
40–59%
20–39%
0–19%
1820–1840
The Industrial Revolution in Europe and North America starts the move from rural areas to urban areas, leading to the beginning of modern urbanization.
1850–1900
The building boom in Chicago triples the population of the city.
1920s
Changes in social structures during World War I encourage many young people to move to cities.
1950s
Just 30 percent of the world's population lives in urban areas.
1970s
Industrialization and more intensive farming facilitate rapid urbanization in low-income countries (LICs), leading to major global urbanization.
1980s
China enters the global economic stage in a major way and sees a rapid growth in its urban population.
2007
More than half of the world's population lives in urban areas.
1820
1840
1860
1880
1900
1920
1940
1960
1980
2000
2020
2040
YEARS
1
2
3
4
5
URBAN POPULATION (in billions)

Global urban population The proportion of people living in urban areas varies between places. In most of Europe and North America, more than 80 percent of the population lives in towns and cities, whereas in India it is between 20 and 39 percent.

▽ **World's rural-to-urban shift** The rate of urbanization has seen a swift growth over time. Until 1960, the rate was slower, but from 1960 onward, global urbanization increased rapidly.

Uneven urban growth

The number of people living in urban areas varies in different parts of the world. High-income countries (HICs) have a higher proportion of people living in urban areas than low-income countries (LICs). More people live in rural areas in LICs and they usually work in agriculture.

KEY
Total urban population
Total rural population

Variations in urban populations ▷
In Africa, only 43 percent of the population lives in towns and cities, whereas in North America, it is 82 percent.

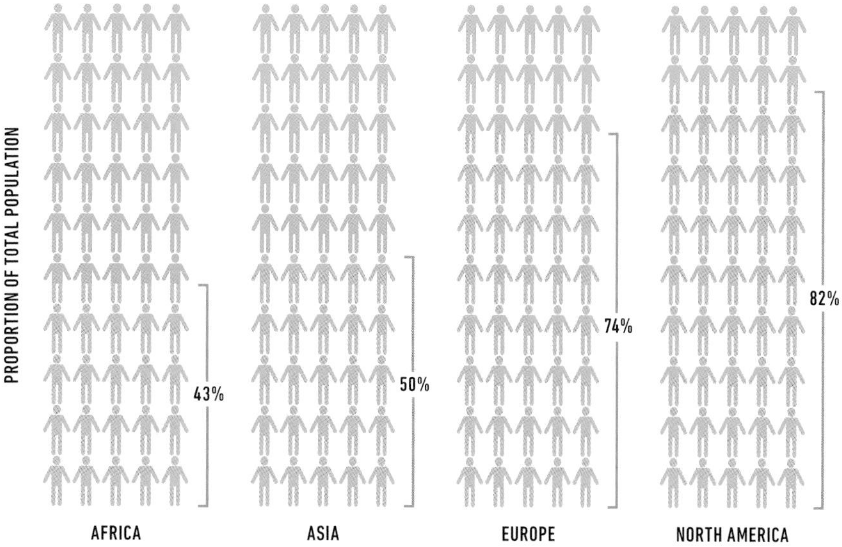
PROPORTION OF TOTAL POPULATION
43%
50%
74%
82%
AFRICA
ASIA
EUROPE
NORTH AMERICA

Push and pull factors

The movement of people from rural to urban areas is caused by push factors such as a lack of job opportunities in the countryside and pull factors such as a better quality of life, which attracts them to a new place. These factors may be stronger in LICs than HICs, and rates for the growth of urbanization are higher in poorer parts of the world.

PUSH

Natural disasters
Natural disasters, such as a tornado, may cause damage in rural areas where help may not be available.

Poor living conditions
The low standard of living in rural areas drives many to the cities in search of a better life.

Crop failure
Poor harvests often cause farmers to migrate to urban areas in search of a livelihood.

Conflict and war
Civil unrest and armed conflict can often destroy villages, forcing people to move to cities for security.

PULL

Higher wages
Jobs in urban areas offer better pay for both skilled and unskilled workers, attracting people to move to the city.

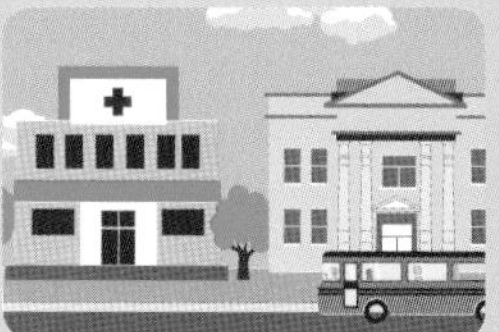

Health care and education
Cities provide better access to healthcare, schools, and other educational opportunities.

More jobs
Cities are perceived to have more job opportunities, which attracts people from rural areas.

Better living conditions
Higher incomes, a wider range of services, and better housing can lead to a higher standard of living in cities.

The spread of cultures

THE SPREAD OF CULTURES THROUGHOUT THE WORLD IS LINKED TO THE MOVEMENT OF BOTH PEOPLE AND IDEAS.

SEE ALSO
❮ 124–125 Migration
Tourism 152–153 ❯
Globalization 162–163 ❯
Global and local interdependence 190–191 ❯

Culture includes language, religion, customs, and art. Cultural heritage is handed down over generations, and carried with people who move far from home.

Colonialism

For many centuries, European nations conquered and occupied distant lands for their resources. They spread their own languages, customs, and religions abroad—often obliterating indigenous cultures in the process—and returned home with products of the cultures they invaded, including food and music. Although most former colonies have now been granted independence, the cultural impact of colonialism continues.

Colonial architecture ▷
Spanish architecture can be seen in churches in countries such as Mexico and Cuba, where Spanish colonialists also made locals practice the Catholic faith.

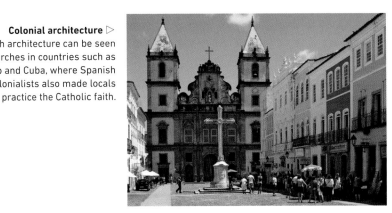

▽ **Distribution of migrant workers, by broad subregion, 2019**
Migrants often move for work, looking for a better life. Most go to wealthier regions, including North America, Europe, and the Arab States.

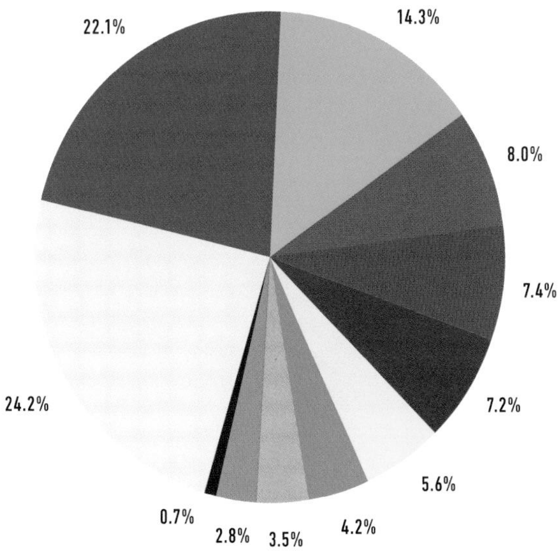
22.1%
14.3%
8.0%
7.4%
7.2%
5.6%
4.2%
3.5%
2.8%
0.7%
24.2%

- Arab States
- Central and Western Asia
- Eastern Asia
- Eastern Europe
- Latin America and the Caribbean
- Northern Africa
- North America
- Northern, Southern, and Western Europe
- Southeastern Asia and the Pacific
- Southern Asia
- Sub-Saharan Africa

Migration

Humans have always moved around the world, spreading ideas and cultures. In recent decades, the most marked migrations have been of people moving from more deprived parts of the world to richer places for work, notably in North America and Europe. These migrants often learn new ways of life but also bring their own customs, helping make the places they move to "multicultural."

Globalization

The spread of cultures through global trade and industry is known as globalization. Goods and services are now traded around the world, with similar products available in almost every country. Movies, TV, music, and the internet have also helped push a globalization of culture. Many people around the world grew up watching American movies or listening to European music, while social-media platforms have created an unprecedented global connection among people everywhere.

McDonald's
麦当劳
24

Shaping the world's taste ▷ Identical fast-food chains are now found in most countries, but their prevalence has led many people to be concerned about the preservation of local cuisines.

Travel and exchange of cultures

Over the last 50 years, long-distance travel, by air, train, and road has become much more affordable and popular with a wide range of people. Most people travel for short trips and vacations, but even so, they often become familiar with some of the local culture. And many people, especially among the younger generation, go for longer periods to study, work, or simply get to know different cultures.

COVID-19 and travel

From 2020, the COVID-19 pandemic dramatically curtailed foreign travel. Around the world, borders were shut, travel bans imposed, and planes grounded. In place of physical travel, people turned to video conferencing and social media to meet and share ideas virtually. It remains to be seen what the long-term impact of the pandemic will be on travel and the spread of cultures.

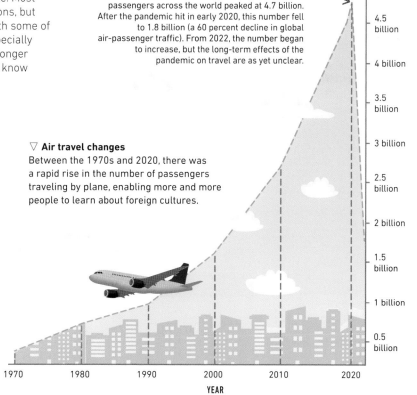

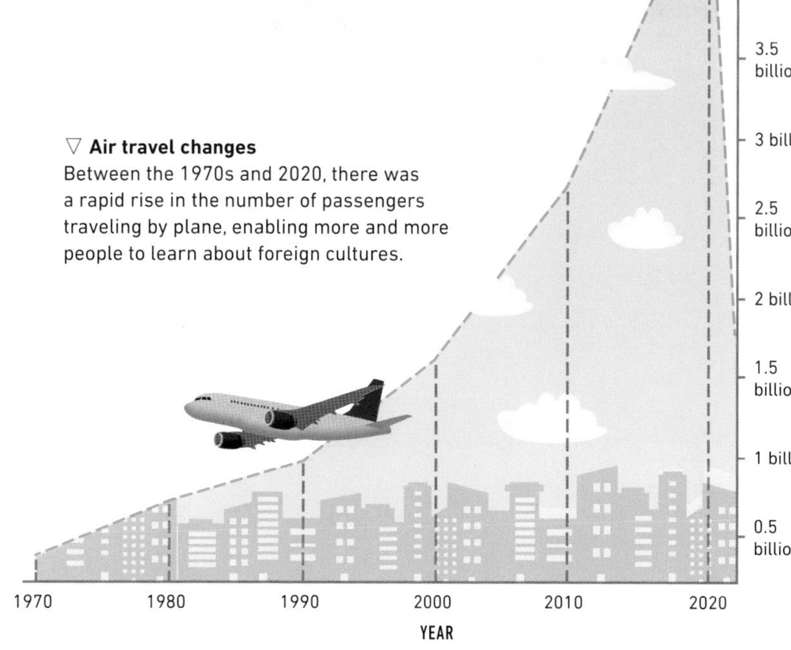
Before COVID-19, the number of scheduled airline passengers across the world peaked at 4.7 billion. After the pandemic hit in early 2020, this number fell to 1.8 billion (a 60 percent decline in global air-passenger traffic). From 2022, the number began to increase, but the long-term effects of the pandemic on travel are as yet unclear.
5 billion
4.5 billion
4 billion
3.5 billion
3 billion
2.5 billion
2 billion
1.5 billion
1 billion
0.5 billion
TOTAL NUMBER OF PASSENGERS TRAVELING BY AIR PER YEAR
1970
1980
1990
2000
2010
2020
YEAR

▽ Air travel changes
Between the 1970s and 2020, there was a rapid rise in the number of passengers traveling by plane, enabling more and more people to learn about foreign cultures.

Health

THROUGHOUT THE WORLD, HEALTH CAN BE LINKED TO SOCIAL AND ECONOMIC WELL-BEING.

SEE ALSO
❮ 136–137 Urbanization
Uneven development 158–161 ❯
Food security 192–193 ❯
Water security 194–195 ❯

Poverty leads to poor health by reducing access to health care, nutritious food, and clean water, and poor health keeps people in poverty as they are unable to work when they are ill.

Global access to health care

The World Health Organization (WHO) believes that access to health care should be a human right throughout the world and that people should not be denied medical treatment and preventive health measures for economic reasons. Many new treatments are expensive and not everyone can afford them, but medicines for problems such as diarrhea are cheap and effective.

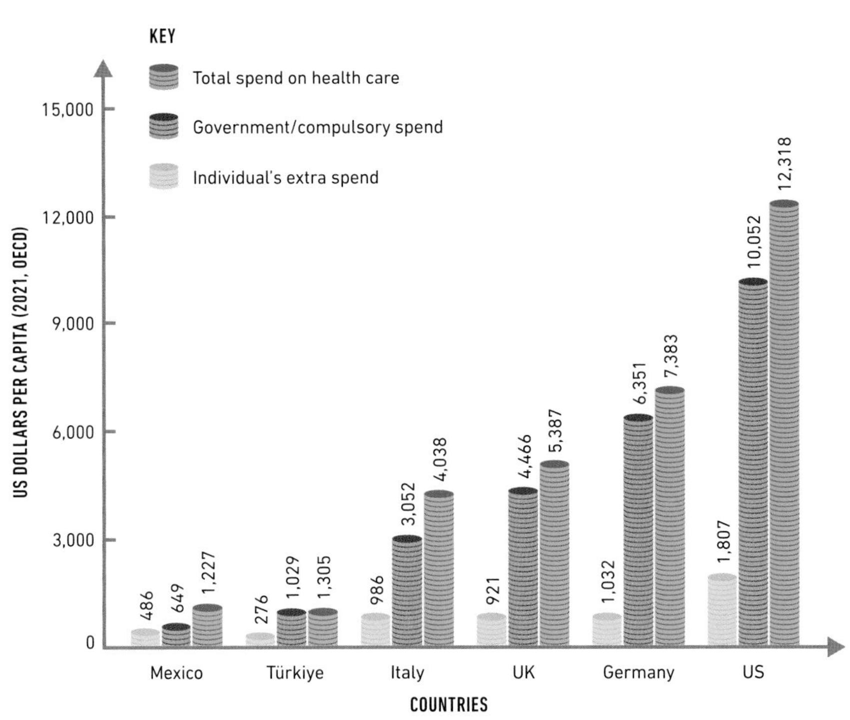
KEY
Total spend on health care
Government/compulsory spend
Individual's extra spend
US DOLLARS PER CAPITA (2021, OECD)
15,000
12,000
9,000
6,000
3,000
0
486
649
1,227
276
1,029
1,305
986
3,052
4,038
921
4,466
5,387
1,032
6,351
7,383
1,807
10,052
12,318
Mexico
Türkiye
Italy
UK
Germany
US
COUNTRIES

Spending on health care ▷ A country's wealth determines the amount of money its citizens and its government spend on health care. This graph shows that high-income countries invest more money in health care.

Equal access to health care

Health care includes access not only to doctors, nurses, hospitals, and medicines but also to clean water, sufficient and nutritious food, and education linked to healthy lifestyles. A healthy population can drive the development of low-income countries (LICs). Ensuring equal access to health care helps bridge the gap between LICs and HICs (high-income countries).

Economic activity
A healthy workforce can support a country's economy, increasing its wealth so that it can pay for health care and other services.

Availability of vaccines
Widespread vaccination of the population prevents diseases and ensures a healthy and productive workforce.

High-quality food supplies
The production of nutritious and high-quality food that is protected from pests and diseases keeps the population well fed and healthy.

Health education
Educating people about good hygiene, diet, and healthy lifestyles helps them make better choices.

Diseases caused by poverty

Infectious diseases may spread from person to person due to poor hygiene and health care. Some diseases, such as avian influenza, are common in LICs and can spread from animals to people. A poor diet can also cause poor health, lead to malnutrition, and make people more vulnerable to preventable disease.

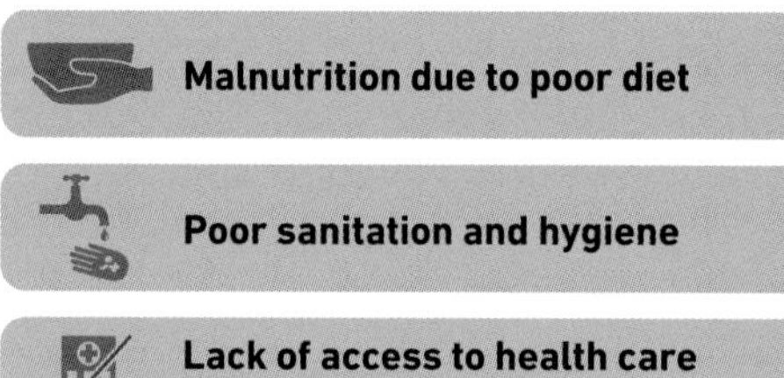

- **Malnutrition due to poor diet**
- **Poor sanitation and hygiene**
- **Lack of access to health care and health advice**

Lead to

- **Diseases such as malaria**
- **HIV/AIDS**
- **Waterborne diseases such as cholera and typhoid**
- **Tuberculosis (TB)**

Diseases caused by affluence

In some HICs, the death rate has recently begun to rise slightly and life expectancy has fallen as diseases linked to increased wealth have become more common. Some of these diseases are connected to each other. For example, obesity can lead to type 2 diabetes and increase the risk of heart disease, as well as some types of cancer.

- **Increased use of cars**
- **Decreased exercise**
- **Easy access to large amounts of low-cost processed food**
- **Lower consumption of fresh and seasonal food**
- **Increased sedentary employment**
- **Greater access to low-cost alcohol and tobacco**

Lead to

- **Type 2 diabetes**
- **Heart disease**
- **Some types of cancer**
- **Alcohol and drug-related issues**
- **Obesity**

Global pandemics

Pandemics are large-scale outbreaks of infectious disease that spread through many countries. Pandemics start when a bacteria or virus mutates to a form that humans can catch. Some spread rapidly but do little damage. Others spread slowly but are highly dangerous. A few spread rapidly and make many who catch it very ill. This was the case with the influenza pandemic of 1918–1919 and the COVID-19 outbreak.

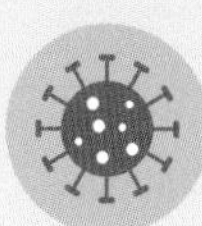

Zoonotic viruses
When habitats are destroyed, people are exposed to zoonotic viruses. These are animal viruses that mutate so they can infect humans.

Stopping the spread
During a pandemic, governments may have to impose strict rules, such as social distancing and wearing masks, to reduce transmission.

Economic activity

THE PROCESSES OF MAKING, SELLING, AND BUYING GOODS AND SERVICES ARE ALL ECONOMIC ACTIVITIES.

Economic activity includes jobs, trade, and industry, and circulates money in a country or region. Money from different types of jobs provides income to people and taxes to the government.

SEE ALSO

Economic sectors

All economic activities or types of industry can be categorized into four sectors. The percentage of income each sector generates in an economy changes over time as the country develops and becomes wealthier.

Primary sector

This sector involves the gathering of raw materials, such as crops and minerals from land and fish and seaweed from the sea.

Mining or quarrying
The processes of mining and quarrying are used to dig coal, iron ore, stone, and other raw materials out of the ground. These materials are then used in the secondary sector.

Agriculture
Farming is used to produce food such as wheat and meat, and grow plants such as cotton and flax, which are processed by the secondary sector.

Fishing
Fish, shellfish, and seaweed collected from seas and rivers can be used as they are for cooking, or be processed to make other products.

Secondary sector

This sector uses raw materials to make finished products, with businesses ranging from small industries in workshops to large factories.

Fabric production
Large looms weave cotton or synthetic fibers into huge rolls of fabric to produce clothes.

Car manufacturing
Making a vehicle involves manufacturing all the parts, including bolts and screws, from raw materials such as metal.

Food processing
Processed foods use raw materials procured by the primary sector. Food items are specially treated in order to preserve them.

Tertiary sector

This sector provides services that can range from personal hairdressing to data processing in a high-tech company.

Retail
Stores offer items made in the secondary sector, and retail associates sell the items to their customers.

Transportation
All forms of public transportation, such as buses and ferries, enable passengers to get to work and travel from one place to another.

Education
Teachers provide education to students, with help from administrators and other employees, in schools and colleges.

Quaternary sector

The latest category of economic activities, this sector provides skills, knowledge, and information to people, and develops new ideas.

Information technology
People working in this industry design and develop software and hardware that can give people access to data and information.

Research and development
This industry carries out research to either develop new products, such as new drugs and driverless cars, or improve existing ones.

Media
Entertainment industries include TV, social media, and video game production.

Employment structure

The distribution of employees across the four sectors of an economy makes up a country's employment structure. Jobs in the primary and secondary sectors tend to rely more on manual labor and physical skills than those in the tertiary and quaternary sectors. The employment structure of an economy changes over time as different sectors develop.

▽ **Distribution patterns in different countries**
While more people work in the tertiary and quaternary sectors in developed countries, more people work in the primary and secondary sectors in developing countries.

KEY

- Employed in primary industries
- Employed in secondary industries
- Employed in tertiary and quaternary industries

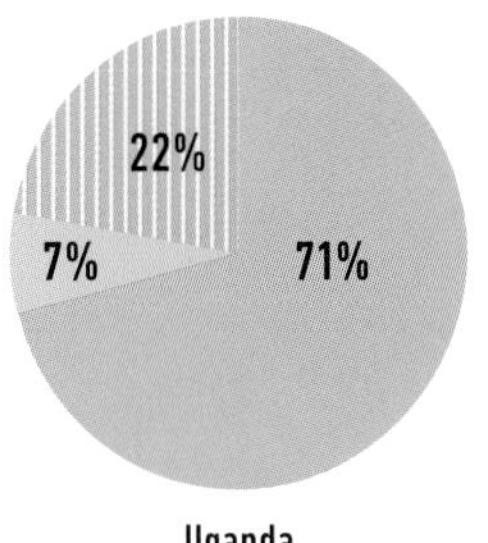
22%
7%
71%
Uganda

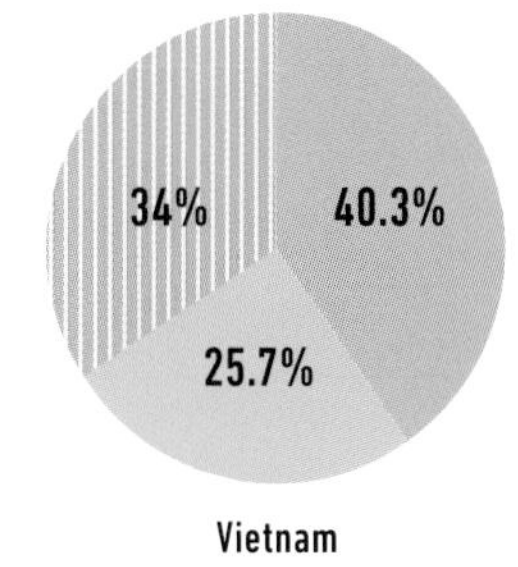
34%
40.3%
25.7%
Vietnam

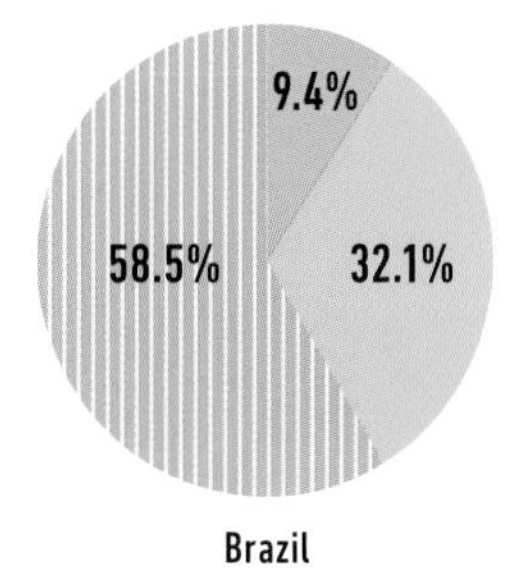
9.4%
58.5%
32.1%
Brazil

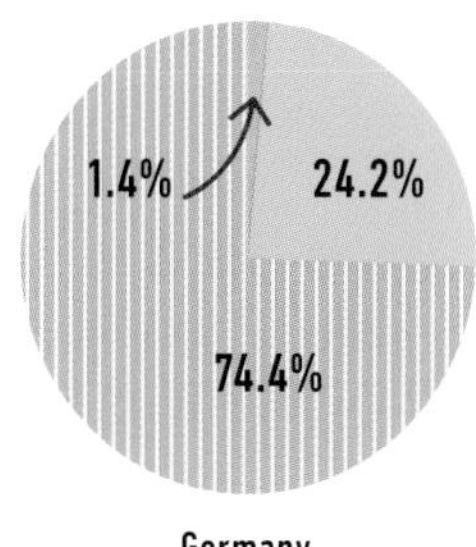
1.4%
24.2%
74.4%
Germany

The industrial system

All industries work as a system with inputs, processes, and outputs. The inputs include everything an industry needs to operate, the processes are what happens to the inputs, and the outputs are the final products. The output can be a finished item, such as a car, or even an idea that a company can sell.

▽ **The linear economic system**
This one-way system for industry, which ends with throwing things away after use, is damaging to the world. More industries are moving toward a "circular economy," designed to recycle as much as possible.

INPUTS

Money invested
All industries need money to buy machinery and raw materials, and to pay their workers. Sources of money may be either investments or profits earned from sales.

Raw materials
Different types of industry require different raw materials. For example, while a bakery uses flour, a car manufacturer uses metal.

Research
All industries try to improve the product they make through research, either undertaken within the company or based on ideas bought from outside.

People
Even the most automated industry requires people to work in it. Labor ranges from cleaners and assistants to directors.

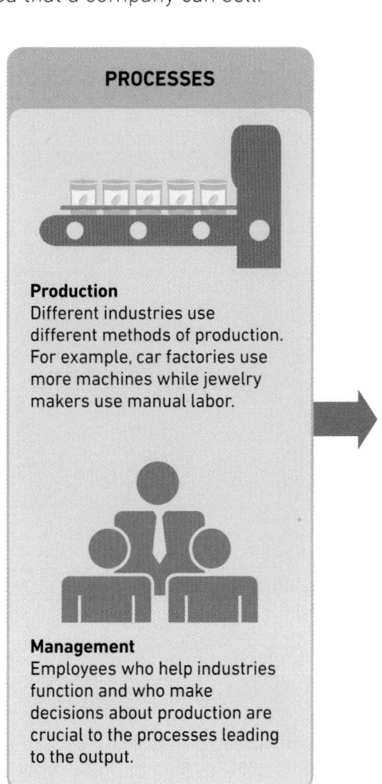

PROCESSES

Production
Different industries use different methods of production. For example, car factories use more machines while jewelry makers use manual labor.

Management
Employees who help industries function and who make decisions about production are crucial to the processes leading to the output.

OUTPUTS

Final product
The output of an industry is the final product it creates—such as a car or a piece of furniture—using the raw materials required.

Profits
The money left over after the final product is sold and all the costs have been recovered is the profit made. Profits may either be invested back in the company, or be claimed by the owners.

Waste
This includes all the waste materials generated during the production process, including packaging materials and factory emissions. It pollutes the environment and can be a problem.

Food and farming

SEE ALSO
❮ 78–79 Climate zones
❮ 140–141 Health
❮ 142–143 Economic activity
Sustainability 184–185 ❯
Food security 192–193 ❯

FARMING IS THE PRODUCTION OF CROPS AND REARING OF ANIMALS FOR CONSUMPTION AND TRADE.

Farming is a primary industry, meaning that the products come direct from the land. Around one-third of the Earth's surface is suitable for farming, although only about 11 percent of that land is actually used.

Types of farming

There are various kinds of farming around the world based on inputs—the amount of money, labor, land, and technology invested; processes—such as milking, animal rearing, and harvesting; and outputs—the final products, which are farmed either for sale or personal consumption.

△ **Arable**
The production of crops, such as wheat, tomatoes, rice, and cotton, on plowed land is called arable farming.

△ **Commercial**
When the farmers grow crops or rear animals to sell in the market, it is called commercial farming.

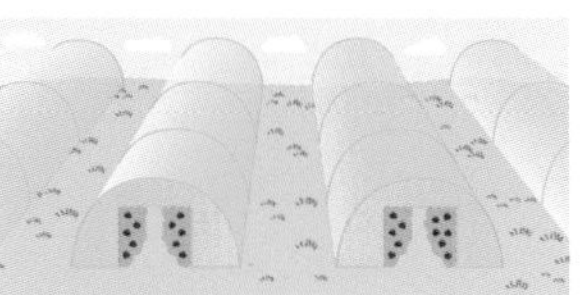

△ **Intensive**
Farming that uses more technology, money, or labor for its size is called intensive farming.

△ **Pastoral**
The production of animals, such as chickens, or animal products, such as milk, eggs, and wool, is called pastoral farming.

△ **Subsistence**
When farmers grow crops and animals for themselves and their families, it is called subsistence farming.

△ **Extensive**
Farming that involves less technology and labor on large farms and brings low yields is called extensive farming.

Factors influencing farming

Physical and human factors together influence farmers' decisions about which crops to grow, which animals to keep, the extent of farming, and the technology and processes to be used.

Physical factors

Climate
The temperature and rainfall of a region influence the crops that can be grown there.

Soil
The fertility and type of soil are important as different crops have different nutritional requirements.

Elevation and relief
Lowlands are good for crops while steep slopes have less soil and hinder the use of machines.

Size
Small patches of land may be used for intensive and large ones for extensive farming.

Human factors

Money available
The money available, to invest in machinery for example, can determine the type of farming that is possible.

Location
Some farmed products need to be processed or sold quickly, so the location of the farm is important.

Government policies
Governments can encourage the production of certain products by offering subsidies to farmers.

Market influence
Demand for certain products usually affects the farmers' choice of crops.

Food miles

The distance the food travels from the producer to the consumer is measured in food miles. Food miles have increased as people demand more exotic foods that cannot be grown in their own country. Food miles have also increased because many consumers want seasonal products, such as tomatoes and strawberries, throughout the year.

▽ **Global supply of ingredients**
The ingredients to make *siu mai*, or pork dumplings, in Hong Kong come from different parts of the world. Some ingredients, such as sesame oil, are sourced from as far away as the US.

NETHERLANDS
GERMANY
ITALY
SPAIN
US
CHINA
SOUTH KOREA
JAPAN
HONG KONG
THAILAND
VIETNAM
MALAYSIA
INDONESIA
NEW ZEALAND
BRAZIL

KEY

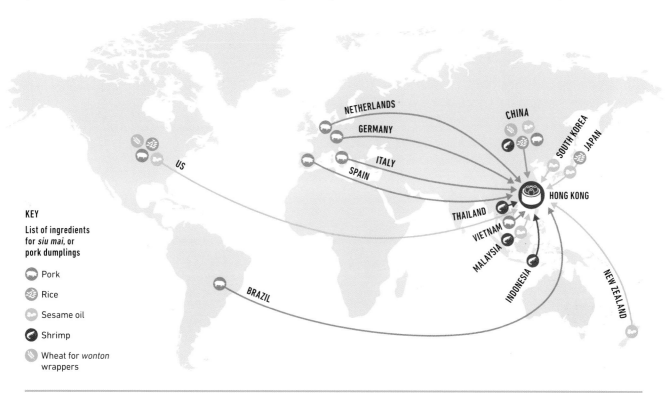

List of ingredients for *siu mai*, or pork dumplings

- Pork
- Rice
- Sesame oil
- Shrimp
- Wheat for *wonton* wrappers

Genetically modified foods

Farmers have always bred plants and animals selectively to increase yield. Global "agribusinesses" can now modify plants and animals in their laboratories to increase yield. Genetically modified crops have been developed to resist pests, weeds, and dry conditions. Rice, cotton, and wheat may be genetically modified, although these modified plants may upset the balance in the ecosystem.

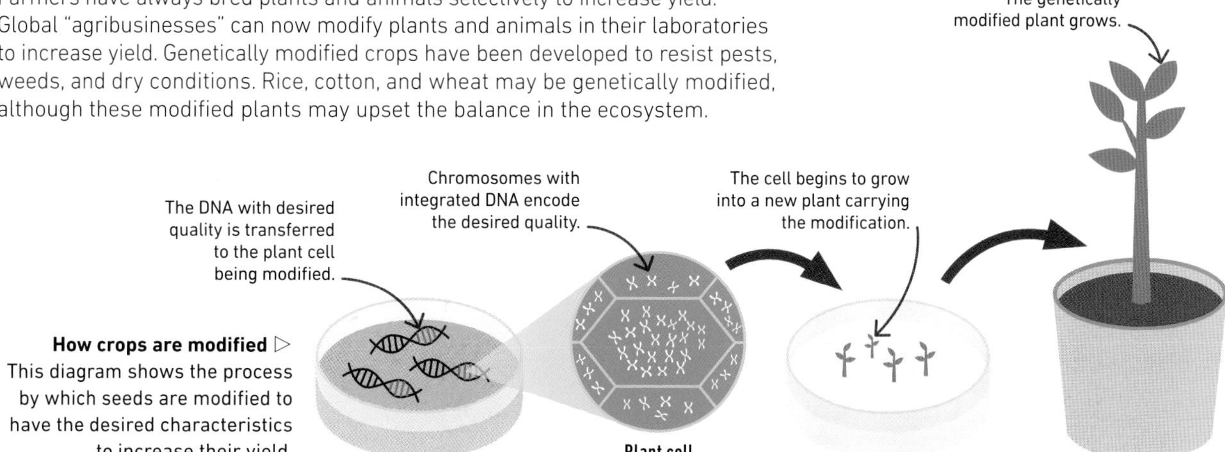
The genetically modified plant grows.
The DNA with desired quality is transferred to the plant cell being modified.
Chromosomes with integrated DNA encode the desired quality.
The cell begins to grow into a new plant carrying the modification.
Plant cell

How crops are modified ▷
This diagram shows the process by which seeds are modified to have the desired characteristics to increase their yield.

Extracting fossil fuels

FORMED OVER MILLIONS OF YEARS, FOSSIL FUELS ARE NONRENEWABLE SOURCES OF ENERGY.

SEE ALSO

Human impact	166–167 ❯
Deforestation	172–174 ❯
Climate change	175–177 ❯
Sustainability	184–185 ❯
Big geographical issues	188–189 ❯

Fossil fuels such as coal, oil, and natural gas are burned to generate electricity. They take millions of years to form, so they are considered nonrenewable resources.

Oil and gas extraction

Formed within layers of sedimentary rock from organic matter (plants and animals that contain carbon and can be burned), oil and gas are often found deep underground or under the seabed. Machines drill through the rock to reach the stored oil and gas. In areas of deep sea or extreme climates, extracting them is expensive and dangerous.

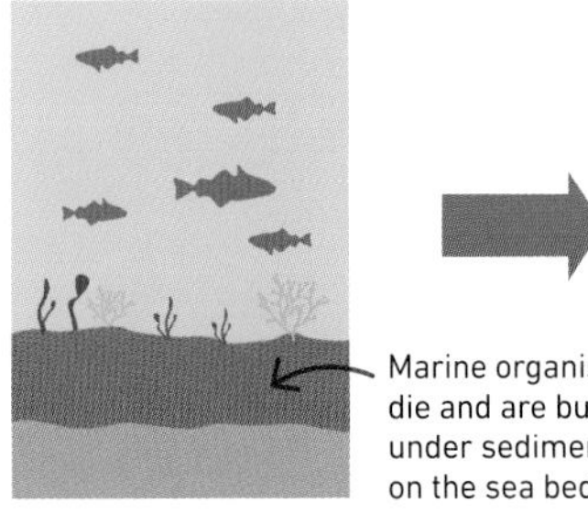

Marine organisms die and are buried under sediment on the sea bed.

1 Decomposing matter Tiny plants and animals such as algae die, fall to the bottom of the sea, and decompose.

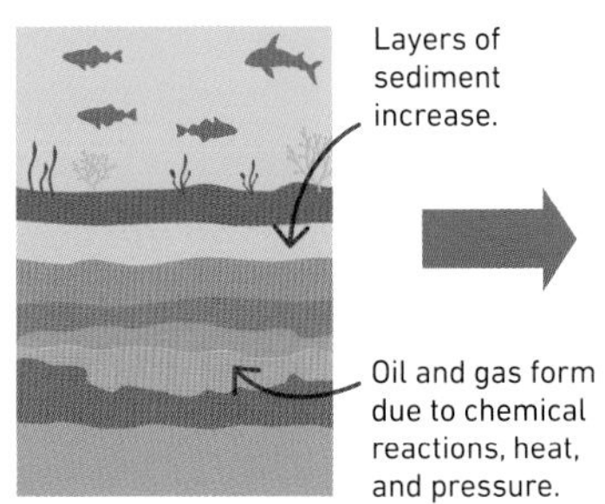

Layers of sediment increase.
Oil and gas form due to chemical reactions, heat, and pressure.

2 Conversion to oil and gas As layers of sediment build up, pressure and heat convert the remains into oil and gas.

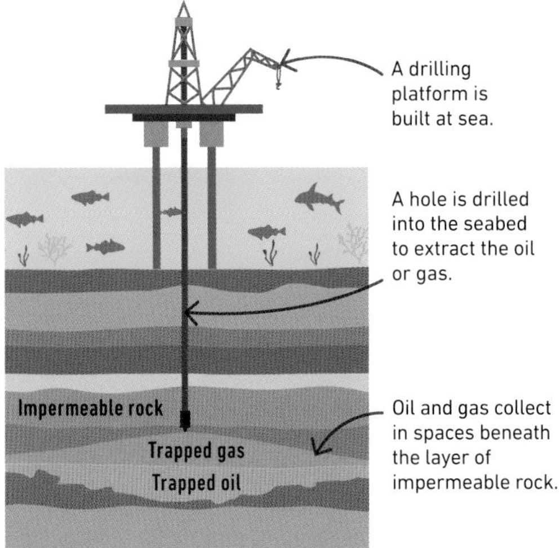

A drilling platform is built at sea.
A hole is drilled into the seabed to extract the oil or gas.
Impermeable rock
Trapped gas
Trapped oil
Oil and gas collect in spaces beneath the layer of impermeable rock.

3 Process of extraction Oil and gas are extracted by a process of drilling down through the layers of rock to reach them.

Shale gas fracking

The gas stores we use are running out, so a new technique called hydraulic fracturing, or fracking, is being developed to extract the gas trapped under land. It is being used in the US and costs less than traditional methods of extraction, though it is an invasive and aggressive process, which makes it controversial.

Fracking process for shale gas ▷
Although it is a cheaper method of extracting gas, fracking is believed to cause water pollution and minor earthquakes.

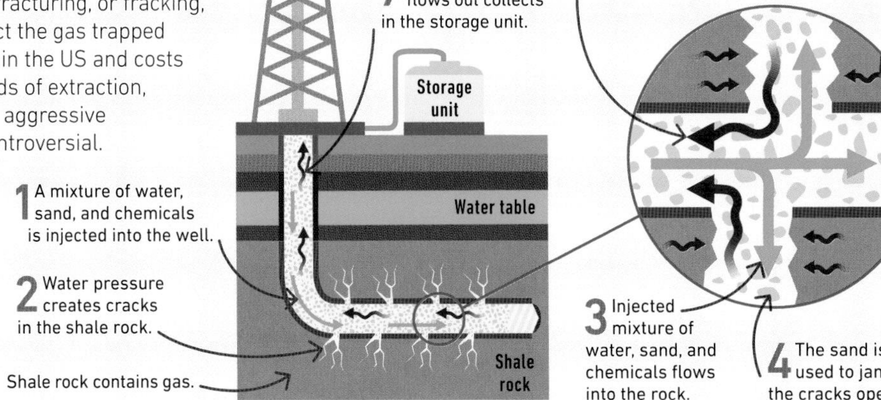

Drilling rig
7 The gas that flows out collects in the storage unit.
Storage unit
Water table
1 A mixture of water, sand, and chemicals is injected into the well.
2 Water pressure creates cracks in the shale rock.
Shale rock contains gas.
Shale rock
6 Natural gas is forced out.
5 Gas flows from the shale rock into the cracks.
3 Injected mixture of water, sand, and chemicals flows into the rock.
4 The sand is used to jam the cracks open.

Coal formation and mining

Tree and plant remains decompose over millions of years underground to form the fossil fuel coal, which is extracted using the process of mining. The open cast method is used to dig coal out of large pits near the surface. If the coal forms very deep under the ground, it is mined by sinking shafts down to the coal layer.

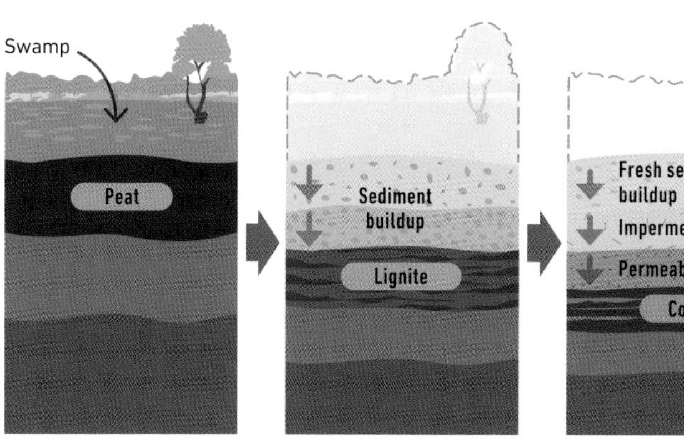
Swamp
Peat
Sediment buildup
Lignite
Fresh sediment buildup
Impermeable rock
Permeable rock
Coal

1 Formation of peat Plant remains sink into the soil in swamps. Owing to the lack of oxygen, they don't decay fully and form brown, spongy peat.

2 Peat to lignite Layers of sediment cover the peat over time, building up pressure. This dries out the peat and forms brown, crumbly lignite.

3 Lignite to coal With further layers of sediment added, pressure and heat continue to build on the lignite layer, turning it into hard, black coal.

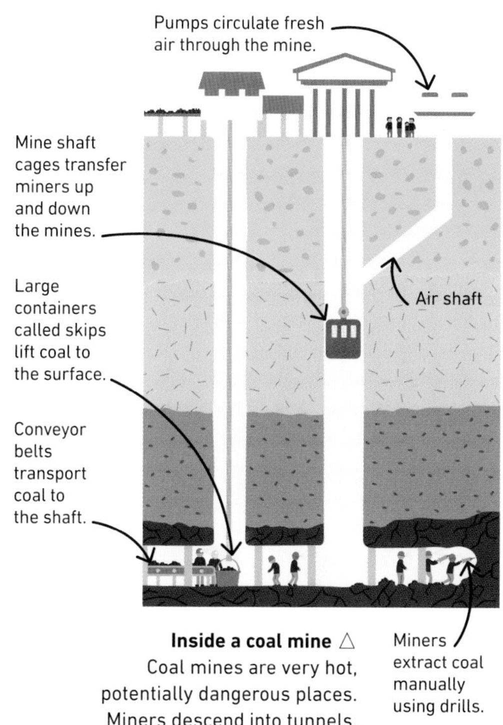
Pumps circulate fresh air through the mine.
Mine shaft cages transfer miners up and down the mines.
Air shaft
Large containers called skips lift coal to the surface.
Conveyor belts transport coal to the shaft.
Miners extract coal manually using drills.

Inside a coal mine △ Coal mines are very hot, potentially dangerous places. Miners descend into tunnels, dig out coal from the rock face, and transport the coal back to the surface.

Disadvantages of fossil fuels

While fossil fuels are convenient stores of energy which are easy to transport, they also have some disadvantages, and are now being phased out. Coal mines, and oil and gas rigs, are dangerous places to work, and the extraction processes and use of the fuels cause great damage to the environment.

Climate change
All fossil fuels produce carbon dioxide when burned, which increases levels of greenhouse gases and leads to climate change.

Damage to the environment
Extracting fossil fuels can lead to areas of wasteland, such as disused mines and spoil heaps, on the landscape.

Limited resources
Many of the most accessible fossil fuel reserves have been extracted, remaining only in areas that are unsuitable for extraction.

Oil spills
Accidents involving large oil tankers can harm seabirds and mammals, devastate coastal areas, and damage entire marine ecosystems.

Hazardous mining conditions
Mining is a dangerous occupation carried out deep underground, exposing miners to dust and gases that can cause health problems.

Future challenge
In the future, fossil fuel extraction will be seen as dirty, inefficient, damaging, and unnecessary.

Manufacturing industry

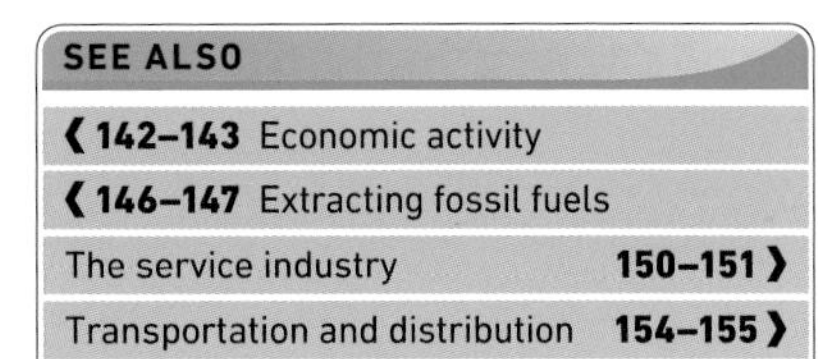
SEE ALSO
❮ 142–143 Economic activity
❮ 146–147 Extracting fossil fuels
The service industry 150–151 ❯
Transportation and distribution 154–155 ❯

THESE INDUSTRIES MAKE GOODS TO SELL, USING RAW MATERIALS, LABOR, AND MACHINES.

Manufacturing industries make new and finished products from raw materials, ranging from heavy industry, such as steel production, to high-tech processes. Manufacturing is a secondary industry.

Industrial location

In the past, industries were located near sources of raw materials and power. With the development of better modes of transportation and communication, the location of an industry is no longer dependent on these factors. Today, incentives offered by local and national governments are more likely to influence the location of an industry.

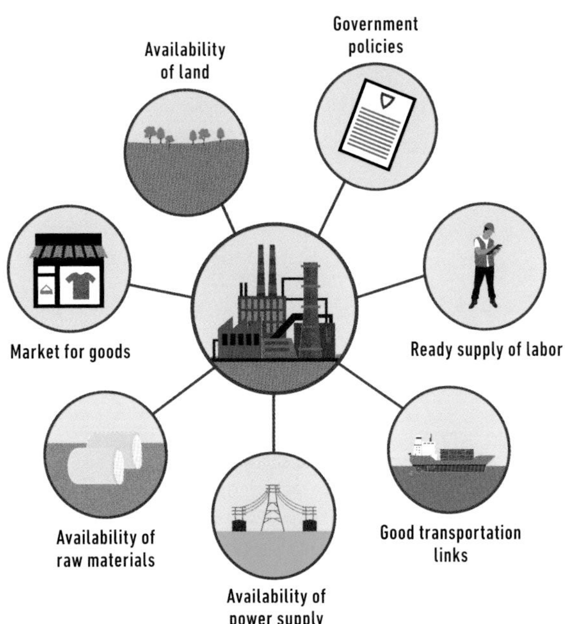
Availability of land
Government policies
Market for goods
Ready supply of labor
Availability of raw materials
Availability of power supply
Good transportation links

Factors affecting the location of industry ▷ Various factors determine the location of an industry. For some industries, proximity to the market, or the place where goods are sold, will be important, while for others it may be access to transportation networks.

How a TNC works

A TNC spreads its operations across the world to reduce the cost of production and increase profits. The TNC headquarters are always located in the company's home country, which has advanced research and development facilities. The production plants are mostly located in LICs, such as China, to benefit from the lower costs of production. It also has sales and marketing offices in the countries where it sells its products.

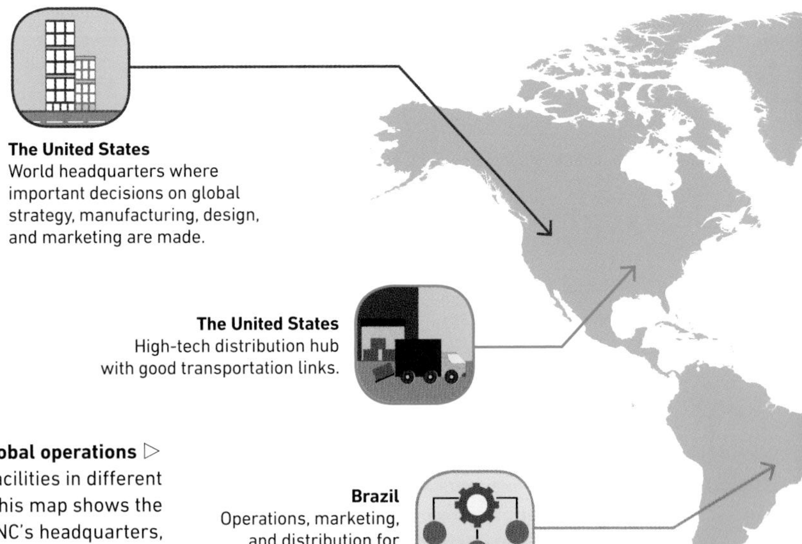
The United States
World headquarters where important decisions on global strategy, manufacturing, design, and marketing are made.
The United States
High-tech distribution hub with good transportation links.
Brazil
Operations, marketing, and distribution for the Central and South American market.

Global operations ▷ A TNC has facilities in different countries. This map shows the spread of a TNC's headquarters, regional offices, and manufacturing plants acround the globe.

Transnational corporation

A transnational corporation (TNC) is a very large company that does business across borders and is able to move its goods (and profits) around the world when it wishes. A smartphone, for example, might be made by a US company in a factory in China that uses materials from all over the world. The transnational corporation can move production to another company at any time it likes.

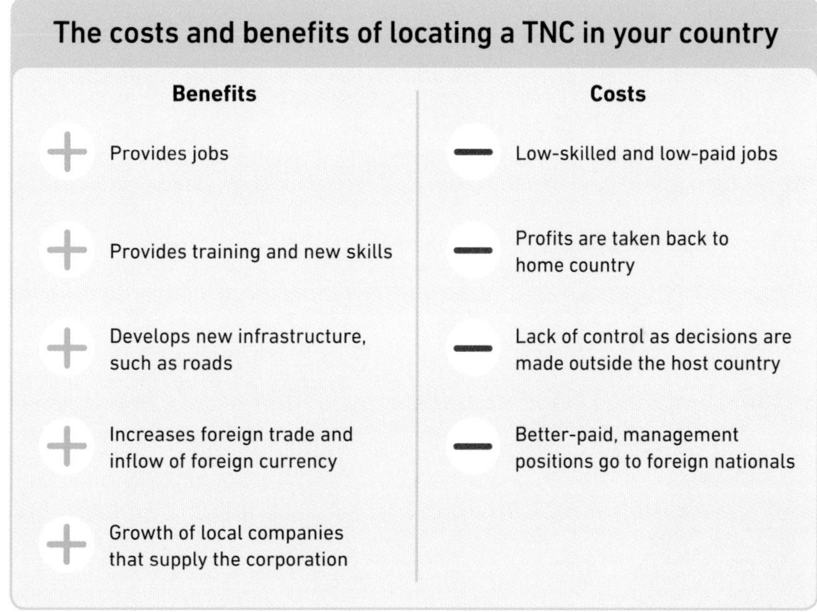

The costs and benefits of locating a TNC in your country

Benefits	Costs
+ Provides jobs	− Low-skilled and low-paid jobs
+ Provides training and new skills	− Profits are taken back to home country
+ Develops new infrastructure, such as roads	− Lack of control as decisions are made outside the host country
+ Increases foreign trade and inflow of foreign currency	− Better-paid, management positions go to foreign nationals
+ Growth of local companies that supply the corporation	

Manufacturing around the world ▷ Setting up a manufacturing plant away from the home country has advantages and disadvantages, both for the company and the host country. In most cases, the advantages outweigh the disadvantages for both.

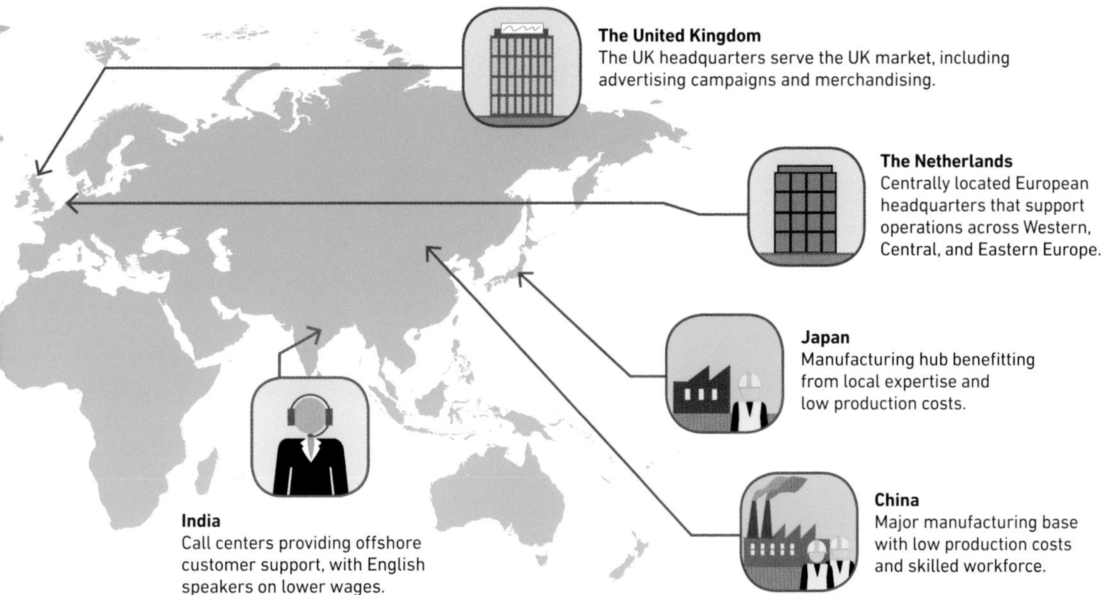
The United Kingdom
The UK headquarters serve the UK market, including advertising campaigns and merchandising.
The Netherlands
Centrally located European headquarters that support operations across Western, Central, and Eastern Europe.
Japan
Manufacturing hub benefitting from local expertise and low production costs.
China
Major manufacturing base with low production costs and skilled workforce.
India
Call centers providing offshore customer support, with English speakers on lower wages.

The service industry

THE GROWTH OF A COUNTRY'S SERVICE SECTOR OFTEN INDICATES ITS GROWING WEALTH AND ECONOMIC DEVELOPMENT.

SEE ALSO

❮ 142–143 Economic activity
❮ 148–149 Manufacturing industry
Tourism 152–153 ❯
Transportation and distribution 154–155 ❯
Globalization 162–163 ❯

The service industry involves the selling of services and skills. Service industries range from hospitals to hairdressers, sports clubs to restaurants, and banking to insurance.

What is the service industry?

Services are activities undertaken by people or businesses for customers. The service industry does not manufacture goods but it sells goods as well as services. It includes areas such as education, shops, and entertainment such as theaters and cinemas.

Service industries are **not involved** in **manufacturing** but **provide a service.**

Provides services to people
Service industries provide wide-ranging services, such as teaching and nursing.

Does not make or produce goods
Service industries do not produce anything but they do use, sell, or buy goods from manufacturing industries.

Location does not rely on raw materials
Service industries do not need to be located near sources of raw materials because they do not make or produce anything.

Does not need to be tied to a location
The service provided may be "virtual" (via the internet) or mobile so the location where the business is based is less important.

Growth of the service industry

In the past 50 years, there has been a huge increase in the growth of the service industry in most countries. This growth is predicted to increase in the future. The main growth is expected in low-income countries (LICs) as their wealth increases, along with rising demand for different services.

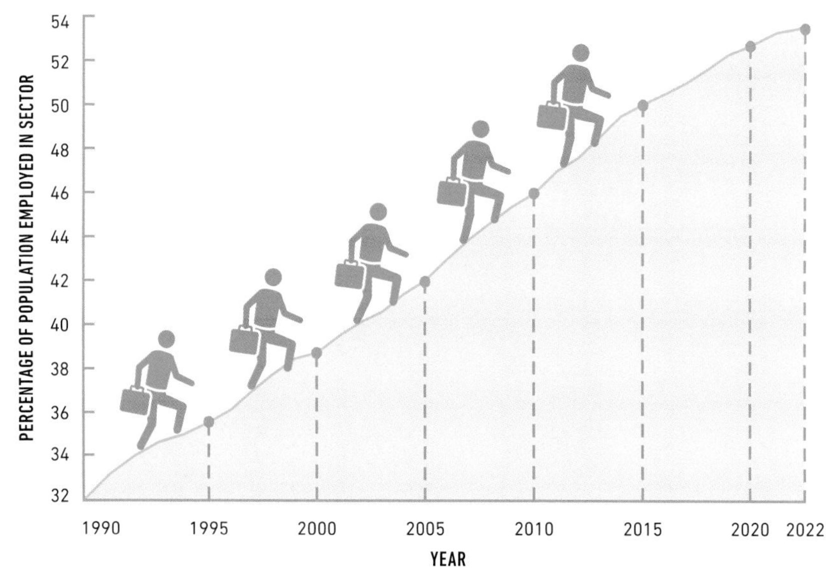

PERCENTAGE OF POPULATION EMPLOYED IN SECTOR
54
52
50
48
46
44
42
40
38
36
34
32
1990
1995
2000
2005
2010
2015
2020
2022
YEAR

Global growth in the sector ▷
There has been a steady rise in employment in the service sector since 1991, and the percentage of people employed looks set to rise.

Factors that affect growth

People in HICs have more time and money to spend on services. This applies to some people in LICs, too. Some service sector jobs are based in LICs, such as in call centers, though many are in HICs where sales and marketing roles tend to be based.

Rise in disposable income
This is the money people have left from their income to spend on services such as entertainment, after they have paid their essential bills.

Development of new technologies
The development of technology such as the internet has led to a demand for new services, such as delivery drivers for internet shopping.

Less dependence on other sectors
As goods are increasingly made in LICs where production costs are lower, most people in LICs are employed in the primary and secondary sectors.

Demographic changes
People are marrying later in life and having fewer children, leaving them more time and money to spend on leisure activities.

The growth of call centers

A call center is an office where people answer telephone inquiries from customers, often to do with banking, cell phone, and media services such as satellite TV. There are many call centers in India and the Philippines, answering queries on behalf of large TNCs.

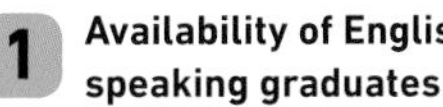

1 Availability of English-speaking graduates
India has a large number of highly educated graduates who speak excellent English and are able to communicate clearly with customers.

2 Fast and cheap modes of communication
The development of fast and cheap methods of communication means calls to India from other countries do not cost the customers more money.

3 Lower salary expectations
Salaries in India are lower than in HICs such as Australia and the US, reducing costs for the TNCs.

4 Flexibility to work shifts and long hours
People work longer hours and shifts to fit in with time differences in the countries where the customers are based.

Service industry statistics across the world

The percentage of people employed in the service sector varies widely in different countries. For a majority of the world's economies, the service sector is the biggest employer. There is a lower percentage of people in the service sector in LICs compared to HICs, but this is starting to change.

LOW-INCOME COUNTRIES (LICs)

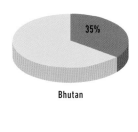

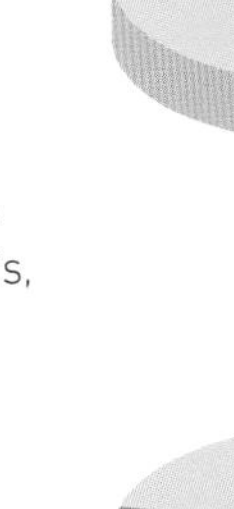

35%
Bhutan

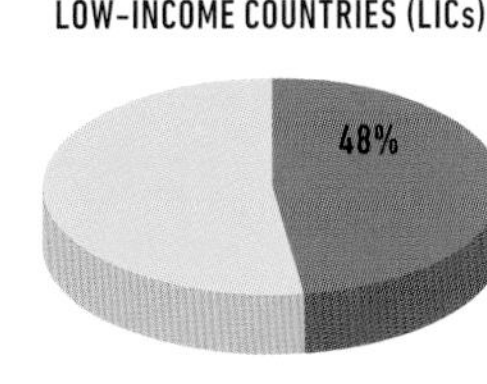

48%
Kenya

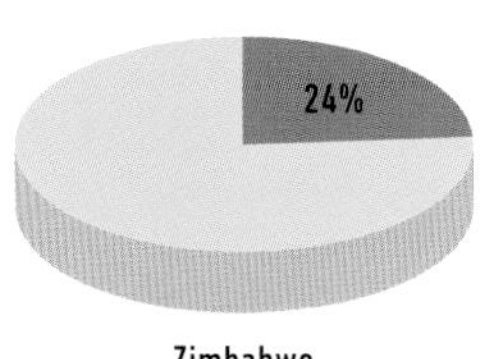

24%
Zimbabwe

HIGH-INCOME COUNTRIES (HICs)

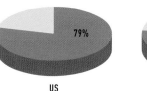

79% US

77% France

71% Japan

KEY

- Population employed in the service sector
- Population employed in other sectors

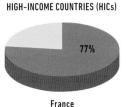

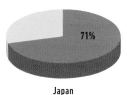

Tourism

THE PHYSICAL AND HUMAN GEOGRAPHY OF A PLACE INFLUENCES ITS POTENTIAL FOR TOURISM.

Tourism refers to people traveling for fun on trips that last for a day or longer, and includes activities such as sightseeing and camping. People who travel in their leisure time are called tourists.

What tourists look for

People have different ideas about what they want from a vacation. While some want adventure and activities, and are attracted to remote and exciting places or those with sports facilities, others are interested in history and culture, and choose to visit places that have historical sites, museums, and art galleries. Those looking for relaxation may head for a beach.

Sightseeing

Culture

Nature

Food

History

Relaxation

Religion

Sports and adventure

Impacts of tourism

Tourism is a very important part of the economy of high-income countries (HICs) as well as low-income countries (LICs). However, while it is a good source of income for LICs, a large number of people arriving as tourists can also have a negative impact on the local environment and culture. It is important to achieve a balance between the positive and negative impacts of tourism.

The **tourism industry** accounts for **10 percent** of the **world's GDP.**

✔ Positive effects

- **More jobs created**
  Local people have access to a wide range of new jobs at hotels and restaurants, or as tour guides, and can learn new skills.
- **More money earned**
  New jobs provide local people with higher wages, which they can use to expand local businesses and, in turn, strengthen the economy.
- **Tax benefits for the government**
  As local people earn more money, they pay more taxes, which the government can use to invest in better infrastructure.
- **Greater cultural awareness**
  People become aware of different cultures as they visit new places. This can improve understanding between countries and reduce conflict.

✖ Negative effects

- **Inconvenience for local people**
  A large number of tourists can cause traffic congestion, parking issues, and litter. The rising demand for vacation homes may increase property prices.
- **Outward flow of profits**
  Many hotels are owned by large transnational corporations (TNCs) that take all profits overseas without benefiting locals in any way.
- **Damage to the local environment**
  New buildings may destroy natural habitats and place stress on water supplies. Littering by tourists can damage the local environment.
- **Cultural clashes**
  Tourists may offend local people if they misunderstand religious or cultural customs, dress inappropriately, or consume alcohol or drugs.

Global growth of tourism

Tourism is one of the world's largest industries and, until 2020, accounted for 10 percent of the global Gross Domestic Product (GDP). The number of tourists and tourist destinations have increased significantly over time. People have been traveling farther and more often, due to various factors.

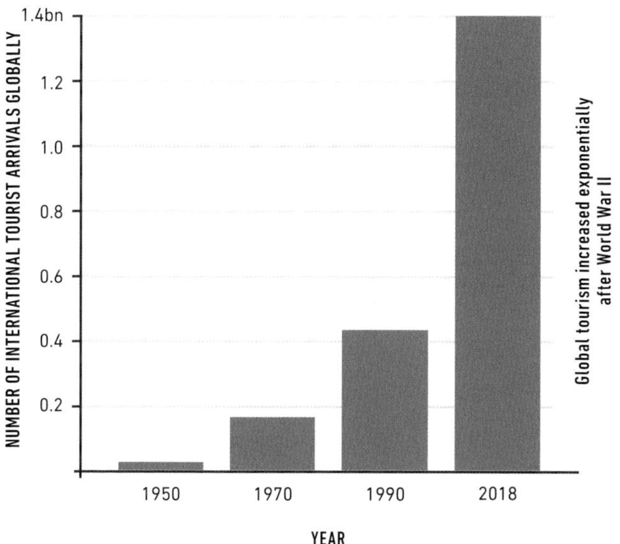
1.4bn
1.2
1.0
0.8
0.6
0.4
0.2
NUMBER OF INTERNATIONAL TOURIST ARRIVALS GLOBALLY
1950
1970
1990
2018
YEAR
Global tourism increased exponentially after World War II

Tourism since 1950

International tourism took off in the 1960s and 1970s, with new passenger planes able to carry large numbers of people. It continued to rise exponentially until 2020, when the COVID-19 pandemic ground most travel plans to a halt. The growth in tourism until 2020 has largely been down to the following factors:

- **Increased wealth**
  People earn more money and can spend more on vacations. More families have multiple incomes because the number of working women has increased.
- **More leisure time**
  The increased amount of paid vacation time and reduced working hours have meant people can travel more.
- **Improvements in technology**
  Improvements in transportation and ease of booking from home via the internet have made traveling easier and cheaper. People also have greater knowledge of the world due to increased access to television and the internet.

The long-term effects of the COVID-19 pandemic on tourism remain to be seen.

Tourism and the environment

While tourism can have a negative impact on the local environment, ecotourism or green tourism tries to minimize this impact while also helping local people benefit from tourist activities. It can provide local people with jobs and enable them to develop new skills.

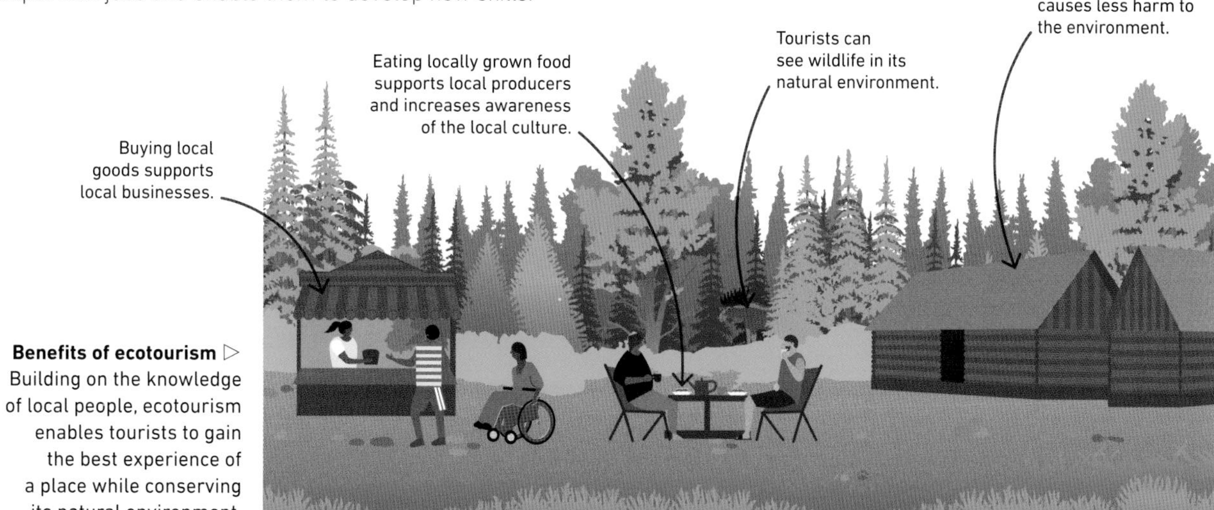
Buying local goods supports local businesses.
Eating locally grown food supports local producers and increases awareness of the local culture.
Tourists can see wildlife in its natural environment.
Accommodation made using local materials causes less harm to the environment.

Benefits of ecotourism ▷ Building on the knowledge of local people, ecotourism enables tourists to gain the best experience of a place while conserving its natural environment.

Transportation and distribution

INDUSTRIES AND SERVICES REQUIRE VARIED AND EFFICIENT TRANSPORTATION OPTIONS.

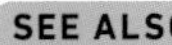

SEE ALSO

❮ 142–143 Economic activity
❮ 148–149 Manufacturing industry
❮ 150–151 The service industry
Globalization 162–163 ❯
Climate change 175–177 ❯

People need to get to work, to school, to the store, to visit friends and family, and to go on vacation. Industries need to transport raw materials to factories and finished goods to customers.

Types of transportation

Public transportation, such as buses or trains, is available for everyone to use and is owned and run by either local or national government or private companies. Private transportation, such as cars and trucks owned by individuals or private companies, is convenient but neither energy efficient nor sustainable.

Cars
This popular form of road transportation allows individuals to get exactly where they want to when they want to.

Buses
Buses transport many people at once, especially between towns or within cities.

Trains
Carrying hundreds of people and tons of goods, trains can travel long distances.

Airplanes
Planes are the fastest mode of transportation between countries and continents.

Motorcycles
If roads are congested or the terrain is unsuitable for cars, motorcycles can be used to travel quickly.

Ships and ferries
Ferries run regular services to places accessible by sea or river. Container ships transport goods overseas.

Trucks
Used to carry raw materials and finished goods, trucks can cover short and long distances.

Bicycles
Bicycles are used to travel shorter distances. Some cities offer bike rental services.

Transportation networks

Transportation hubs are places where many forms of transportation connect. Examples are airports with train and bus connections to the city center and ports with rail links that enable shipped goods to be transported over land to stores, warehouses, and consumers. Companies also have distribution centers such as warehouses, which allow them to store and transport goods.

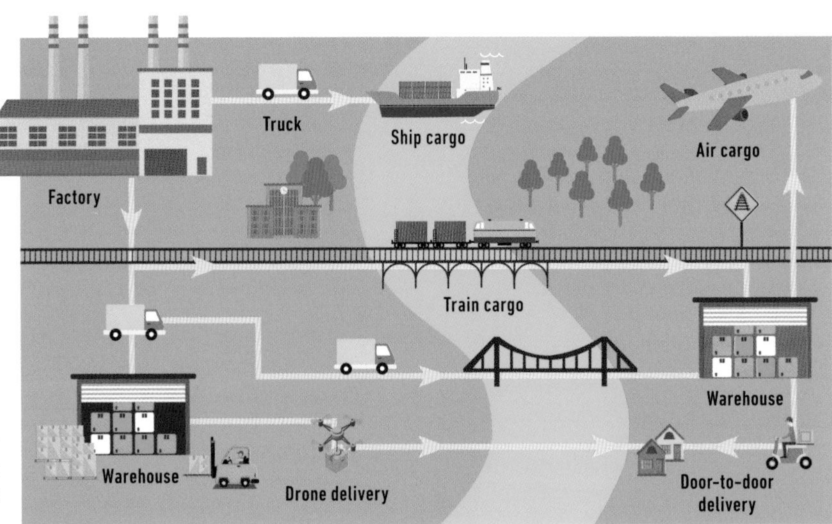
Factory
Truck
Ship cargo
Air cargo
Train cargo
Warehouse
Warehouse
Drone delivery
Door-to-door delivery

The distribution network ▷
Goods are transported through a distribution network, from the factory to a warehouse and from the warehouse to customers.

Transportation problems

Manufacturers need to transport raw materials to factories and goods to their customers. However, transportation presents several problems. Road transportation causes pollution and congestion, air transportation is expensive, and rail transportation has a limited range.

Air transportation involves high costs linked to fuel and the development and maintenance of infrastructure such as airports.

Container ships are a cheaper but slower mode of transportation.

Ports, airports, and parking lots for trucks and cars take up large areas of land, reducing green space and affecting wildlife habitats.

Emissions from vehicles include greenhouse gases, which contribute to climate change as well as air pollution, which causes health issues.

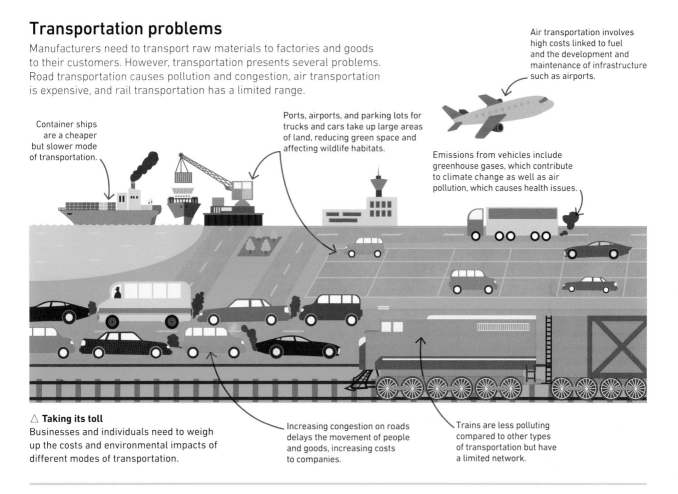

△ **Taking its toll**
Businesses and individuals need to weigh up the costs and environmental impacts of different modes of transportation.

Increasing congestion on roads delays the movement of people and goods, increasing costs to companies.

Trains are less polluting compared to other types of transportation but have a limited network.

Solving transportation problems

A move to more sustainable transportation would reduce congestion on roads as well as air pollution. Many countries are improving their distribution facilities for goods by increasing freight train capacity, which reduces the number of trucks on the roads. Some countries are imposing regulations on night flying and noise and also on emissions from ships, planes, and cars.

Increasing connections between bus and metro routes makes public transportation convenient.

Car sharing or carpooling to and from work reduces daily traffic and emissions and saves fuel.

Bike rental services and designated cycle paths encourage the use of eco-friendly bikes.

Electric cars with cheap and widespread charging points reduce the use of polluting vehicles.

Rental car facilities allow people to use cars when needed without having to buy their own.

Congestion charges levied on vehicles that enter cities or polluted areas act as a deterrent.

Park-and-ride facilities encourage people to use public transportation in towns, reducing traffic.

Drones are being developed to deliver packages to reduce traffic on roads.

Technology

TECHNOLOGY IS THE USE OF SCIENCE TO HELP SOLVE PROBLEMS AND IMPROVE THE WAY WE DO THINGS.

SEE ALSO
❮ 142–143 Economic activity
❮ 148–149 Manufacturing industry
Globalization 162–163 ❯
Sources of energy 182–183 ❯

From the invention of the wheel to the development of computers, technology has introduced new processes and products to our lives. We use some sort of technology for almost everything we do.

Technological change

Technological innovation is often driven by the need to solve a problem, or to save time or money. Discoveries and inventions are researched and then need to be developed into products that must be tested, improved, and perfected. As human needs continually change, technology must continue to develop.

▽ **Pace of technological change**
The number of inventions and new technologies has increased hugely in the last 200 years. Although growth in the power of computers—especially in the last 20 years—has driven great technological progress, there now seems to be a slowdown.

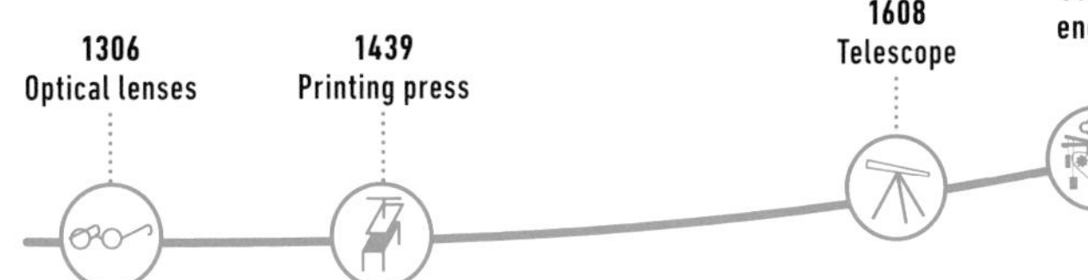
1306
Optical lenses
1439
Printing press
1608
Telescope

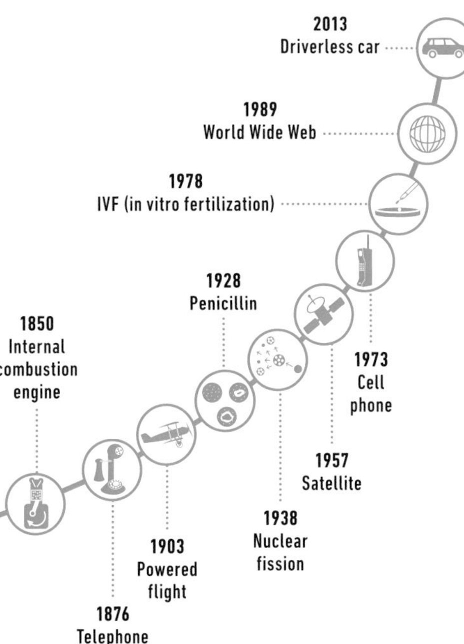
1698
Steam engine
1752
Electricity
1850
Internal combustion engine
1876
Telephone
1903
Powered flight
1928
Penicillin
1938
Nuclear fission
1957
Satellite
1973
Cell phone
1978
IVF (in vitro fertilization)
1989
World Wide Web
2013
Driverless car

Types of technology

We use technology in almost all parts of our daily lives. Technology has provided fast, safe transportation and useful materials such as plastic. It has also enabled quick and cheap worldwide communications, brought us lifesaving medical inventions such as pacemakers, and given us time-saving appliances like washing machines.

Information
The use of computers to process data (information technology, or IT) has made businesses more efficient. Many tasks that were previously manual can now be done automatically.

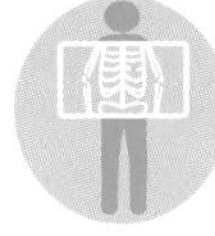

Medical
X-rays, contact lenses, and drugs such as antibiotics and painkillers are all forms of medical technology that have been invented and developed to improve and extend human lives.

Transportation
Transportation technology has enabled people and goods to travel quickly and cheaply across the globe. As well as trains, planes, and cars, it includes infrastructure such as roads and bridges.

Energy
Technology has brought constant development in the energy industry, from fossil fuel extraction to renewable sources such as water (hydropower), solar panels, and wind turbines.

Communications
Computers have enabled us to send information across the globe instantly. Phones, satellites, television, and the internet are all forms of communications technology.

Manufacturing
Automated production and the use of robots have changed manufacturing, reducing the need for factory workers to do repetitive and sometimes dangerous tasks.

Advantages and disadvantages

There are differing opinions over whether technology has improved life or made it worse. Some forms of technology have both advantages and disadvantages. For example, automated harvesting machines have released people from poorly paid, exhausting work but also taken away their source of income by removing their jobs.

Advantages

Technological innovation saves time and money, and leads to greater efficiency and productivity.

New communication technologies provide quicker and easier access to information, and improve education opportunities.

Medical breakthroughs have led to improvements in health across the world.

Technology helps countries develop their economies more quickly.

Disadvantages

People have lost their jobs as more work in industry and agriculture is done by machines.

New technologies have been developed that can cause potential harm, or even destruction, to our planet, such as plastics and nuclear weapons.

The increasing dependence on technology leads to the loss of traditional skills.

The rise of the internet and social media have led to concerns over privacy, the spread of disinformation, and cyberbullying.

Appropriate technology

The application of technology on a small, manageable scale is called appropriate technology. It uses local materials, limited amounts of energy, and the knowledge and skills of local people. Because it is suitable to the economic and social conditions of an area, it can be employed to help reduce the development gap between HICs and LICs. It can be applied to both agriculture and industry.

Small scale
Affordable, adaptable, and matches the needs of the community.

Simple
Equipment can be easily repaired and maintained. Less can go wrong.

Sustainable
Uses local resources without harming the environment.

Community driven
Local problems are solved using local skills, benefiting the local community.

REAL WORLD

Drip irrigation, Kenya

In drought-prone, semi-arid regions of Kenya, a successful smallholder irrigation system applies the ideas of appropriate technology. Crops such as corn are grown using "drip irrigation," a simple system that minimizes evaporation by allowing water to drip directly onto the plants' roots via a network of tubes. Equipped and trained locally, farmers have been able to improve crop yields while saving water and reducing their workloads.

Uneven development

DEVELOPMENT IS THE PROCESS BY WHICH A COUNTRY IMPROVES THE LIVES OF ITS INHABITANTS.

SEE ALSO
❮ 124–125 Migration
❮ 126–127 Population change
❮ 142–143 Economic activity
Big geographical issues 188–189 ❯
Conflict and resolution 196–197 ❯

Although people's quality of life is improving around the world, countries do not develop at an even rate. While some countries are developing quickly, others are falling behind.

Indicators of development

The greatest challenge in studying development is measuring it. Some measures ("indicators") are economic, but development is not about economics alone. The other big issue is justice—the extent to which people have a chance to improve their lives.

Economic indicators

Wealth
How much money a country has is usually a good indicator of its quality of life.

Employment structure
This indicator categorizes a country's development by how its economy splits into primary, secondary, tertiary, and quaternary sectors.

Trade
Less-developed countries often trade raw materials, which are generally cheaper than manufactured goods, and thus bring in less money from trade.

Levels of debt
Wealthier nations lend money to poorer nations, who then have to pay it back, sometimes with interest. This can lead to spiraling debt for poor nations.

Social indicators

Literacy rates
The percentage of people able to read and write is a good indicator of a country's education system.

Birth and death rates
High birth and death rates indicate a low level of development. As countries develop, both rates fall.

Life expectancy
A measure of a country's level of health care is the average age that its people can expect to reach.

Infant mortality
Less-developed countries often have high levels of infant mortality, measured in terms of the number of children who die before the age of one.

Wealth and development

Development is usually closely linked to wealth, commonly measured using Gross Domestic Product and Gross National Income. Using a single measure can be misleading, though, as wealth is never shared equally. Therefore, the United Nations (UN) has developed a "composite" measure, combining several indicators, to provide a more accurate picture. It is called the Human Development Index.

Gross Domestic Product (GDP)

GDP is the total value of goods and services produced in a country. It is always given in US$ to allow easy comparison between countries.

Gross National Income (GNI)

GNI is the total value of goods and services produced, plus the income earned from investments overseas made by its people and businesses.

Human Development Index (HDI)

The UN's Human Development Index combines life expectancy, income per head, and education (measured in years of schooling), and calculates a value for each country on a scale from 0 (lowest) to 1 (highest). It allows more accurate comparisons of standards of living than wealth alone.

Measuring wealth inequality

The World Bank uses GNI per capita (Gross National Income divided by population) to calculate the average income of a country's citizens and then classifies each country according to whether it has a high, middle, or low income. Some low- and middle-income countries are rapidly becoming wealthier as their economies move from primary to secondary and tertiary industries. These countries are called Newly Emerging Economies (NEEs) and include Russia, India, and China.

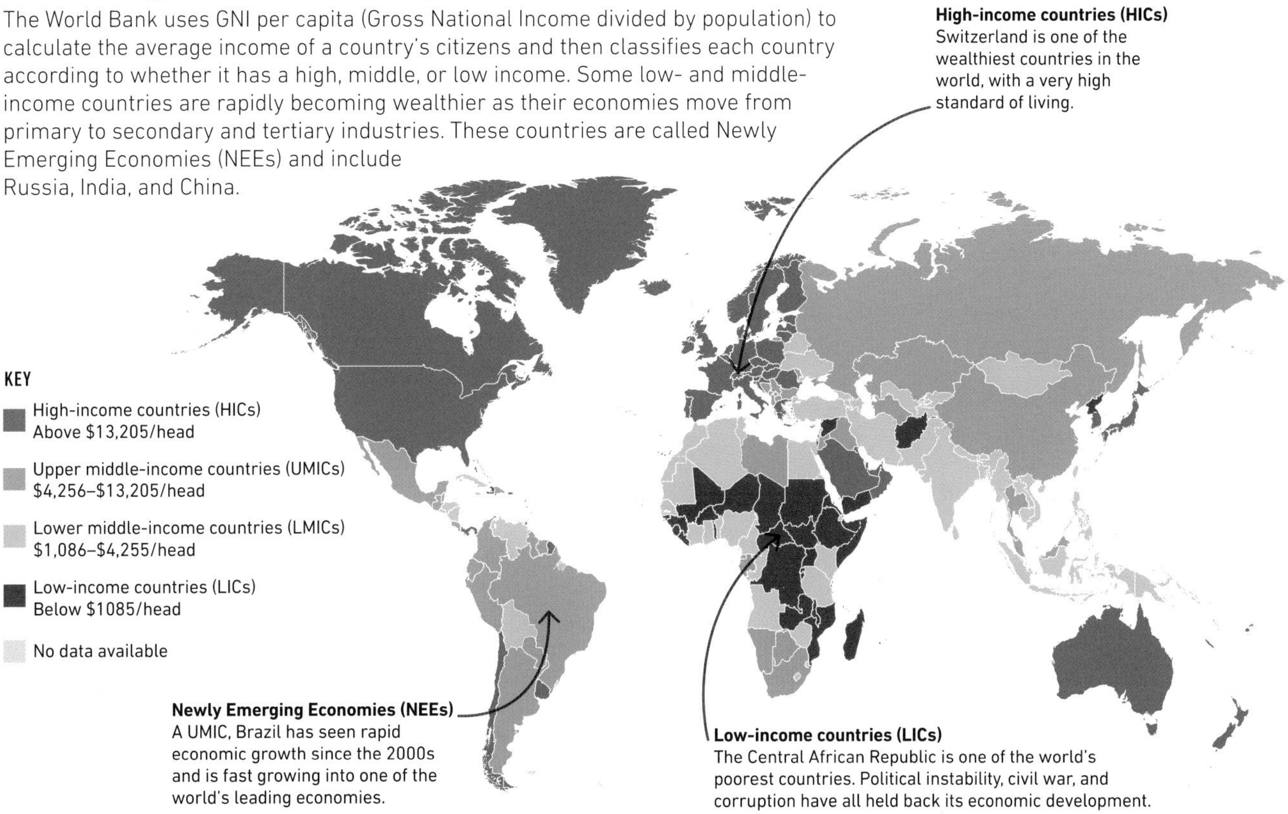

High-income countries (HICs)
Switzerland is one of the wealthiest countries in the world, with a very high standard of living.

KEY

- High-income countries (HICs) Above $13,205/head
- Upper middle-income countries (UMICs) $4,256–$13,205/head
- Lower middle-income countries (LMICs) $1,086–$4,255/head
- Low-income countries (LICs) Below $1085/head
- No data available

Newly Emerging Economies (NEEs)
A UMIC, Brazil has seen rapid economic growth since the 2000s and is fast growing into one of the world's leading economies.

Low-income countries (LICs)
The Central African Republic is one of the world's poorest countries. Political instability, civil war, and corruption have all held back its economic development.

Comparing wealth and development

As a country's wealth increases, its level of development usually also improves. Increased wealth enables more money to be spent on education and health care, for example, which helps improve the quality of life of its population. However, there are variations. Improvements in living standards often depend on how evenly wealth is distributed within a country and how a government chooses to spend its money.

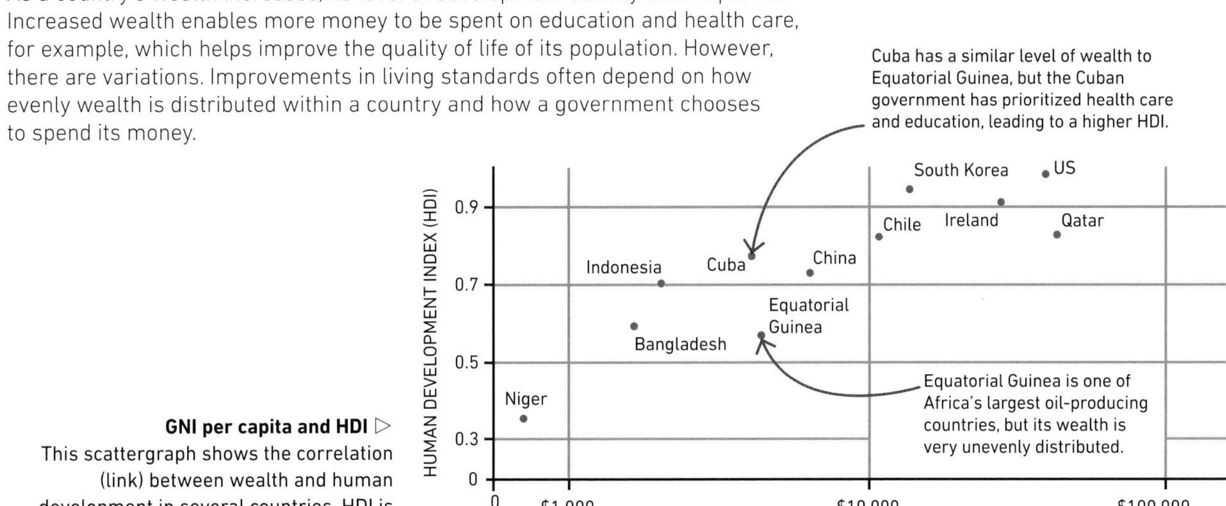
Cuba has a similar level of wealth to Equatorial Guinea, but the Cuban government has prioritized health care and education, leading to a higher HDI.
HUMAN DEVELOPMENT INDEX (HDI)
0.9
0.7
0.5
0.3
0
South Korea
US
Chile
Ireland
Qatar
China
Indonesia
Cuba
Equatorial Guinea
Bangladesh
Niger
Equatorial Guinea is one of Africa's largest oil-producing countries, but its wealth is very unevenly distributed.
0
$1,000
$10,000
$100,000
GNI PER CAPITA

GNI per capita and HDI ▷
This scattergraph shows the correlation (link) between wealth and human development in several countries. HDI is measured on a scale from 0 to 1.

The Development Gap

The UN's Human Development Index (HDI) shows that—with some exceptions—countries in the northern hemisphere enjoy a higher level of development than those in the southern hemisphere. This north–south divide is known as the Development Gap. In some ways, this gap has narrowed in recent years, but in other ways the inequalities have grown.

Half of the world's wealth belongs to just **1 percent** of its population.

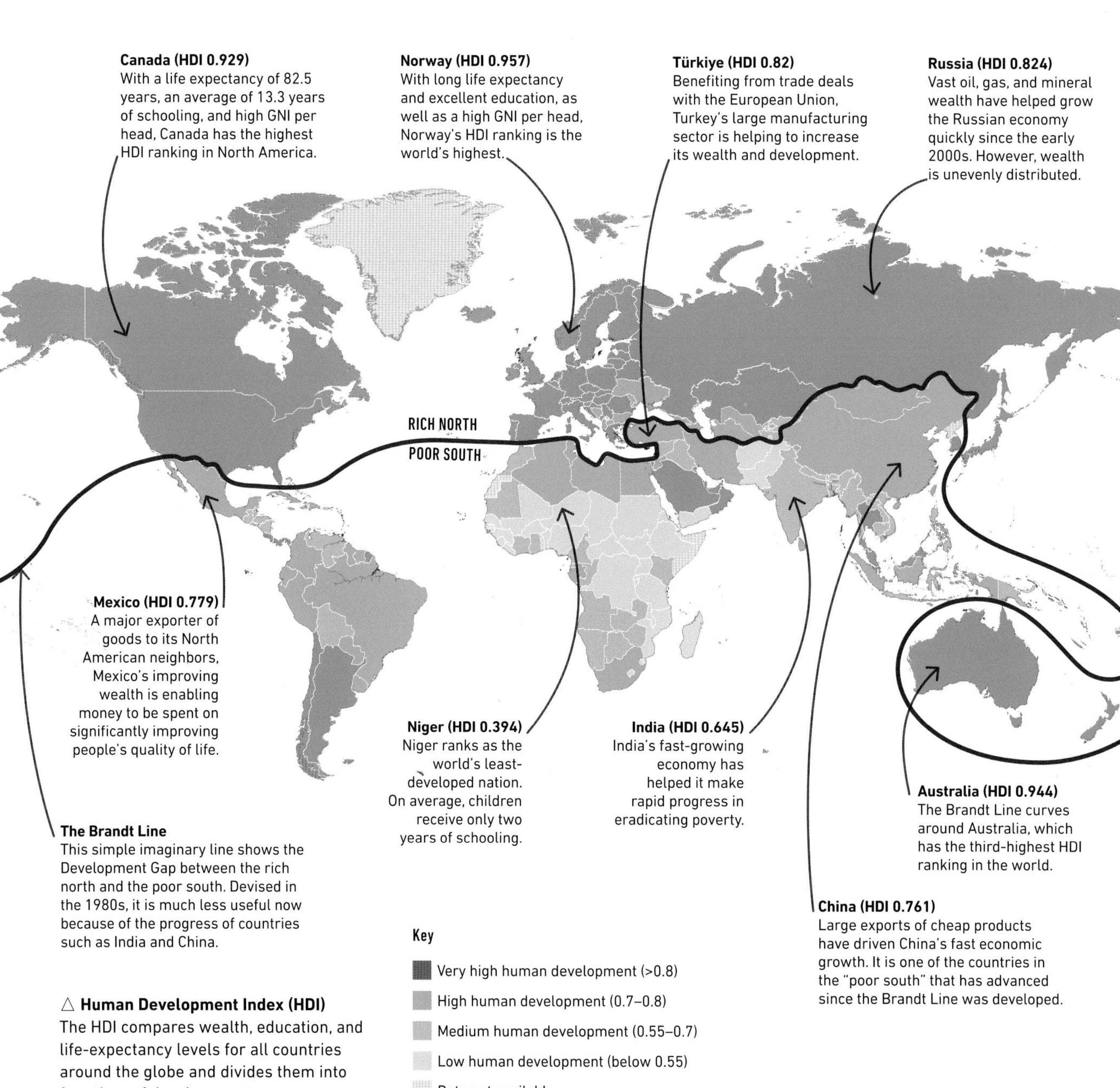
Canada (HDI 0.929)
With a life expectancy of 82.5 years, an average of 13.3 years of schooling, and high GNI per head, Canada has the highest HDI ranking in North America.
Norway (HDI 0.957)
With long life expectancy and excellent education, as well as a high GNI per head, Norway's HDI ranking is the world's highest.
Türkiye (HDI 0.82)
Benefiting from trade deals with the European Union, Turkey's large manufacturing sector is helping to increase its wealth and development.
Russia (HDI 0.824)
Vast oil, gas, and mineral wealth have helped grow the Russian economy quickly since the early 2000s. However, wealth is unevenly distributed.
RICH NORTH
POOR SOUTH
Mexico (HDI 0.779)
A major exporter of goods to its North American neighbors, Mexico's improving wealth is enabling money to be spent on significantly improving people's quality of life.
Niger (HDI 0.394)
Niger ranks as the world's least-developed nation. On average, children receive only two years of schooling.
India (HDI 0.645)
India's fast-growing economy has helped it make rapid progress in eradicating poverty.
Australia (HDI 0.944)
The Brandt Line curves around Australia, which has the third-highest HDI ranking in the world.
The Brandt Line
This simple imaginary line shows the Development Gap between the rich north and the poor south. Devised in the 1980s, it is much less useful now because of the progress of countries such as India and China.
China (HDI 0.761)
Large exports of cheap products have driven China's fast economic growth. It is one of the countries in the "poor south" that has advanced since the Brandt Line was developed.
Key
Very high human development (>0.8)
High human development (0.7–0.8)
Medium human development (0.55–0.7)
Low human development (below 0.55)
Data not available

△ **Human Development Index (HDI)**
The HDI compares wealth, education, and life-expectancy levels for all countries around the globe and divides them into four tiers of development.

Inequality within countries

One drawback of the HDI is that it measures each country's overall development without revealing inequalities within it. All countries have richer and poorer areas, and—especially in LICs—people who live in cities often have better access to services than those in rural areas.

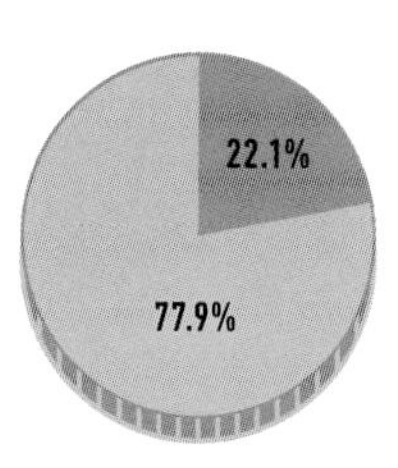
22.1%
77.9%

◁ **United States: 728 billionaires** The US has a population of 325 million people, yet almost a quarter of its wealth belongs to just 728 individuals.

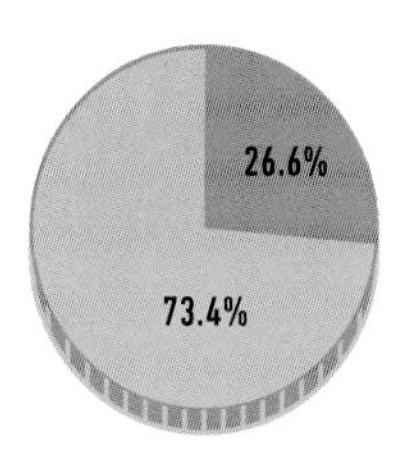
26.6%
73.4%

◁ **India: 100 billionaires** More than 200 million still live in poverty in India, but just 100 people own more than 26 percent of its wealth.

KEY
Gross Domestic Product (GDP)
Billionaires' net worth as percentage of national GDP

Causes of uneven development

Uneven development around the globe is caused by a combination of many factors. Natural hazards and an unfavorable climate can affect a country's ability to develop, but political and economic factors—such as war and trade—also often have a significant impact.

Historical and political factors

Colonization
Most LICs were colonized in the 18th and 19th centuries. European powers exploited their colonies for their raw materials, and sometimes their people, which has held back their development to the present day.

Conflict
Development in many LICs has been slowed down by wars. More money spent on armies and weapons means that less is available for development.

Political instability
Many LICs do not have stable governments. Unstable countries fail to attract foreign investment to develop their industries.

Economic factors

Poor trade links
LICs tend to trade with fewer countries than HICs, many of which are part of large trading organizations such as the European Union. They therefore make less money through exports.

Heavy debt
LICs sometimes have to borrow money from HICs, for example, to help rebuild after a natural disaster. This money sometimes has to be repaid with interest.

Emphasis on raw materials
The cost of primary products, such as raw materials, is much lower than the cost of secondary products, such as manufactured goods. The economies of many LICs are based on primary products, which are less profitable than exported goods.

Physical factors

Natural disasters
Development in LICs can be held back by earthquakes, floods, and other natural disasters. The rebuilding they require can cost a lot of money.

Climate
Countries that have low levels of wealth and technology have difficulty in growing food in climates with long droughts and extreme temperatures. Climate change can make these difficulties worse.

Poor farming land
If a country has poor soil, or is very mountainous, it may not be able to produce a lot of food.

Lack of natural resources
Countries that lack natural resources will not be able to make products they can sell.

Consequences of uneven development

Uneven development leads to global inequality. Hundreds of millions of people are forced to live in poverty. Many have inadequate access to health care, education, sanitation, food supplies, and other daily necessities. It is also a cause of international migration and can lead to conflict.

Wealth inequality
Uneven development can cause vast differences in wealth between rich and poor countries, and greater inequality within countries.

Inadequate health care
Many people in LICs die young due to poor health care. Life expectancy in Chad is 44 years, compared to 75 in Japan.

International migration
People in LICs often seek to move to HICs as a result of war or natural disasters, or for better job opportunities and a better quality of life.

Globalization

MANY COMPANIES OPERATE ON A GLOBAL SCALE WITH FACTORIES AND OFFICES AROUND THE WORLD.

SEE ALSO
❮ 124–125 Migration
❮ 138–139 The spread of cultures
❮ 148–149 Manufacturing industry
❮ 152–153 Tourism
Global and local interdependence 190–191 ❯

Globalization is not new, but as transportation, trade links, and communications have improved, economic globalization has sped up. Many firms now have global operations.

What is globalization?

The idea of a rapidly shrinking world is often used to explain globalization. Wherever people go, they are likely to find the same fast food at the same outlets, and even watch the same TV programs. Food is exported around the world and people travel across continents to find work or for vacations.

Video-sharing websites are accessed by about 2 billion users around the world.

Almost 1.6 billion people travel internationally every year.

TV programs are exported around the world.

Major fast food companies operate in more than 110 countries.

Migrant workers have filled the demand for labor created by the rapidly growing economies of the world.

FIFA, the governing body of football (soccer), consists of 211 member nations from across the globe.

Fruits that grow only in certain climates, like the tropical banana, are now available worldwide.

The production and export of coffee have played a pivotal role in the ongoing development of Brazil.

How the world comes closer

Advancements in technology have led to the quick and easy transportation of goods. The internet has brought about new ways to connect with others and it has played a big role in the spread of media and culture.

Transnational companies (TNCs)
Large TNCs that have offices and factories in many countries offer the same products and services across the world.

Tourism
Cheaper and faster flights allow more people to travel the world, visit other countries, and experience different cultures.

Sports
International sporting events with worldwide participation, such as the Olympics and the soccer World Cup, are enjoyed by people around the world.

Exports
Crops that grow in specific places and require certain climatic conditions are now available to people across the globe.

Culture
Technological advancements and the growth in tourism have increased awareness of music, movies, and traditions across the world.

Migrant workers
Workers migrate to other countries for better job opportunities and pay. They often take with them their culture and way of life.

Factors that lead to globalization

Globalization is helped by the development of new forms of technology that help connect places. Changes in the laws of a country may also facilitate foreign investment, and in turn globalization.

Improved transportation
Increased access to international travel allows both tourists and businesses to reach previously inaccessible locations.

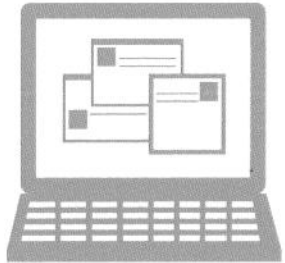

Advanced technology
The internet has sped up the transfer of ideas and images around the world, enabling faster communication.

Better communication
Satellite technology has enabled the transmission of "live" information, which connects distant places to each other.

Changes in domestic laws
Many countries have changed their trade regulations to allow TNCs to invest there, encouraging global businesses.

Advantages of globalization

Globalization can have many positive impacts on individuals and the world's economies. Ever-widening networks and increased communication have led to a free flow of goods, services, and information, which strengthens connections between countries.

More jobs created
Professionals in one country may be able to find opportunities for work and better pay in other countries.

Wider range of goods
Faster means of transport and cheaper production costs make a variety of goods available across the world.

Reduced costs of production
TNCs are able to produce goods in countries where wages are lower, keeping their production costs low.

Variety of vacation destinations
Media can bring attention and tourism-related income to new and lesser-known vacation destinations.

Increased cultural exchange
Greater awareness and appreciation of other people's cultures increases understanding between countries.

Increased trade
The demand for raw materials and the sales of finished products by TNCs promote trade, which brings in wealth.

Problems caused by globalization

While there are many advantages to becoming a global community, globalization also has its disadvantages. Powerful high-income countries (HICs) may dominate world trade and exert pressure on smaller, less influential, low-income countries (LICs), which can lose their own values and traditions in the process.

Outsourcing and unemployment
As TNCs move production to LICs where labor is cheaper, unemployment among primary and secondary sector workers in HICs increases.

Loss of local and national identity
Local traditions, values, and languages are under threat from the more dominant cultures of richer countries.

Manufacturing in HICs
As industries move to LICs where production is cheaper, manufacturing jobs in HICs start to decline.

Urban housing

HOUSING IN CITIES REFLECTS THE SOCIAL, CULTURAL, AND ECONOMIC CHARACTERISTICS OF THE PEOPLE LIVING THERE.

SEE ALSO
❮ 124–125 Migration
❮ 130–131 Cities
❮ 136–137 Urbanization

Cities have many functions, including providing space for people to live. They usually have residential zones with different types of homes for people from a range of social and economic backgrounds.

Home ownership

Houses can be owned by individuals, property companies, or local authorities. Privately owned homes are normally occupied by the owner but can be rented out to other people. Property companies also rent out the housing they own but the rents may be higher. Housing owned by local authorities, or social housing, is rented to people who meet the criteria for need.

Private ownership
This is when a person owns a house or apartment and lives in it, or rents it out to other people and becomes the landlord.

Social housing
This is owned by the local authority or housing associations and rented to low-income families who cannot afford the rent in the private market.

Private-rented sector
This housing is owned by individuals or property companies and rented out to make a profit.

Redevelopment

The renewal and renovation of run-down or derelict urban areas is called redevelopment. Properties may be demolished and replaced with new housing, or buildings such as warehouses and factories may be converted into flats, turning previously industrial areas into residential areas. They are close to the city center with great access to jobs and services.

Before

Dilapidated warehouses
Industrial buildings that are no longer in use fall into disrepair and are left empty, so they become derelict.

After

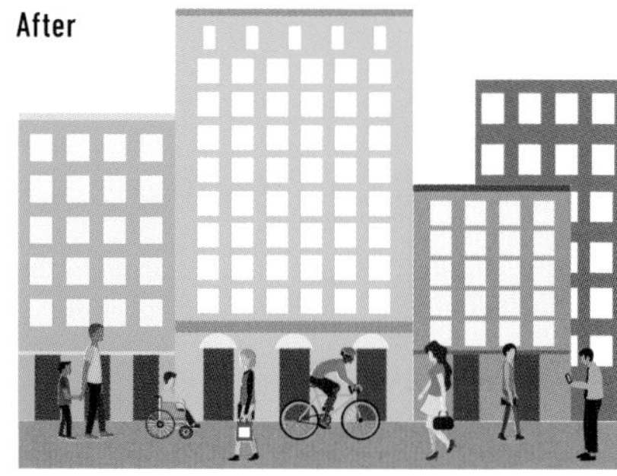

Apartments converted from warehouses
Warehouses converted into apartments are popular, as they are close to the city center with access to jobs and entertainment.

Suburbanization

As cities grow, the demand for housing increases and new housing developments are built on the edge of the city. The land here is cheaper, so they often contain larger detached houses with yards or communal spaces. These areas are called the suburbs and are often some distance from the city center, requiring people to travel to work in the city.

Move toward counterurbanization ▷
In high-income countries (HICs), there has been a trend to move away from city centers and into suburban areas for a better quality of life. This is known as counterurbanization.

Gentrification

In inner cities, the process of improving housing in neighborhoods that are normally associated with lower-income groups is known as gentrification. With a rise in the number of affluent people, there is a demand for better housing, which is met by improving and converting older housing. Gentrification causes lower-income groups to be forced out due to increased rents and house prices.

Derelict building
This housing, previously used by low-income groups, has been abandoned and has fallen into disrepair.

Popular café
Older housing, which is cheap to rent or buy, may be taken over by people who want to set up a new local business.

Renovated shop
Now that the area has become more popular, properties have become more desirable and businesses are doing well.

Upmarket branded shop
As the area attracts more affluent people, upmarket shops move into the area to serve the demand from the new well-off residents.

Human impact

OVER THE LAST TWO CENTURIES, HUMAN ACTIVITY HAS BECOME THE DOMINANT INFLUENCE ON THE ENVIRONMENT.

SEE ALSO

Deforestation	172–174 ›
Climate change	175–177 ›
Plastic pollution	186–187 ›
Big geographical issues	188–189 ›
Food security	192–193 ›
Water security	194–195 ›

With population growth and technological advancement, humans are having a growing impact on the planet. Human activity has changed our landscape and oceans and is changing our climate.

Plastic pollution

Plastic is a wonder substance: a human invention with many uses. However, plastic pollution is an increasing concern. Some plastic waste is recycled but most ends up in landfill sites or—worse—the oceans, where it can take centuries to break down. Often, marine animals get tangled up in items such as fishing nets or confuse toxic plastic particles for food. Plastic also poses a risk to humans: as it is broken down, it creates smaller, harmful microplastics, which enter the food chain and, eventually, the body. More positively, environmental campaigns have forced governments and businesses to begin to tackle these problems.

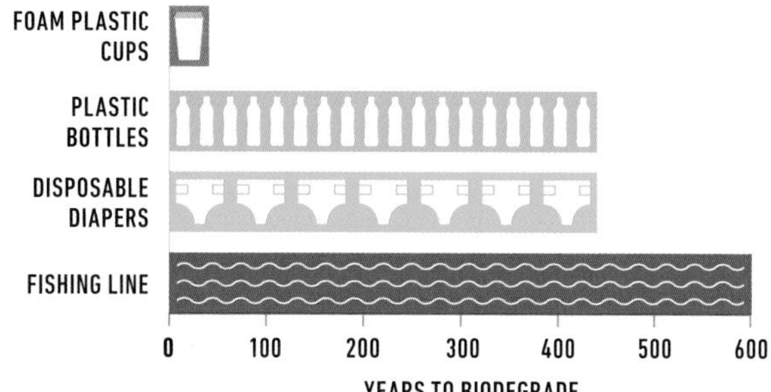

FOAM PLASTIC CUPS
PLASTIC BOTTLES
DISPOSABLE DIAPERS
FISHING LINE
0
100
200
300
400
500
600
YEARS TO BIODEGRADE

Effects of farming

People will always need food; therefore, they will always need farming of some kind. Ever since early humans first chopped down the forests to clear land for farming, they have impacted the environment. Over time, farming has become more intensive, with more land cleared, and plants and insects destroyed.

Deforestation
Clearing forests makes more land available for farming. However, it destroys habitats and helps cause climate change as trees absorb carbon dioxide, a greenhouse gas.

Irrigation
Excessive irrigation can deplete water resources. Lakes, rivers, and groundwater sources may dry up or drop to levels that can no longer sustain ecosystems.

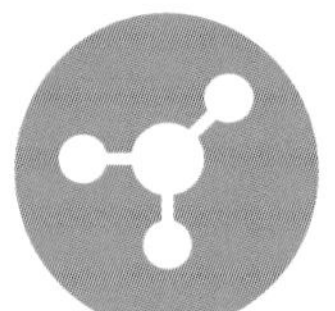

Pollution
Pesticides and fertilizers help increase production, but they cause polluting agricultural runoff, when the chemicals they contain "run off" the land into waterways.

Genetic modification (GM)
Crops can be genetically engineered to improve yields or become resistant to pests. However, there are multiple risks, including the contamination of wild plants.

Soil degradation
Intensive cultivation and poor farming practices can lead to a decline in soil quality and soil erosion. Ultimately, the soil may turn into desert, unable to sustain life.

Methane emissions
Livestock, particularly cattle, produce large amounts of methane. This potent greenhouse gas traps heat in the atmosphere and thus contributes significantly to global warming.

Overfishing

The rise in the global population has contributed to a significant increase in the demand for fish over recent decades. Attempting to meet this demand has led to overfishing, where fish are caught at a faster rate than they can reproduce. Many species are now endangered or being pushed to the point of extinction. An increasing proportion of the fish we now eat is produced in fish farms, where they are reared in tanks and fed artificially with food pellets.

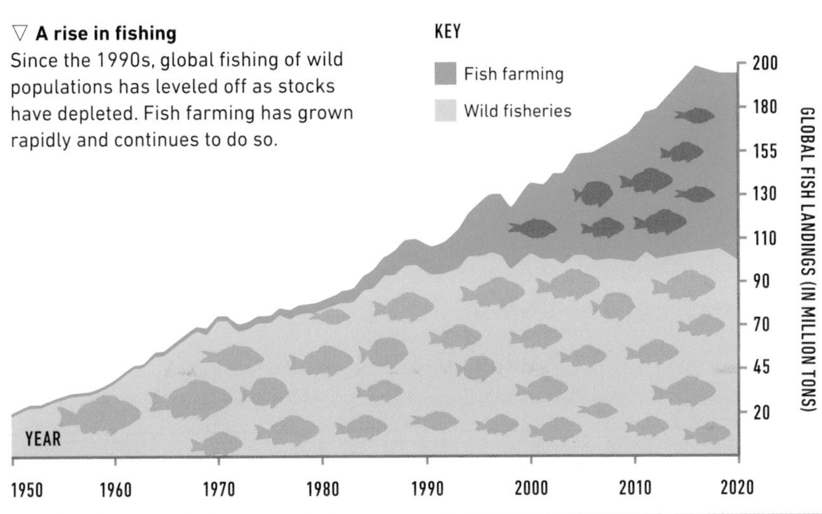
KEY
Fish farming
Wild fisheries
YEAR
1950
1960
1970
1980
1990
2000
2010
2020
200
180
155
130
110
90
70
45
20
GLOBAL FISH LANDINGS (IN MILLION TONS)

▽ **A rise in fishing**
Since the 1990s, global fishing of wild populations has leveled off as stocks have depleted. Fish farming has grown rapidly and continues to do so.

REAL WORLD

Oil spills

When crude oil accidentally spills into the ocean, it can have a devastating effect on marine life. Oil spills destroy habitats, block sunlight, damage the fur and plumage of birds and mammals (making them susceptible to hypothermia), and introduce toxic chemicals to the food chain. It can take many years for ecosystems to recover. Large-scale spillages occur when offshore drilling operations go wrong, when big oil tankers sink, or when pipelines break.

Anthropocene epoch

Geologists divide Earth's vast history into eons, eras, periods, and epochs, identified by characteristic fossils found in rocks. But human activity is changing the world. Natural habitats are vanishing, landscapes are being remodeled, and even the climate is being altered by the way we burn fossil fuels. The most recent geological period, the Holocene, began with the end of the Ice Age 11,500 years ago. Scientists believe the impact of humans on the world has been so severe that we are now living in a new epoch, the Anthropocene (the "human epoch").

Markers of the new epoch

- Fallout from the first nuclear tests in the 1950s
- Airborne soot
- Raised levels of carbon dioxide in the air causing climate change
- Buildup of nitrogen and phosphorus in the soil from fertilizers
- Chicken bones in landfill, since chickens are the world's most numerous birds
- Plastic pollution, especially in the oceans
- Mass extinction of species

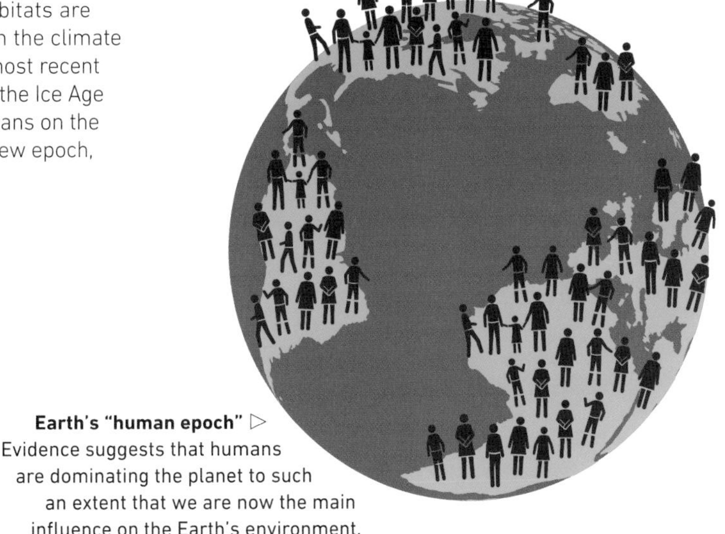

Earth's "human epoch" ▷
Evidence suggests that humans are dominating the planet to such an extent that we are now the main influence on the Earth's environment.

Pollution

SEE ALSO
❮ 148–149 Manufacturing industry
❮ 166–167 Human impact
Plastic pollution 186–187 ❯
Big geographical issues 188–189 ❯

POLLUTION IS THE DAMAGE TO THE ENVIRONMENT CAUSED BY HARMFUL SUBSTANCES IN WATER, THE AIR, OR THE GROUND.

There are three major types of pollution: air (atmospheric), land (terrestrial), and water (hydrospheric). Pollution can be chemical and biological, but it can also be caused by light and noise.

Air pollution

Air pollution occurs when harmful gases or particles are emitted into the atmosphere. In high concentrations, these pollutants can cause damage to health, especially for those with breathing or heart problems. Causes of air pollution include burning fossil fuels in power stations, and traffic fumes.

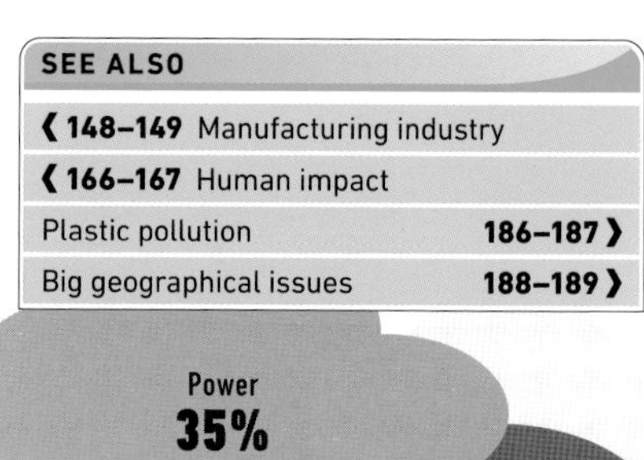
Power
35%

Transportation
11%
Fuel supply
7%

Sulfur dioxide ▷
A major polluting gas, sulfur dioxide is a principal component of acid rain. Around 75 percent of sulfur dioxide emissions are caused by burning fossil fuels for industry and power.

Polluting gases, carried by the wind, dissolve in clouds, forming acid rain. This damages buildings and trees, and harms life in rivers and lakes.

Coal is burned in industrial plants and power stations. Acid gases (sulfur dioxide and nitrogen oxide) are released into the atmosphere.

Land pollution

Overuse of chemicals, damaging mining techniques, and the dumping of waste all contribute to terrestrial pollution. A large proportion of trash is buried in landfill sites. As the trash decomposes, it emits methane, a powerful greenhouse gas.

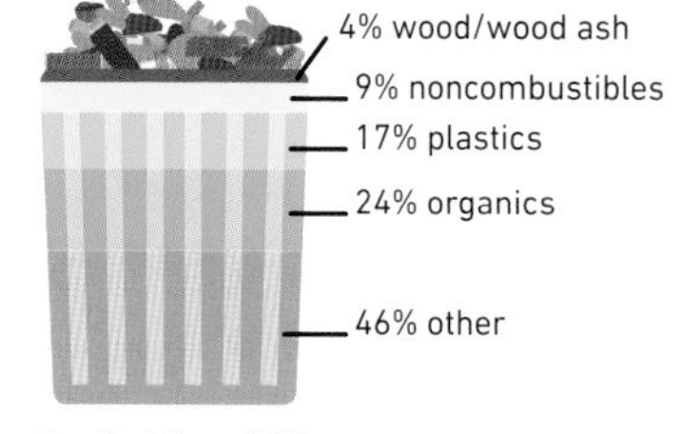
4% wood/wood ash
9% noncombustibles
17% plastics
24% organics
46% other

New York State (HIC)
2¾ lb (1.3 kg) per person per day

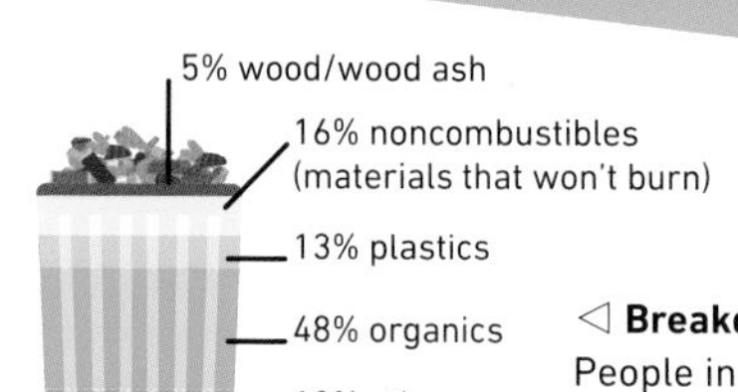
5% wood/wood ash
16% noncombustibles (materials that won't burn)
13% plastics
48% organics
18% other

Lagos, Nigeria (LIC)
1¼ lb (0.6 kg) per person per day

◁ Breakdown of waste
People in HICs produce far more waste than those in LICs, and much more that cannot be recycled or will not biodegrade.

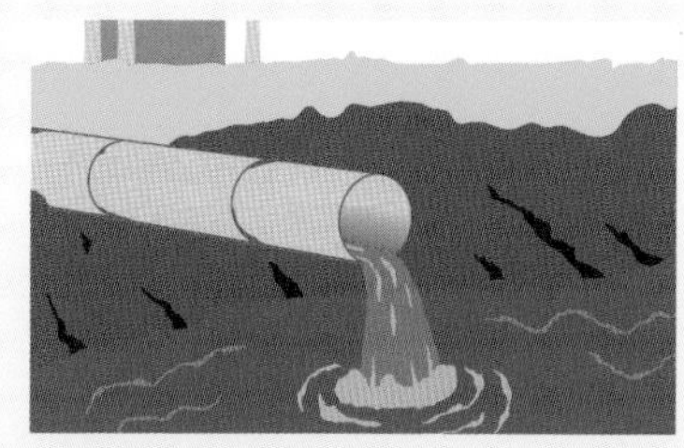

Point sources

A point source is a single, identifiable source of pollution. A common point source of water pollution, for example, is a pipe or a drain from a factory or sewage treatment works. Nonpoint source pollution cannot be traced to a single point of origin, such as agricultural run-off, which can accumulate from multiple sources.

Water pollution

Contamination of water bodies including aquifers, rivers, and oceans is called water pollution. Fertilizers, used by farmers to improve crop yields, can be carried into waterways from fields during rainfall (agricultural run-off). Toxic chemicals such as lead and arsenic are released illegally into water bodies as the by-product of industries, particularly textile production and mining. Raw sewage can also be a major water pollutant, causing serious health risks if consumed.

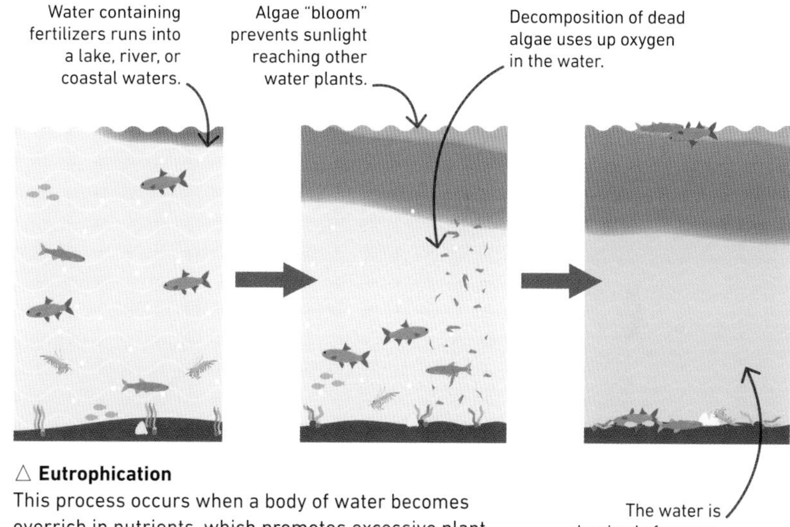
Water containing fertilizers runs into a lake, river, or coastal waters.
Algae "bloom" prevents sunlight reaching other water plants.
Decomposition of dead algae uses up oxygen in the water.
The water is deprived of oxygen, causing fish to die.

△ **Eutrophication**
This process occurs when a body of water becomes overrich in nutrients, which promotes excessive plant growth. A major cause is agricultural run-off.

Cars are responsible for the emission of greenhouse gases and harmful black carbon.
Noise pollution (caused by loud transportation, music, and construction work) and light pollution (caused by artificial light in the night sky) occur in built-up areas, disrupting behavior patterns in animals and humans.
Liquid drains or "leaches" from landfill sites, causing leachate, which runs off into the soil.
Pesticides and fertilizers from farms run off into rivers.
Approximately 8 million tons of plastic are thrown into the ocean each year. Plastic can take thousands of years to biodegrade.
Cruise liners burn damaging fuel oil, which releases harmful sulfur dioxide, and dump more than a million liters of waste into the ocean per day.

▽ **Pollution**
Pollution can have many diverse sources. It can also be carried far away from its original source, meaning it can damage the environment of more than one country.

The changing landscape

LANDSCAPES ARE DYNAMIC AND ALWAYS EVOLVING.

Landscapes can evolve naturally, or they can evolve as a result of human activity. There are very few landscapes left in the world that have not been shaped by humans in some way.

SEE ALSO
❮ 44–46 Weathering and erosion
❮ 72–73 Erosion in deserts
❮ 166–167 Human impact
Deforestation 172–174 ❯
Climate change 175–177 ❯
Big geographical issues 188–189 ❯

Natural changes

Usually, changes to the landscape happen gradually over a very long period of time. These slow processes of erosion are caused by moving air (wind) or the movement of water (in rivers or even as ice). Moving water can also result in very sudden and catastrophic landscape changes—causing landslides, for example.

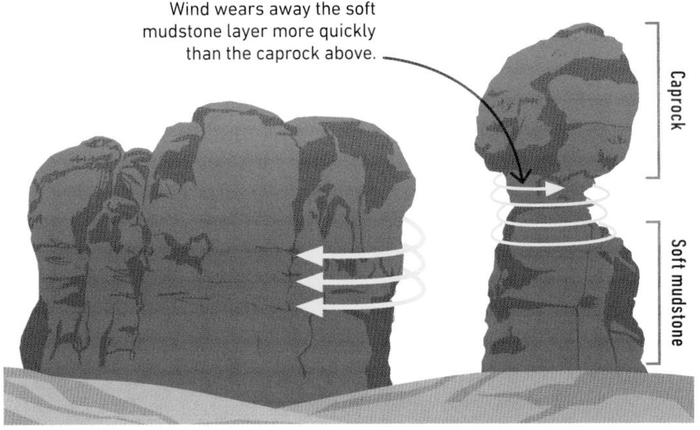
Wind wears away the soft mudstone layer more quickly than the caprock above.
Caprock
Soft mudstone

Wind erosion ▷
Particles carried by the wind grind away rock surfaces. The rock is worn away into new, sometimes unusual, shapes. Wind erosion is a significant driver of landscape change in dry desert regions.

Human changes

Landscape changes caused by humans can be dramatic. Since the 1800s, the rate at which humans are changing the planet has accelerated. Human activities that can cause landscape change include urbanization, deforestation, and agriculture. These changes can also cause natural processes to speed up.

▽ Quarrying and deforestation
The ground beneath these trees holds important resources. In cutting down trees to extract these resources, humans have a major impact on the landscape.

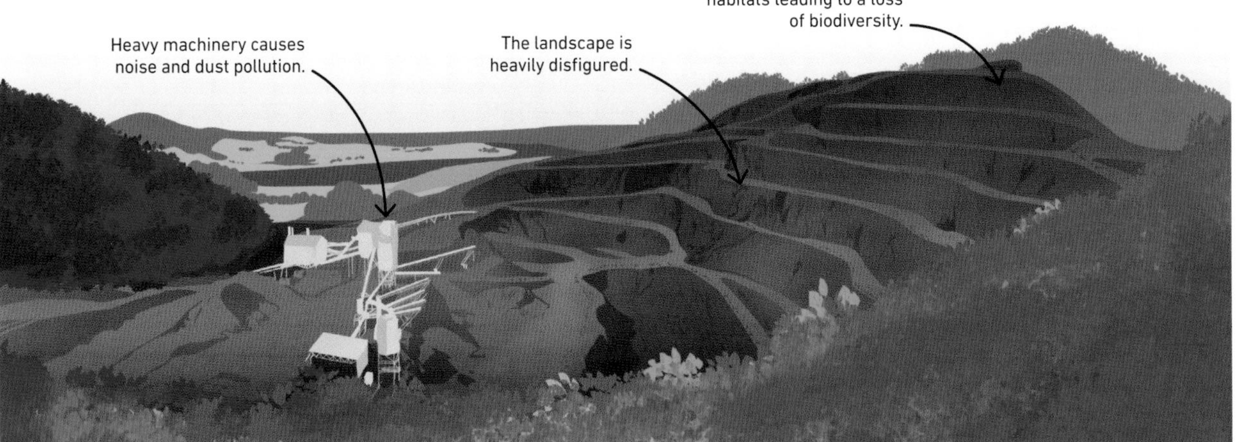
Quarrying destroys wildlife habitats leading to a loss of biodiversity.
Heavy machinery causes noise and dust pollution.
The landscape is heavily disfigured.

Climate change

Natural processes causing landscape change have been influenced by human-induced climate change. The release of greenhouse gases into the atmosphere has led to global warming. Across the world, sea levels are rising as a result of ice melting, and because the water is getting warmer and therefore expanding. Many coastal locations and port cities are facing inundation. In addition, climate change causes more frequent extreme weather events such as storms and hurricanes, which can cause flooding, landslides, and soil loss from unprotected slopes.

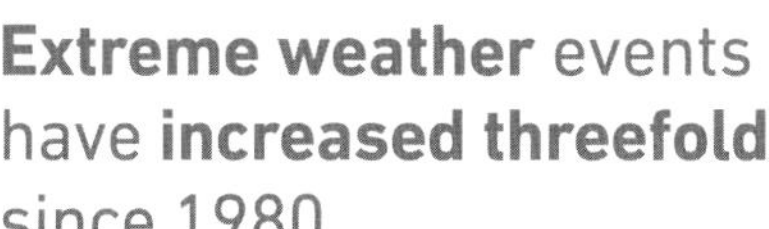

Extreme weather events have **increased threefold** since 1980.

1890s
The growth of factories increased the amount of greenhouse gases being released into the air, warming the planet.

Present day
Melting global ice cover is contributing to a rise in sea levels. This means that low-lying areas are at risk of flooding.

2050
With temperatures 5.4°F (3°C) warmer than today, melting ice caps may cause widespread flooding.

Desertification

Desertification is the process of land slowly transforming into desert as the quality of the soil declines over time. It can be caused by human activity such as deforestation and overgrazing, which exposes the soil and increases the likelihood of erosion. It is also made worse by climate change: as desert areas become hotter and drier, the soil can be carried away by the wind more easily.

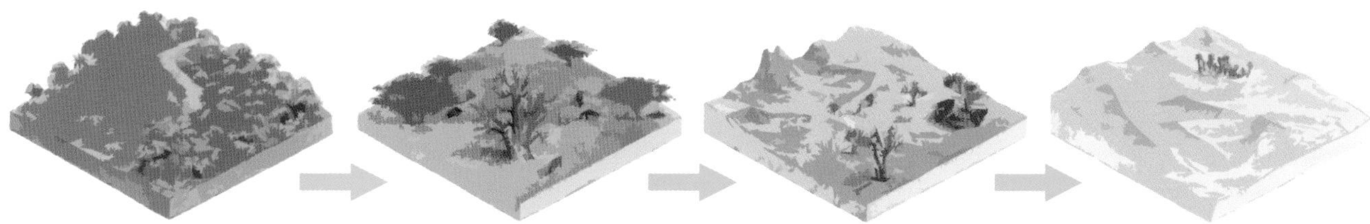

Lush vegetation
Existing vegetation keeps the soil healthy, and the land appears lush. However, with little rainfall, vegetation has only just enough water to survive.

Vegetation clearance
Farming, overgrazing, and tree-felling mean there is less vegetation to hold the soil in place. Soil is exposed and erodes more easily.

Desertification intensifies
Owing to more land overuse, grasses die and are replaced by shrubs. The ground receives fewer nutrients, and more soil is eroded.

The area becomes desert
With little plant life covering the soil, rain evaporates quickly. The wind carries away the topsoil that is left, leaving a barren wasteland.

Deforestation

DEFORESTATION IS THE PERMANENT REMOVAL OF TREES SO THAT LAND CAN BE USED FOR OTHER PURPOSES.

For thousands of years, humans have cut down small areas of woodland in order to build houses or plant crops. But the rate of deforestation is now much faster, with devastating consequences.

SEE ALSO
❮ 54–55 Soil
❮ 102–103 Ecosystems
❮ 104–105 Tropical grasslands and rainforests
❮ 108–109 Temperate forests and grasslands
❮ 110–111 Boreal forests and tundra
❮ 144–145 Food and farming
Climate change 175–177 ❯

Importance of forests

Forests can be grouped into three broad categories, according to the latitudes at which they grow: tropical, temperate, and boreal. Due to photosynthesis, forests play a key role in the water cycle and provide a natural defense against climate change. They are therefore vital to the planet's health.

Water supply
Trees release water from their leaves during photosynthesis, which helps form rain clouds.
Carbon storage
During photosynthesis, trees remove carbon from the atmosphere and store it in their leaves, helping reduce the effects of greenhouse gases.
Ecotourism
Areas of natural forest are beautiful and attract tourists. Sustainable ecotourism can earn income for local communities.
Biodiversity
About 70 percent of the world's land-dwelling species live in forests. Loss of habitat can lead to extinction.

△ **Life in the rainforest**
Home to a huge variety of plants and animals, rainforests teem with life. There are ways in which humans can earn income from forests without destroying them.

Clearing the rainforest

Rainforests grow in a narrow zone around the equator, where it is hot and wet all year round. Just like the forests in more temperate zones (such as in Europe) that were largely cleared in previous centuries, these forests are under extreme threat of deforestation. The main threat is from commercial interests, including agriculture, logging, and mining. These industries provide profits for companies, and jobs and income for local people. However, clearing these forests destroys the richest source of biodiversity on the planet. It also removes one of the Earth's most important carbon stores, which exacerbates climate change.

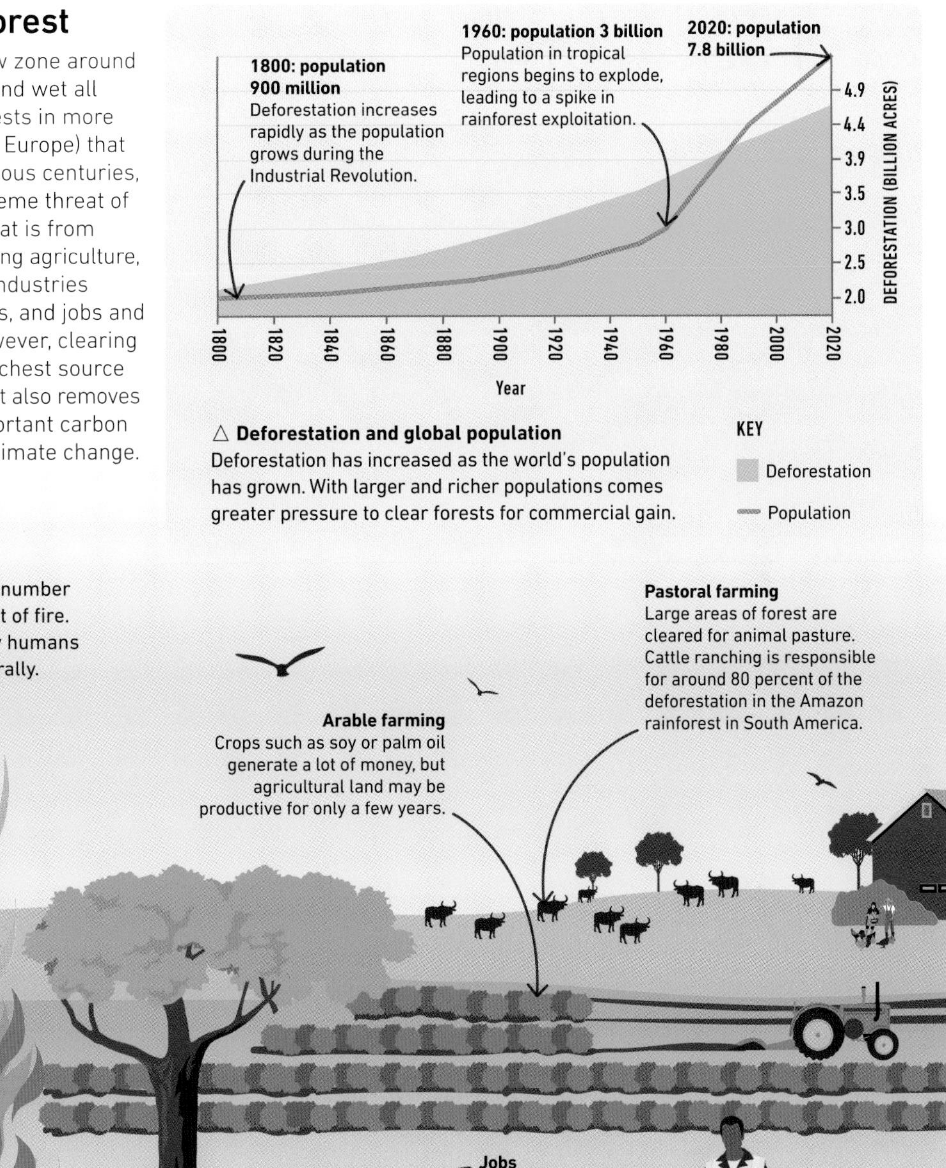

1800: population 900 million
Deforestation increases rapidly as the population grows during the Industrial Revolution.
1960: population 3 billion
Population in tropical regions begins to explode, leading to a spike in rainforest exploitation.
2020: population 7.8 billion
4.9
4.4
3.9
3.5
3.0
2.5
2.0
DEFORESTATION (BILLION ACRES)
1800
1820
1840
1860
1880
1900
1920
1940
1960
1980
2000
2020
Year
KEY
Deforestation
Population

△ **Deforestation and global population**
Deforestation has increased as the world's population has grown. With larger and richer populations comes greater pressure to clear forests for commercial gain.

▽ **Causes of deforestation**
Deforestation can happen for a number of reasons but is mainly a result of fire. Most deforestation is caused by humans but sometimes it happens naturally.

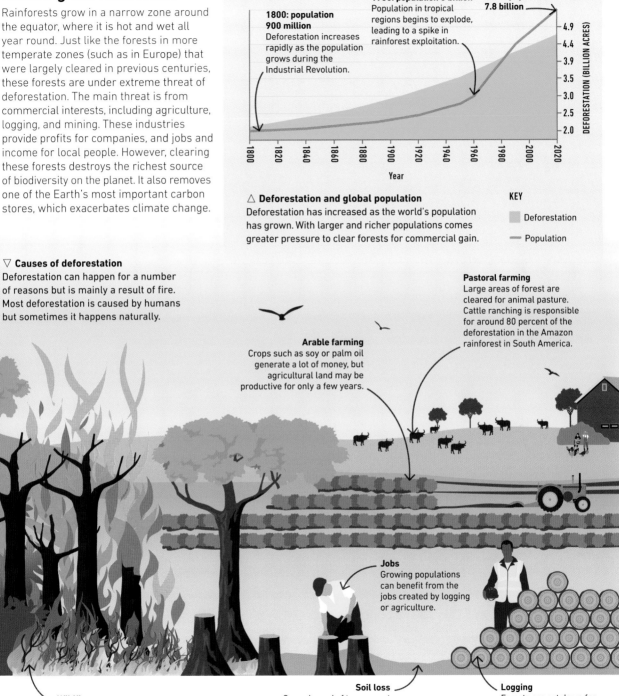

Arable farming
Crops such as soy or palm oil generate a lot of money, but agricultural land may be productive for only a few years.
Pastoral farming
Large areas of forest are cleared for animal pasture. Cattle ranching is responsible for around 80 percent of the deforestation in the Amazon rainforest in South America.
Jobs
Growing populations can benefit from the jobs created by logging or agriculture.
Wildfires
Though some have natural causes, such as lightning strikes, most wildfires are caused by humans. They can have a devastating effect on forests.
Soil loss
Once cleared of trees, most forest soils dry out. Soil nutrients are washed away by rain, leaving topsoils dry. The soil then becomes infertile and unable to grow anything.
Logging
Forests are cut down for timber or paper products. However, much of the world's logging, particularly in tropical rainforests, is illegal.

Rate of deforestation

Rising population figures combined with improved forest clearance techniques have increased the global rate of deforestation. More than 1 billion acres of forest have been lost since 1990. From the mid-20th century onward, the highest rate of deforestation has occurred in the tropical rainforests.

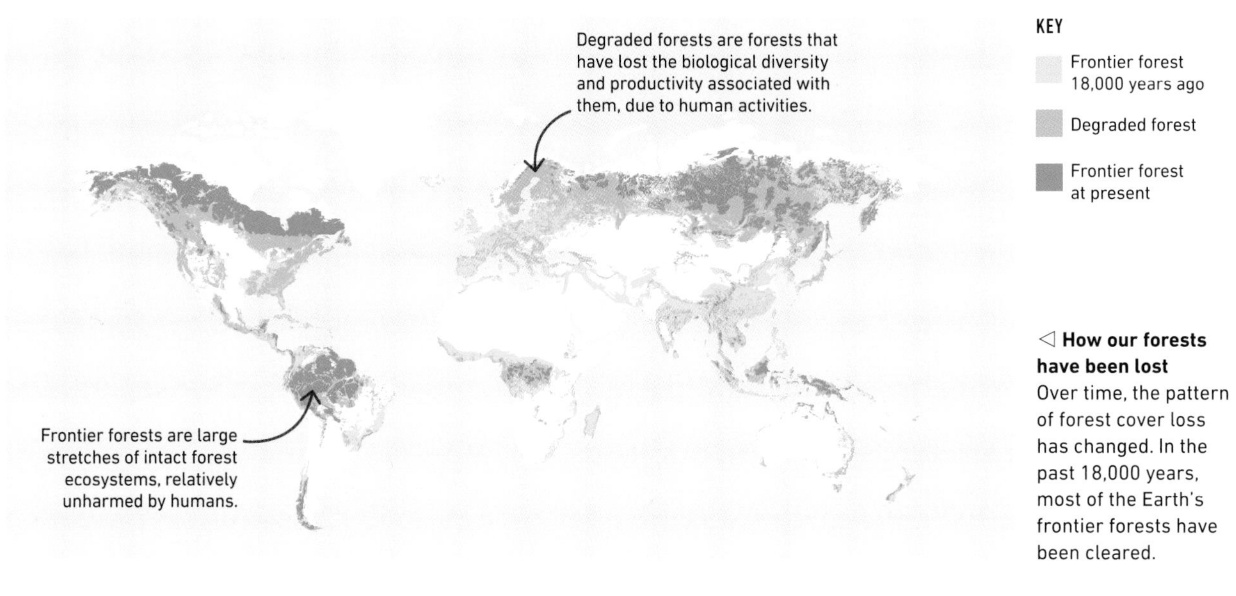
Degraded forests are forests that have lost the biological diversity and productivity associated with them, due to human activities.
Frontier forests are large stretches of intact forest ecosystems, relatively unharmed by humans.
KEY
Frontier forest 18,000 years ago
Degraded forest
Frontier forest at present

◁ **How our forests have been lost**
Over time, the pattern of forest cover loss has changed. In the past 18,000 years, most of the Earth's frontier forests have been cleared.

Soil loss

As trees are cleared, there is less canopy cover to intercept rainfall and fewer roots to hold the soil together. So, when it rains, soil is easily washed away along with nutrients from the plants. This eventually dries up the topsoil, which is then carried away by the wind, resulting in severe erosion.

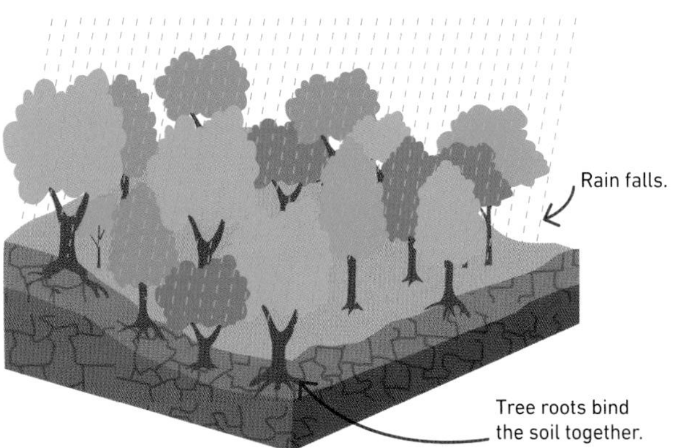
Rain falls.
Tree roots bind the soil together.

△ **How trees protect soil**
The tree canopy intercepts rainfall, which means raindrops hit the soil with less force, causing less erosion. The roots also bind the soil together and protect it from erosion.

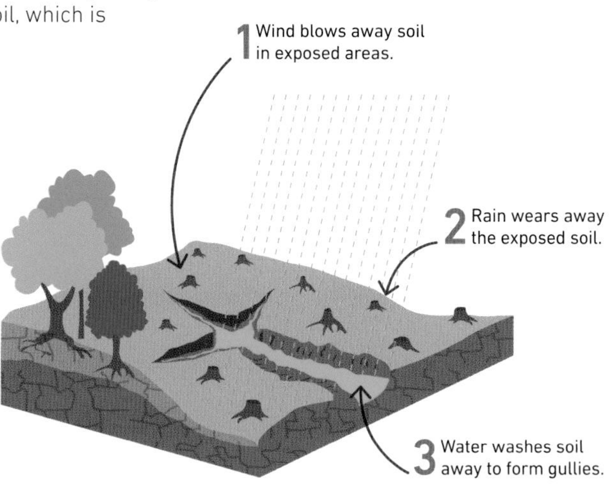
1 Wind blows away soil in exposed areas.
2 Rain wears away the exposed soil.
3 Water washes soil away to form gullies.

△ **Impact of felling trees**
Tree roots hold the soil in place. Felling or chopping down trees exposes the soil and makes it vulnerable to erosion by water and wind.

Climate change

SEE ALSO
❮ 166–167 Human impact
❮ 170–171 The changing landscape
Food security 192–193 ❯
Water security 194–195 ❯

CHANGE IS NORMAL AND NATURAL. "CLIMATE CHANGE" IS A PHRASE THAT REFERS TO VERY RAPID, HUMAN-MADE CHANGE.

Over cycles of thousands of years, the average global temperature can change. This is natural climate change. However, in recent decades, we can see that very rapid change is now happening as a result of human activities.

Natural causes of climate change

Over millions of years, the Earth's climate has swung between hot and cold periods. Long-term climate change can occur due to several natural causes, such as volcanic activities, variations in the amount of heat emitted by the sun, and cyclical variations of the Earth's orbit around the sun, also known as Milankovitch cycles.

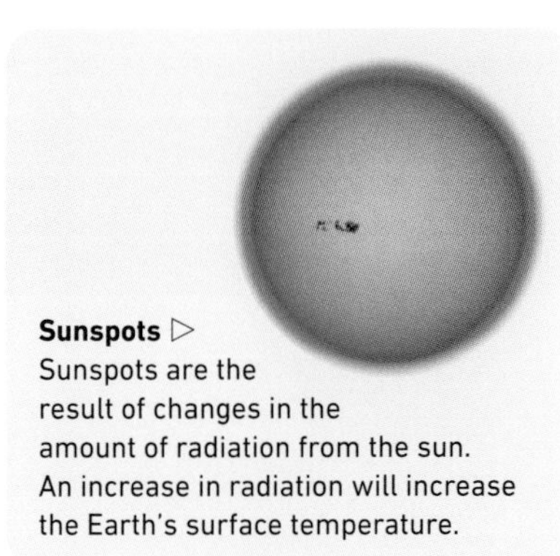

Sunspots ▷
Sunspots are the result of changes in the amount of radiation from the sun. An increase in radiation will increase the Earth's surface temperature.

△ Volcanic activity
Volcanic eruptions can push ash, dust, and gas into the atmosphere, blocking the sun's energy and cooling the Earth. Prolonged eruptions can have a deeper impact on the climate.

Milankovitch cycles ▷
The variations in the tilt and orbit of the Earth around the sun are known as Milankovitch cycles, and these can affect the Earth's climate. There are three cycles—eccentricity, obliquity, and precession.

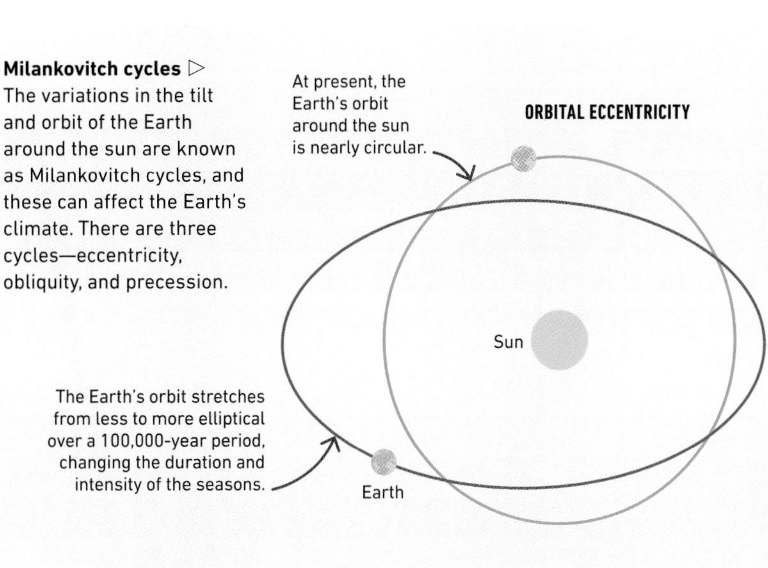
ORBITAL ECCENTRICITY
At present, the Earth's orbit around the sun is nearly circular.
Sun
Earth
The Earth's orbit stretches from less to more elliptical over a 100,000-year period, changing the duration and intensity of the seasons.

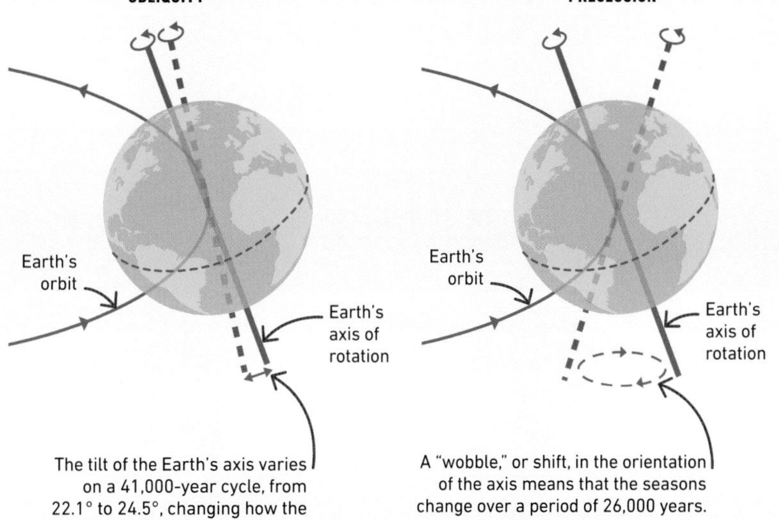
OBLIQUITY
Earth's orbit
Earth's axis of rotation
The tilt of the Earth's axis varies on a 41,000-year cycle, from 22.1° to 24.5°, changing how the sun's heat falls on the Earth.
PRECESSION
Earth's orbit
Earth's axis of rotation
A "wobble," or shift, in the orientation of the axis means that the seasons change over a period of 26,000 years.

Human causes of climate change

Over the last 150 years, various human activities have become the main cause of climate change. Greenhouse gas emissions from the increased burning of fossil fuels are warming the Earth. According to the Intergovernmental Panel on Climate Change (IPCC), global temperatures are predicted to rise by 5.4–9 °F (0.3–4.6°C) by the end of the century.

▽ **Accelerated greenhouse effect**
Greenhouse gases act as a "blanket," trapping just the right amount of heat for life to survive on the Earth. However, the extra CO_2 in the atmosphere helps thicken this blanket and traps heat around our planet.

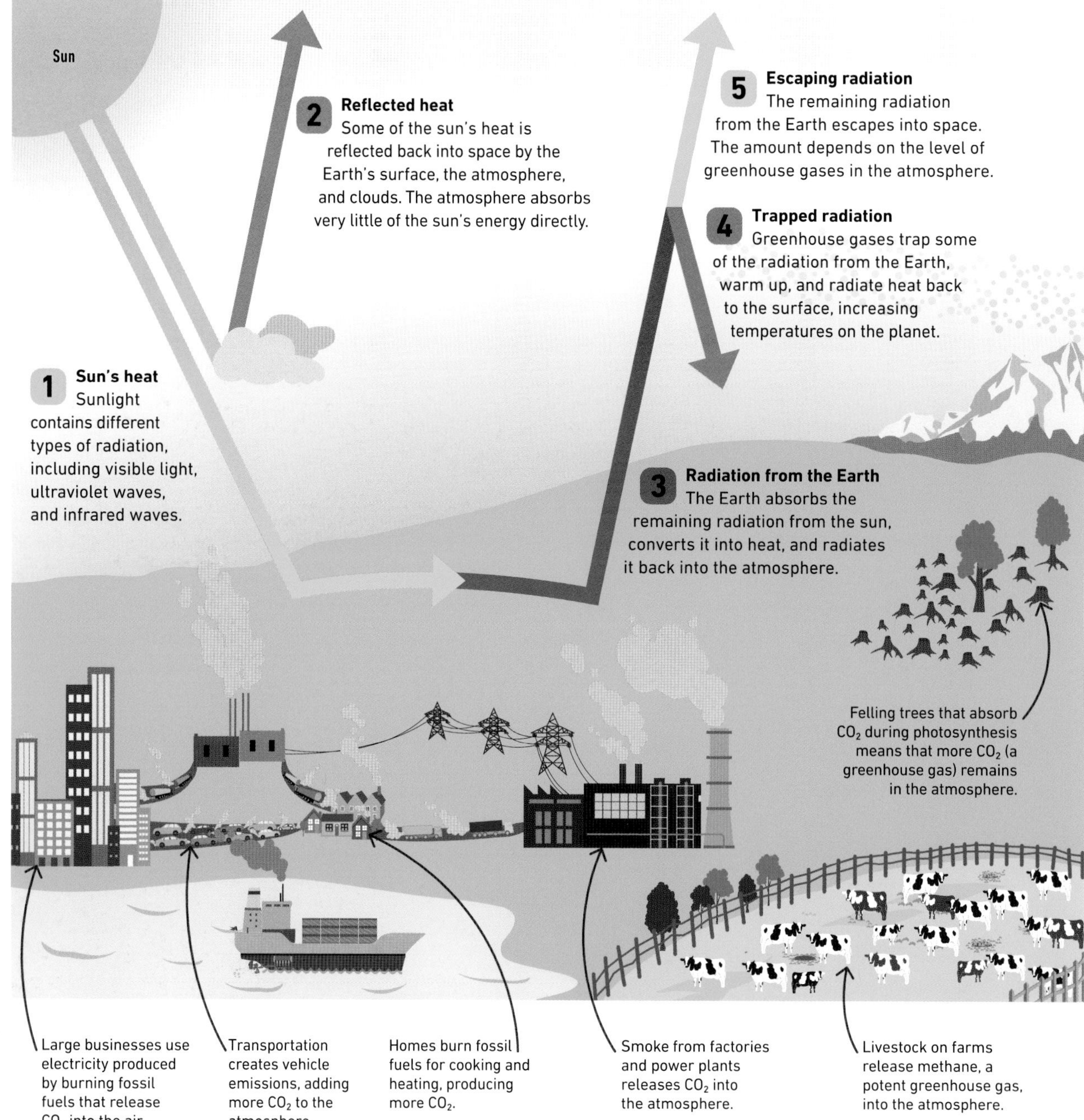
Sun
2 Reflected heat
Some of the sun's heat is reflected back into space by the Earth's surface, the atmosphere, and clouds. The atmosphere absorbs very little of the sun's energy directly.
5 Escaping radiation
The remaining radiation from the Earth escapes into space. The amount depends on the level of greenhouse gases in the atmosphere.
4 Trapped radiation
Greenhouse gases trap some of the radiation from the Earth, warm up, and radiate heat back to the surface, increasing temperatures on the planet.
1 Sun's heat
Sunlight contains different types of radiation, including visible light, ultraviolet waves, and infrared waves.
3 Radiation from the Earth
The Earth absorbs the remaining radiation from the sun, converts it into heat, and radiates it back into the atmosphere.
Felling trees that absorb CO_2 during photosynthesis means that more CO_2 (a greenhouse gas) remains in the atmosphere.
Large businesses use electricity produced by burning fossil fuels that release CO_2 into the air.
Transportation creates vehicle emissions, adding more CO_2 to the atmosphere.
Homes burn fossil fuels for cooking and heating, producing more CO_2.
Smoke from factories and power plants releases CO_2 into the atmosphere.
Livestock on farms release methane, a potent greenhouse gas, into the atmosphere.

Effects on the environment

A rise in global temperature will have an adverse impact on the entire planet. Some effects of warmer temperatures are ocean acidification, rising sea levels, changing weather patterns, and the loss of ice cover and habitats. The warmer our planet gets, the more severe the impact on the environment will be.

△ **Ocean warming**
Higher temperatures will mean warmer oceans. This will increase the evaporation of water and also impact marine species and ecosystems.

△ **Tropical storms**
Higher evaporation will lead to the formation of more clouds. In the tropics, this can increase the frequency and intensity of tropical storms.

△ **Loss of habitat**
Warmer waters will melt the sea ice, resulting in sea levels rising and the loss of habitat for Arctic wildlife, such as polar bears.

Effects on human life

Climate change has been described as the "challenge of our generation." It poses a serious threat to lives, habitats, and businesses across the world. The less-developed regions and poorer people are likely to be more vulnerable and will be most affected by climate change.

Rising water levels and flooding
As sea levels rise, low-lying coastal areas will be flooded, forcing people to move out of their homes, creating environmental refugees.

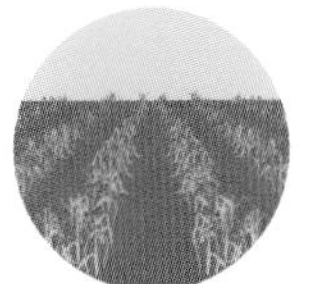

Reduced food supply
Changes in weather patterns will reduce crop yields in various parts of the world, leading to higher food prices and an increase in food insecurity.

Reduced water supply
Higher temperatures, along with reduced rainfall, will deplete the water in rivers and reservoirs, resulting in water scarcity.

Change in seasons
Plant growth and crop yields can suffer due to unseasonal weather. A change in the seasons can also allow disease-carrying insects longer periods to breed and spread.

Damage to fishing industry
Warmer oceans will cause damage to coral reefs and threaten marine habitats, reducing fish numbers. This can have a significant impact on the fishing industry.

Increase in wildfires
Warmer temperatures and less rainfall can dry up areas of land, increasing the chances of forest fires and the destruction of lives, homes, and infrastructure.

Action on climate change

In 1988, an international group of scientists, the IPCC, was established to keep track of the climate crisis. Politicians began to set goals for cutting carbon emissions, including the Paris Agreement in 2015. But progress has been slow, and targets have been missed. Most scientists now think carbon emissions must be cut to zero by 2030 to avoid disaster. In 2018, Swedish schoolgirl Greta Thunberg led a global school strike to demand action from politicians. World leaders continue to meet every year to assess progress, discuss climate-change policy, and update goals.

Preservation and conservation

SEE ALSO
❮ 98–99 Biomes
❮ 102–103 Ecosystems
❮ 152–153 Tourism
❮ 166–167 Human impact

THE PROTECTION OF WILDLIFE AND NATURAL HABITATS IS CHALLENGED BY CONFLICTING HUMAN NEEDS.

Preservation and conservation are two similar ideas about protecting the environment and are often used interchangeably, but there are subtle differences between the two.

Preservation

Preservation is the idea that the areas of the Earth so far untouched by humans need to be maintained in their current condition, or "preserved." Human activity in these places must be restricted to a minimum so that the landscape remains untouched.

The Great Bear Rainforest, Canada ▷
This vast rainforest covers 12,000 sq miles (32,000 sq km). In February 2016, the Canadian government signed an agreement to ban commercial logging in 85 percent of the forest.

In-situ vs. ex-situ

In-situ and ex-situ conservation are methods of protecting endangered wildlife, at risk of extinction. In-situ means conservation "in place," so the animal is protected within its natural habitat. Ex-situ means conservation "out of place," so the animal is taken outside of its natural habitat, for example to a zoo, or even relocated elsewhere in the wild.

Plant species can also be **endangered**. **Seed banks** can help **preserve** them.

△ **In-situ**
This Alaskan brown bear is seen fishing in the McNeil River Sanctuary in Alaska where bear hunting is prohibited and the number of visitors is regulated.

△ **Ex-situ**
In the Chengdu Research Base of Giant Panda Breeding in China, the natural habitat of pandas is recreated to encourage them to reproduce.

Conservation

Conservation is an active form of management that balances the need to protect landscapes and their ecosystems while also acknowledging the demands of human activity. It is different to the "hands off" approach of preservation and provides a managed environment for visitors, such as a national park.

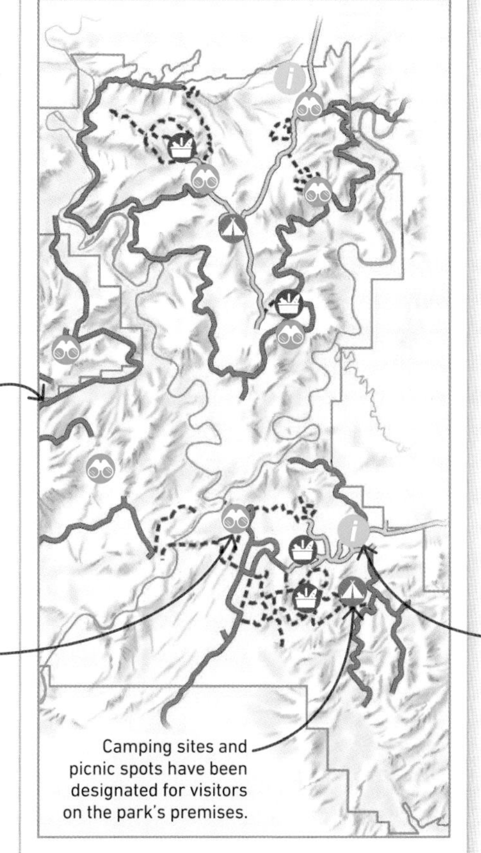
Camping sites and picnic spots have been designated for visitors on the park's premises.

Roads for four-wheel drive vehicles and hiking trails let visitors explore different parts of the park.

Elevated viewing platforms allow visitors to observe wildlife.

Information centers provide visitors with all the necessary details of the park.

KEY

- Camping site
- Viewing platform
- Picnic spot
- Information center
- Highway
- Four-wheel drive vehicles only
- Hiking trail
- Park border

◁ **Canyonlands National Park**
In 2019, this national park in the US had nearly 750,000 visitors. The area has been developed so that tourists can access it more easily, but it is still home to mule deer and coyotes.

REAL WORLD

Tigers in Nepal

In 2009, there were just 135 tigers left in the wild in Nepal. Currently, there are 235 tigers living across five national parks due to successful conservation efforts by the government, such as expanding the boundaries of the existing national parks, and increasing patrols.

Challenges

Conservation and preservation can cause conflicts between those who want to protect the land and those who want to use it for its resources. These conflicts will become more common as the world becomes more populated. The rapidly growing demand for energy resources and land for human use are among the most severe challenges.

△ **Oil drilling**
The global demand for oil is growing, but supply is limited. When new oil supplies are discovered under the ocean or land, there is pressure to drill for oil.

△ **Poaching**
Poaching is the illegal hunting of wildlife, often for pelts, horns, or ivory. It poses a challenge for conservationists because habitats have to be guarded.

△ **Tourism**
Overtourism can cause damage to ecosystems and wildlife. Delicate habitats, such as coral, can get damaged just by tourists touching it.

Managing natural hazards

SEE ALSO
❮ 28–30 Earthquakes and tsunamis
❮ 34–37 Volcanoes and hot springs
❮ 96–97 Hurricanes and tornadoes

BEING PREPARED FOR NATURAL DISASTERS IS ESSENTIAL TO REDUCE THE DISRUPTION AND LOSS OF LIFE THEY CAUSE.

Natural events that can endanger people's lives, cause damage to property, and disrupt economic activity are known as natural hazards. The chances of their occurring vary from place to place.

Types of natural hazards

Natural hazards can be either climatic or tectonic. The weather causes climatic hazards, and movements of tectonic plates, which are pieces of the Earth's outer layer or crust, cause tectonic hazards. Natural hazards can occur all over the world, but some areas are more likely to experience them than others. Sometimes, human activities, such as deforestation, can increase the chances of natural hazards such as landslides.

Tectonic hazards

Volcanic eruption
Plate movements can cause hot molten rock deep down in the Earth to rise to the surface and erupt as lava.

Earthquake
The sudden release of pressure at the margin of a tectonic plate can cause the ground to shake.

Tsunami
Earthquakes under the sea can displace a large amount of water, causing a giant wave called a tsunami.

Climatic hazards

Tropical storms
In tropical areas, warm ocean water evaporates, causing low-pressure areas where storms develop.

Flood
Sea or river water flowing above its normal level causes a flood. Heavy rain can also cause floods.

Drought
Low rainfall or lack of rainfall in some arid areas can lead to a severe shortage of water.

Tornado
A column of rapidly rotating wind called a tornado can cause loss of life and damage to property.

Restless Earth ▷
Weather events and movements of tectonic plates can give rise to various natural hazards that can be extremely destructive.

Vulnerability to disaster

A community is said to be vulnerable to natural disasters if a natural hazard is likely to occur in its location and cause destruction. The level of development of the community also influences vulnerability—poorer communities have fewer resources to prepare for natural hazards and are most affected by them.

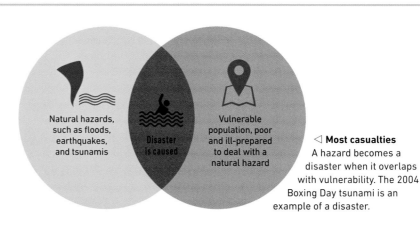

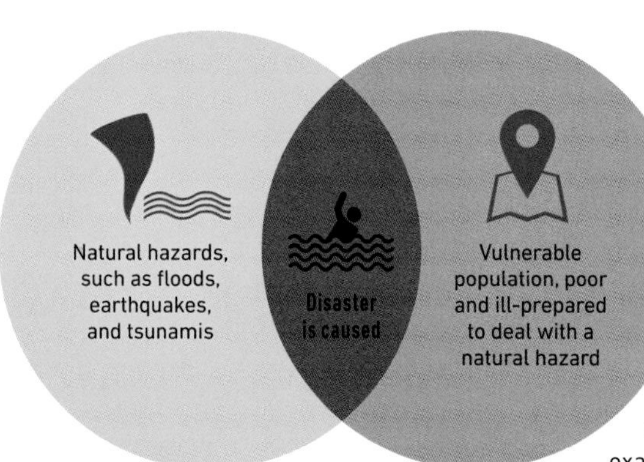
Natural hazards, such as floods, earthquakes, and tsunamis
Disaster is caused
Vulnerable population, poor and ill-prepared to deal with a natural hazard

◁ Most casualties
A hazard becomes a disaster when it overlaps with vulnerability. The 2004 Boxing Day tsunami is an example of a disaster.

Engineering and planning

Man-made structures, such as earthquake-resistant buildings, flood defense systems, and storm drains, built using methods of hard engineering, can reduce the impact of specific natural hazards and also make communities less vulnerable to them. They can be very effective but are very expensive to build.

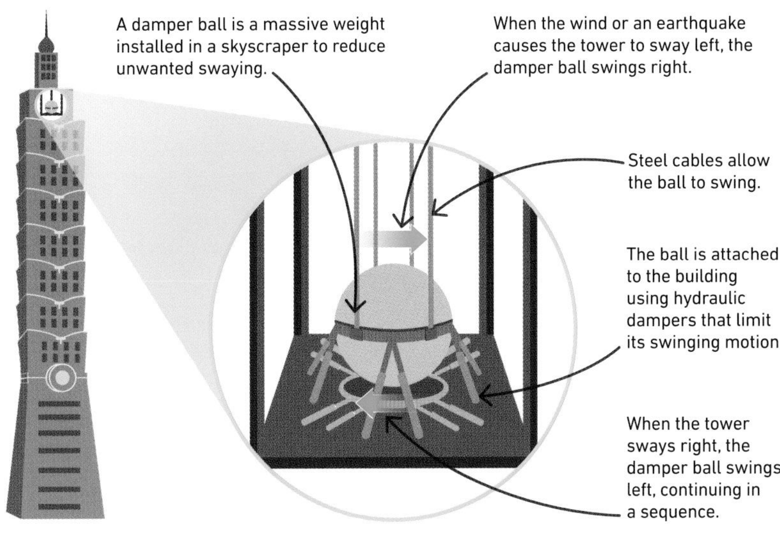

A damper ball is a massive weight installed in a skyscraper to reduce unwanted swaying.
When the wind or an earthquake causes the tower to sway left, the damper ball swings right.
Steel cables allow the ball to swing.
The ball is attached to the building using hydraulic dampers that limit its swinging motion.
When the tower sways right, the damper ball swings left, continuing in a sequence.

Earthquake resistance ▷
One of the world's tallest buildings, Taipei 101 in Taiwan has very deep foundations, a steel frame, and a gigantic damper ball that counters its movement during earthquakes.

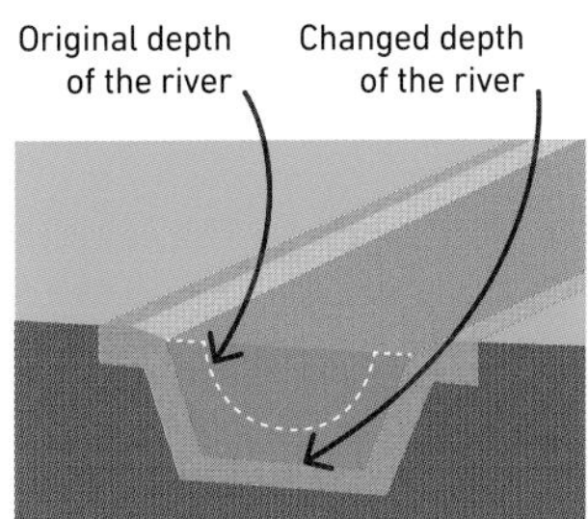

Original depth of the river
Changed depth of the river

◁ Altering a river's course
Making a river deeper so it can carry more water and making its path straighter so the water flows faster can reduce the risk of a river flood. However, this may lead to flooding downstream.

◁ Dams and reservoirs
A dam built on a river can trap water, forming a reservoir, and release it in a controlled way. This prevents the river from flooding its banks.

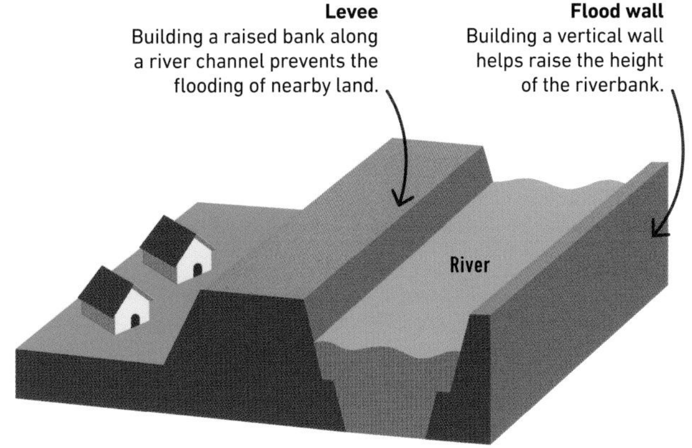

Levee
Building a raised bank along a river channel prevents the flooding of nearby land.
Flood wall
Building a vertical wall helps raise the height of the riverbank.
River

△ Flood defenses
Structures like levees and flood walls can protect communities from floods. Levees can also prevent seawater from flooding coastal areas.

Being prepared

A good way to reduce vulnerability to natural disasters is to be prepared for them. Improving people's preparedness is known as the soft management of hazards because it does not involve the building of any physical structures. It includes increasing awareness through education and conducting activities, such as practice drills and first-aid training.

△ Earthquake drills
Earthquakes are common in Japan, so many primary schools organize drills teaching children what to do when the ground begins to shake.

△ Flood barriers
Volunteers help stack sandbags in Torgau, Germany, to build a barrier against flooding if the water level in the Elbe River rises.

Sources of energy

ENERGY IS NECESSARY FOR INDUSTRY AND TO HEAT OUR HOMES AND COOK OUR FOOD.

SEE ALSO
❮ 146–147 Extracting fossil fuels
❮ 166–167 Human impact
❮ 175–177 Climate change
Sustainability 184–185 ❯

The sun is the main source of energy, but the Earth also produces its own energy, which comes from its molten core. Humans have harnessed both energy sources for their needs.

Fossil fuels

Coal, oil, and natural gas are the three forms of fossil fuel, which store carbon made by plants using the sun's energy over millions of years. This carbon is burned as a fuel and has become the dominant source of global electricity production. Coal alone produces 40 percent of the world's electricity. This is not sustainable because the carbon dioxide being returned to the atmosphere is a greenhouse gas.

KEY
Natural gas
Oil
Coal

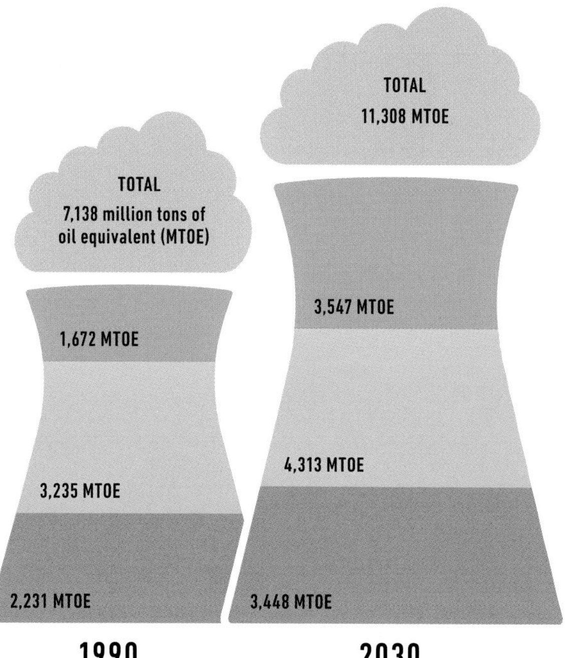
TOTAL
7,138 million tons of oil equivalent (MTOE)
1,672 MTOE
3,235 MTOE
2,231 MTOE
1990
TOTAL
11,308 MTOE
3,547 MTOE
4,313 MTOE
3,448 MTOE
2030

Rising demand ▷ Large-scale energy production is measured by a unit called MTOE. By 2030, projections show there will be demand for 11,308 MTOE from fossil fuels. In order to avert the climate emergency, it will be necessary to reverse the projections shown here.

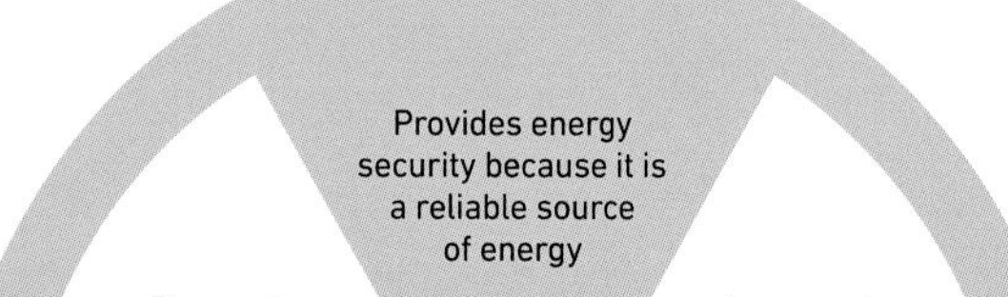
Provides energy security because it is a reliable source of energy
Is expensive to set up
Can provide the energy for extremely destructive weapons
Produces radioactive waste, which is difficult to manage in the long term
Can be dangerous if the power station is damaged in any way
Does not add carbon dioxide to the environment

◁ Destructive resource Nuclear energy has a number of advantages, but the Chernobyl disaster in 1986 in the former Soviet Union, when radioactivity was released into the atmosphere, made this source of energy highly controversial.

Nuclear energy

In nuclear power stations, atoms are split apart using a process called nuclear fission to release their vast amounts of energy and produce power. The fuel used for this reaction is usually uranium or plutonium, which are limited resources, and so nuclear power is also a nonrenewable source of energy.

A **nuclear power** plant generates **as much energy** as **1,500 wind turbines.**

Renewable sources of energy

Energy sources that are not limited and do not deplete when used, such as sunlight and wind, are called renewable. They can be harnessed directly, for example, by a wind turbine and transformed into usable forms of energy, such as electricity. Most renewable energy comes from the sun, but there are other forms: hydrogen fuel produced by electrolysis of water, tidal energy harnessed from tides in the ocean, and geothermal energy harnessed from the hot water and steam that lie deep under the Earth's surface.

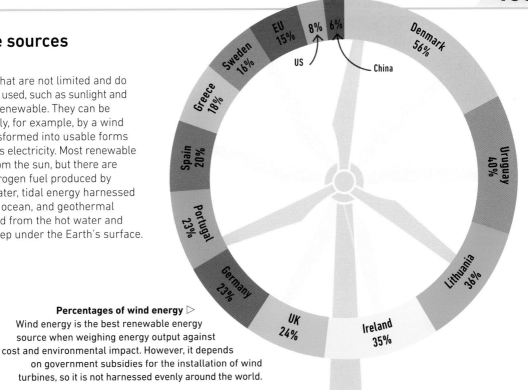

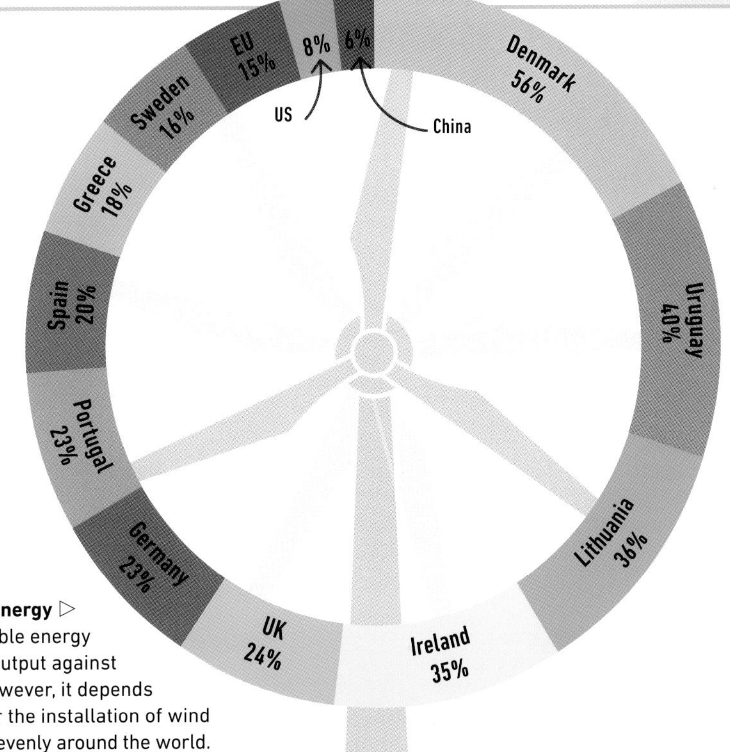
Denmark 56%
Uruguay 40%
Lithuania 36%
Ireland 35%
UK 24%
Germany 23%
Portugal 23%
Spain 20%
Greece 18%
Sweden 16%
EU 15%
8%
6%
US
China

Percentages of wind energy ▷
Wind energy is the best renewable energy source when weighing energy output against cost and environmental impact. However, it depends on government subsidies for the installation of wind turbines, so it is not harnessed evenly around the world.

The four main sources of renewable energy

	Biopower	Hydropower	Wind power	Solar power
Summary	• Includes biomass burned to replace gas and coal, and biofuel to replace fossil fuels • Creates 50 percent of renewable energy	• Generated from flowing water, such as rivers and hydroelectric dams • Creates 31 percent of renewable energy	• Harnessed by wind turbines • Creates 9 percent of renewable energy	• Includes solar thermal energy and energy generated by solar panels • Creates 8 percent of renewable energy
Pros	• Widely available • Cheap to produce • Reduces waste in landfill	• Provides clean energy • Power output flexible due to adjustable water flow • Very reliable	• Clean energy source, with no emissions • Once installed, turbines are cheap to run	• Clean energy resource • Cheap to run • Solar technology is constantly improving
Cons	• Relies on combustion, so not emission-free • Inefficient compared to other fuel types	• Hydroelectric dams are expensive to build • Electricity production is affected by droughts	• Installation costs of turbines are high • Dependent on weather conditions	• Dependent on weather conditions • Electrical output may not justify installation cost

Sustainability

THE WAY WE USE THE PLANET'S RESOURCES TODAY WILL SHAPE THE LIVES OF FUTURE GENERATIONS.

SEE ALSO
❮ 166–167 Human impact
❮ 168–169 Pollution
❮ 175–177 Climate change
❮ 178–179 Preservation and conservation

Sustainability is the process of using resources such as food, energy, and housing responsibly, ensuring that enough is preserved or created to meet the needs of future generations.

Sustainable development

A balance between the needs of the environment, society, and the economy is needed to achieve sustainability. For example, a country that lowers its environmental standards to bring about growth would be seen as unsustainable. Similarly, a country with strong environmental laws but unequal opportunities in society would also be considered unsustainable.

Social factors

Economic factors

Environmental factors

Sustainable development

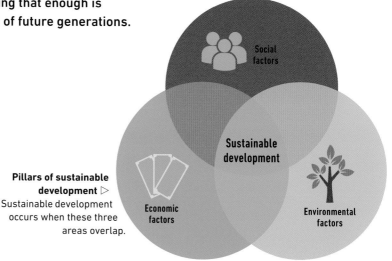

Pillars of sustainable development ▷ Sustainable development occurs when these three areas overlap.

UN goals for sustainable development

The Sustainable Development Goals (SDGs) are a set of 17 global targets that combine social and environmental actions to be met by 2030. They were established by the United Nations (UN) in 2015. They cover eliminating hunger and achieving clean water worldwide as well as taking action on climate change and wildlife conservation.

▽ New global goals
These 17 Sustainable Development Goals (SDGs) have been adopted by all 193 countries that are members of the United Nations General Assembly.

Climate action

Responsible consumption and production

Sustainable cities and communities

Zero hunger

Quality education

No poverty

Good health

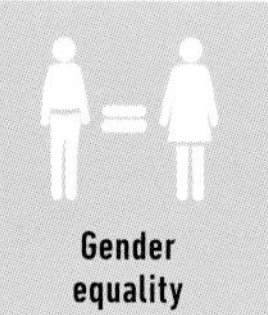
Gender equality

Clean water and sanitation

Clean and affordable energy

Reduced inequalities

Industry, innovation, and infrastructure

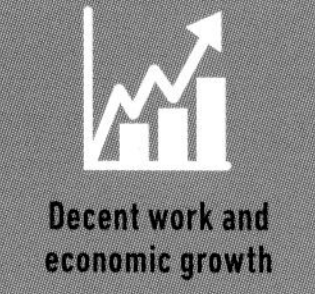
Decent work and economic growth

Global partnership

Peace and justice

Conservation of life on land

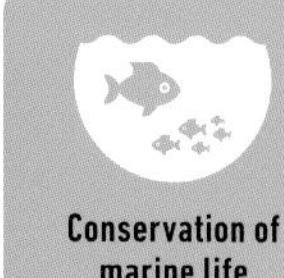
Conservation of marine life

Carbon footprint

The phrase "carbon footprint" refers to the amount of carbon dioxide (CO_2) released into the atmosphere directly or indirectly by an individual, organization, community, or product. The greater the carbon emissions, the larger the carbon footprint and, by extension, the bigger the contribution to climate change.

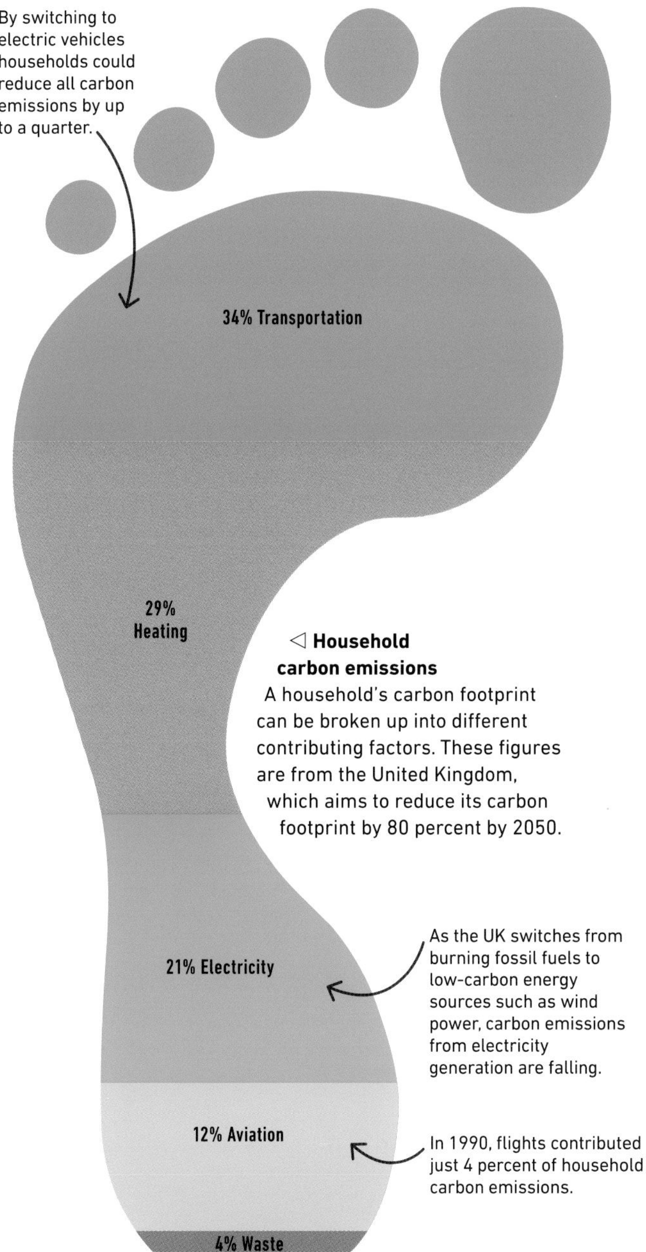
By switching to electric vehicles households could reduce all carbon emissions by up to a quarter.
34% Transportation
29% Heating
21% Electricity
As the UK switches from burning fossil fuels to low-carbon energy sources such as wind power, carbon emissions from electricity generation are falling.
12% Aviation
In 1990, flights contributed just 4 percent of household carbon emissions.
4% Waste

◁ **Household carbon emissions**
A household's carbon footprint can be broken up into different contributing factors. These figures are from the United Kingdom, which aims to reduce its carbon footprint by 80 percent by 2050.

Sustainable future

Technological solutions may help us bring about a sustainable future where all three pillars of sustainable development are met. Changes in our behavior can also ensure sustainability. The examples below show sustainable measures that have already been adopted.

△ **Ecotourism**
Many eco-lodges around the world produce their own electricity, recycle water, and support local efforts to conserve the environment.

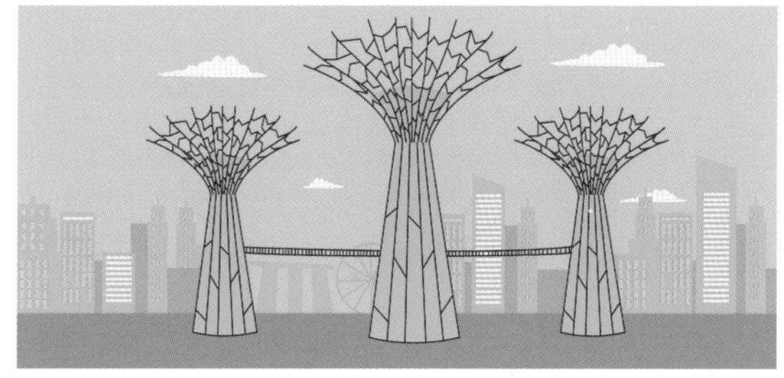

△ **Sustainable structures**
More than 200 species of plants are grown within the steel "Supertrees" at Singapore's Gardens by the Bay. The trees produce their own electricity and collect rainwater for reuse.

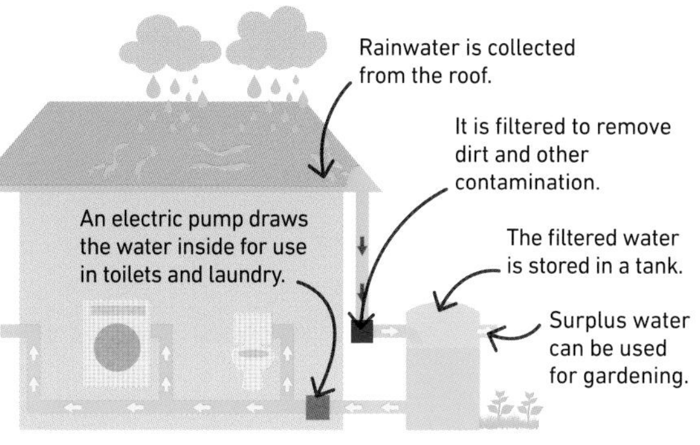
Rainwater is collected from the roof.
It is filtered to remove dirt and other contamination.
An electric pump draws the water inside for use in toilets and laundry.
The filtered water is stored in a tank.
Surplus water can be used for gardening.

△ **Rainwater recycling**
Reusing rainwater for household purposes such as flushing the toilets and watering the garden can help us waste less water.

Plastic pollution

PLASTICS ARE MADE OF FOSSIL FUELS AND DON'T BIODEGRADE, LEADING TO ENVIRONMENTAL PROBLEMS.

SEE ALSO
❮ 84–86 Ocean currents
❮ 102–103 Ecosystems
❮ 166–167 Human impact
❮ 168–169 Pollution

Plastic is everywhere and in almost everything we use. Most plastics take a very long time to break down, which has led to plastic pollution becoming an important environmental challenge.

The rise of plastic

Plastic, as we know it, was first formulated and manufactured in 1907 using fossil fuels. It was a revolutionary invention, marketed as "the material of a thousand uses." During World War II, the production of plastic for weapons increased rapidly because it was cheap and durable. After the war, producers switched to making consumer goods and plastic soon became hugely popular.

Global plastic production

Every year, 405 million tons of plastic are produced, about 500 billion plastic bags are used, and in the US alone, more than 100 billion plastic bottles are sold. A lot of plastic is now produced in emerging economies such as China, which accounts for more than 25 percent of the total global production.

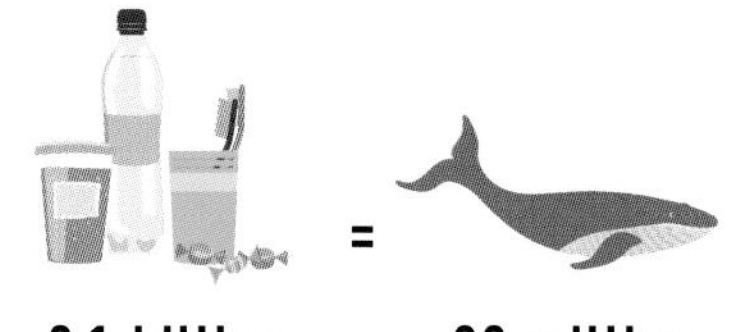
=
9.1 billion
tons of plastic
80 million
blue whales

◁ **All the plastic in the world**
Since plastic was artificially synthesized in 1907, 9.1 billion tons have been produced. That's equal to 80 million blue whales!

Polluting the planet ▽
Plastic waste is dumped in the oceans by various countries. It is carried by circular currents, called gyres, to form huge, floating waste patches, the largest of which, the Great Pacific patch, is three times the size of France.

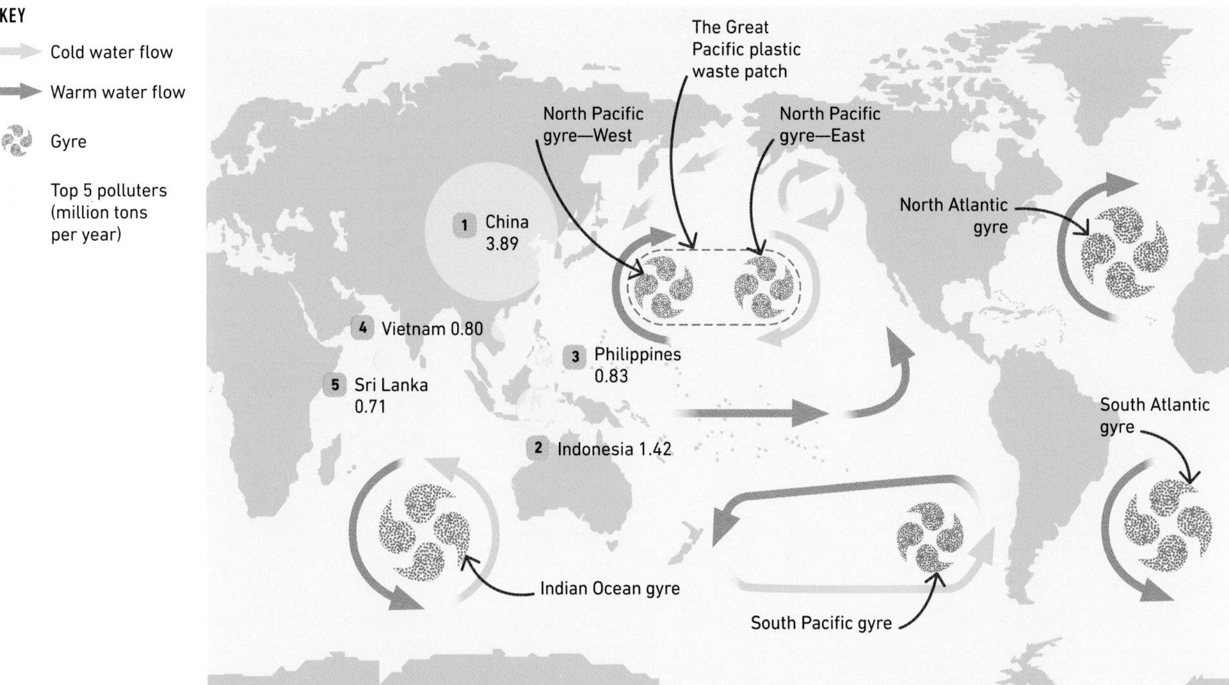
KEY
Cold water flow
Warm water flow
Gyre
Top 5 polluters (million tons per year)
The Great Pacific plastic waste patch
North Pacific gyre—West
North Pacific gyre—East
North Atlantic gyre
1 China 3.89
4 Vietnam 0.80
3 Philippines 0.83
5 Sri Lanka 0.71
2 Indonesia 1.42
South Atlantic gyre
Indian Ocean gyre
South Pacific gyre

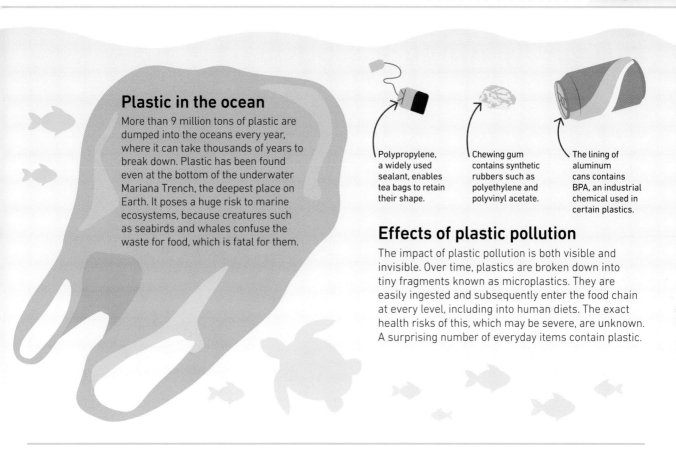

Plastic in the ocean

More than 9 million tons of plastic are dumped into the oceans every year, where it can take thousands of years to break down. Plastic has been found even at the bottom of the underwater Mariana Trench, the deepest place on Earth. It poses a huge risk to marine ecosystems, because creatures such as seabirds and whales confuse the waste for food, which is fatal for them.

Polypropylene, a widely used sealant, enables tea bags to retain their shape.

Chewing gum contains synthetic rubbers such as polyethylene and polyvinyl acetate.

The lining of aluminum cans contains BPA, an industrial chemical used in certain plastics.

Effects of plastic pollution

The impact of plastic pollution is both visible and invisible. Over time, plastics are broken down into tiny fragments known as microplastics. They are easily ingested and subsequently enter the food chain at every level, including into human diets. The exact health risks of this, which may be severe, are unknown. A surprising number of everyday items contain plastic.

What we can do

There is a growing awareness of the harm that plastic is doing to the planet. However, its usage is so widespread that it can be hard to avoid. Real solutions to plastic pollution require individual actions, such as using a "bag for life" and more importantly, government regulations to reduce plastic usage and waste.

▽ **Living with less plastic**
There are a number of actions that each individual can take to reduce the amount of plastic waste that is produced.

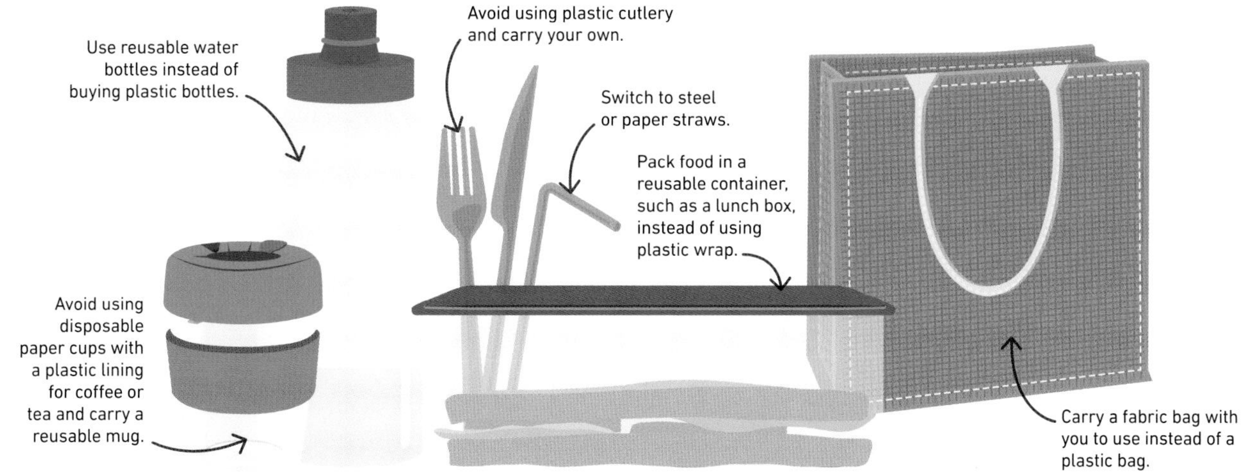
Use reusable water bottles instead of buying plastic bottles.
Avoid using plastic cutlery and carry your own.
Switch to steel or paper straws.
Pack food in a reusable container, such as a lunch box, instead of using plastic wrap.
Avoid using disposable paper cups with a plastic lining for coffee or tea and carry a reusable mug.
Carry a fabric bag with you to use instead of a plastic bag.

Big geographical issues

THE WORLD IS FACING MANY GEOGRAPHICAL CHALLENGES THAT NEED TO BE ADDRESSED URGENTLY.

SEE ALSO
❮ 124–125 Migration
❮ 162–163 Globalization
❮ 175–177 Climate change
❮ 184–185 Sustainability
Water security 194–195 ❯

Geography is the study of places and the relationship between people and their immediate environment. Geography is an important factor in many of the challenges the world is facing today.

Uneven development

High-income countries (HICs) can afford more infrastructure such as housing, transportation, and public services than low-income countries (LICs). Uneven development can occur within the same country, too, with wealthier regions attracting more investment.

Drastic differences ▷
Uneven development can affect people's access to health care, education, and other vital services such as sanitation and clean water.

Migration

Uneven economic development and the gross inequalities this causes in incomes and opportunities leads to migration both within nations and internationally. Human migrations have always taken place and are usually to the benefit of everyone, but there is concern about the scale of current movements.

△ **Finding new lands**
People migrate due to push factors that make them decide to leave an area, and pull factors that make them choose a new area as their home. These factors can be economic or political.

Climate change

The Earth's climate is changing at a fast pace due to greenhouse gas emissions caused by human activity. Climate change is one of the biggest challenges that the planet is facing today because it affects people, animals, plant life, and ecosystems, as well as weather.

△ **Natural disasters**
It is believed that climate change will cause the world's ice caps to melt and sea levels to rise, leading to flooding. It will also result in more frequent tropical storms and other extreme weather events.

Biodiversity loss

The impact of humanity is destroying the diversity of the natural world. So many thousands of wild species go extinct each year that scientists are calling it a mass extinction. Since 1970, wild animal vertebrate populations—birds, reptiles, mammals, and amphibians—have declined by 60 percent. Farm animals now account for 60 percent of all the biomass (weight) of mammals on Earth, while we humans account for a further 36 percent, leaving just 4 percent to wild animals.

◁ **Lost habitats**
When habitats are destroyed, many species die out.

Globalization

The process by which the world is becoming more interconnected through trade, technology, and cultural exchange is called globalization. Distances between places and people are shrinking because of the ease and speed with which people, messages, and goods now travel around the world.

◁ **Breaking down borders**
Businesses and consumers are no longer dependent on their local area. The buying and selling of goods and services can now take place across international borders.

Nationalism vs internationalism

A nation's desire to be independent from the rule or influence of other nations is known as nationalism, while internationalism is the belief that countries can achieve more together through political, economic, and cultural alliances such as the European Union (EU).

◁ **International boundaries**
Borders are imaginary lines that mark the boundaries between nations. They are defined on the basis of politics or war.

Water wars

Essential for all life on Earth, water is not distributed evenly across the planet, which means that access to water is an important political issue. Water-rich regions, with an abundant water supply, have a lot of advantages over water-deficient regions. This can cause water-related disputes, often heightened by climate change.

△ **Shared resources**
Rivers often cross political boundaries and infrastructure such as dams in one region can limit access to water in another, leading to conflict.

Sustainability

The process of fulfilling the needs of the present generation for resources such as food, water, and shelter, without sacrificing the needs of future generations, is known as sustainability. The population of the world is estimated to increase by over 2 billion by 2050 and, therefore, sustainable practices are becoming increasingly important.

△ **The non-meat way of life**
Not eating meat is the best way to reduce one's impact on the Earth, as meat and dairy farming use 83 percent of farmland and produce 60 percent of agricultural greenhouse emissions.

Global and local interdependence

PEOPLE RELY ON OTHER PEOPLE FROM ALL OVER THE WORLD FOR WORK, GOODS, AND CULTURAL INTERESTS.

SEE ALSO
❮ 142–143 Economic activity
❮ 144–145 Food and farming
❮ 162–163 Globalization
❮ 184–185 Sustainability

Interdependence describes a situation in which people (or groups of people) rely or depend on each other. This interdependence can exist in the smallest community as well as on a global scale. Even small local economies are part of the global economic system.

Globalization means that the world is becoming more and more **interdependent.**

Local interdependence

People who live in small communities, such as isolated villages, rely on each other for many of the most important aspects of their daily lives. For example, when villagers buy essential food and services from local businesses, the business owner makes a profit. In recent years, local economies have become an important part of the global economic system.

Local businesses rely on farmers for produce and on the local community to buy their products.
Local business
Local people may work on farms or in businesses and buy locally produced goods.
Local producer
Local community

A local economy ▷ Local relationships are two-way. The farmer sells produce to local businesses, who provide him or her with money in exchange.

Shrinking world

Improvements in technology over the past two centuries have led to information, people, and goods moving around the world at ever-faster speeds. This has made it feel like the world is "shrinking." This shrinking world has allowed the development of interdependence on a global scale, as people around the world build connections with one another effortlessly.

1800

3 months
At the start of the 19th century, a letter to Australia would be sent by sailing ship. The trip took several months.

1850

1 month
By the mid-19th century, the steamship had reduced the amount of time a letter took to reach Australia to one month.

Global interdependence

Countries depend on each other for different things. Imports and exports are key in creating webs of global interdependence. Typically, less-developed countries depend on more-developed countries for manufactured goods, aid, or income from tourism. Richer countries, meanwhile, depend on poorer countries for primary products such as coffee, cocoa, and iron.

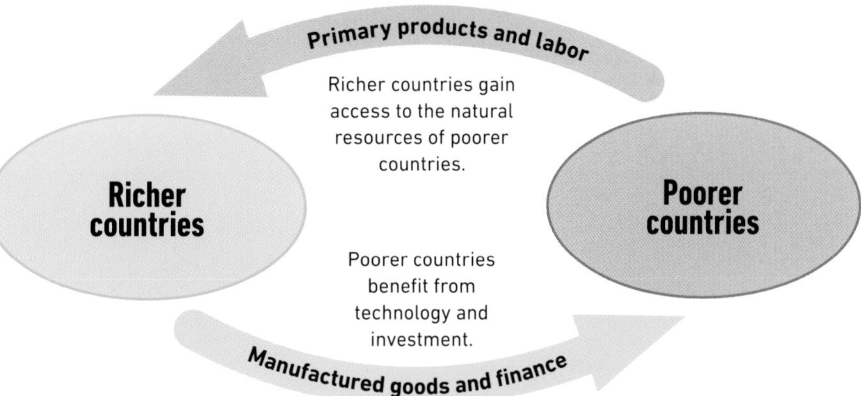
Primary products and labor
Richer countries gain access to the natural resources of poorer countries.
Richer countries
Poorer countries
Poorer countries benefit from technology and investment.
Manufactured goods and finance

Banking

Banking is an example of an industry that is highly interdependent on a global scale. Money can be transferred or shares traded instantly from anywhere in the world by anyone with access to the internet. This has created a complicated web of connections between people, organizations, and countries. However, this interdependence means that financial problems are quickly felt all over the world.

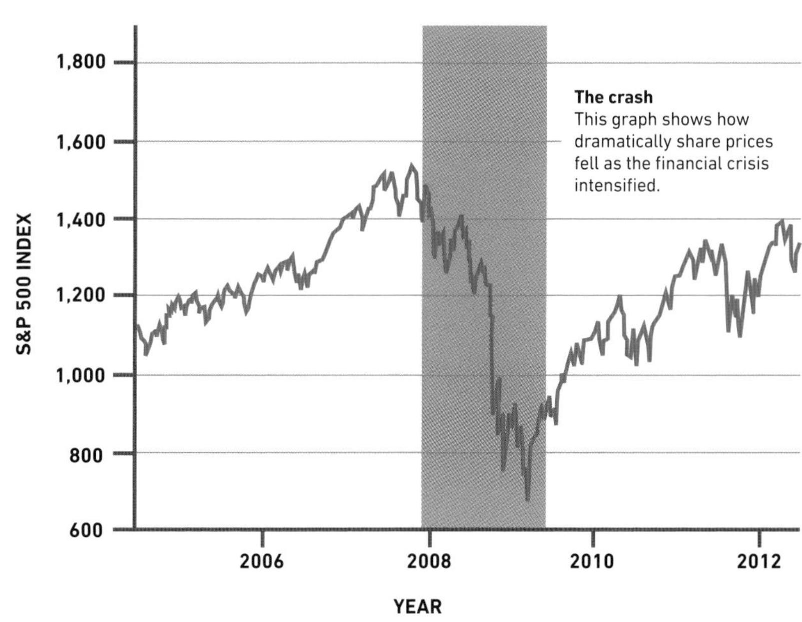
1,800
1,600
1,400
1,200
1,000
800
600
S&P 500 INDEX
2006
2008
2010
2012
YEAR
The crash
This graph shows how dramatically share prices fell as the financial crisis intensified.

The 2008 crisis ▷
In 2008, a major bank in the US failed. This triggered the beginning of the worst financial crash since the 1930s. The consequences of this crash have been felt all over the world.

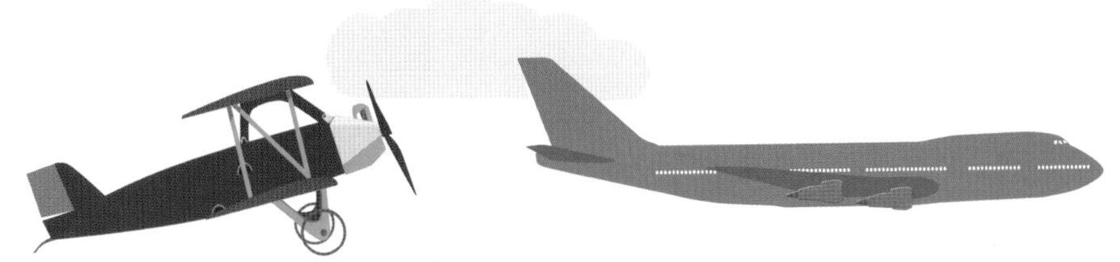

1940

4 Days
The development of air travel meant that taking a letter to the other side of the world was possible in a matter of days.

1960

1 Day
The jet engine made air travel even faster. Mail, plus goods and people, could now cross the globe in a single day.

1995

Seconds
The internet has revolutionized global communications, allowing people to receive a message almost instantaneously.

Food security

THE ABILITY OF A COUNTRY TO FEED ITS POPULATION IS A CONCERN TO ALL COUNTRIES.

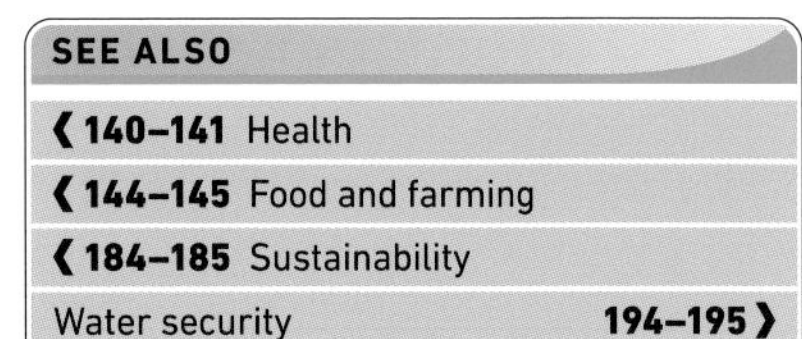
SEE ALSO
❮ 140–141 Health
❮ 144–145 Food and farming
❮ 184–185 Sustainability
Water security 194–195 ❯

The amount of nutritious food available and how well it is distributed determine an area's food security. If an area has inadequate food resources, it is food insecure.

Global food consumption

Rapidly growing populations, increasing wealth in large parts of the world, and the mechanization of farming, which increases yields and lowers the cost of food production, have all contributed to an enormous increase in the global production and consumption of food over the last century. However, the distribution of food around the world is very uneven.

KEY
Daily calorie intake per person

3,270–3,770
1,890–2,390
2,850–3,270
Less than 1,890

2,390–2,850
No data

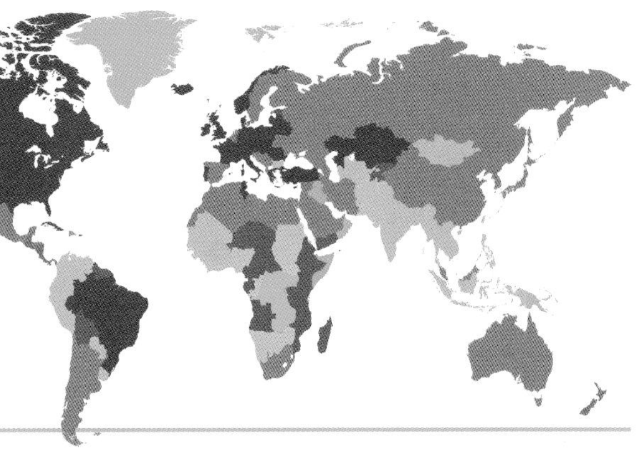

Global calorie intake ▷
On average, the amount of food consumed (and wasted) by a person per day in a more wealthy country is much higher than in a poor country where some people may not be able to afford good food.

Increasing global demand

The United Nations (UN) has predicted that the global demand for food will increase by 70 percent by 2050, partly due to the increase in population and wealth. People in HICs tend to demand richer foods, such as meat and dairy, in greater quantities.

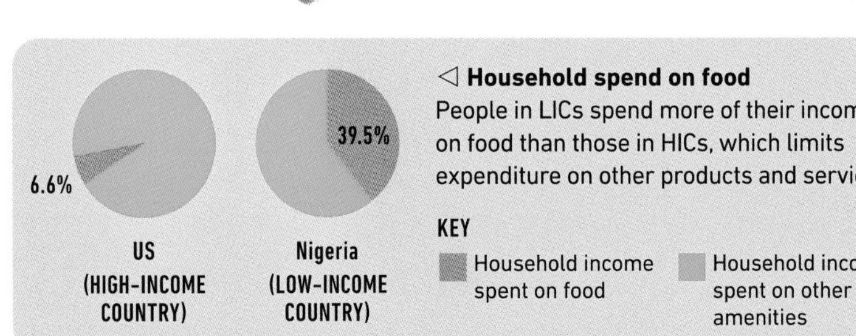
6.6%
US
(HIGH-INCOME COUNTRY)
39.5%
Nigeria
(LOW-INCOME COUNTRY)
KEY
Household income spent on food
Household income spent on other amenities

◁ Household spend on food
People in LICs spend more of their income on food than those in HICs, which limits expenditure on other products and services.

▽ Rise in grain production
Technological developments have helped ensure an adequate supply of grain, such as wheat and corn, for the growing population of the world.

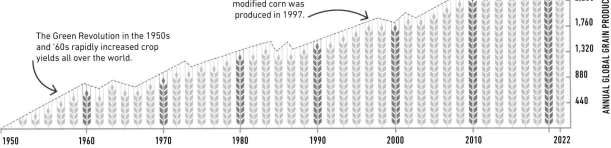

The Green Revolution in the 1950s and '60s rapidly increased crop yields all over the world.

The first genetically modified corn was produced in 1997.

By 2012, China, the US, and India produced nearly half of the world's grain.

3,060
2,650
2,205
1,760
1,320
880
440
ANNUAL GLOBAL GRAIN PRODUCTION (MILLION TONS)

1950 1960 1970 1980 1990 2000 2010 2022
YEAR

Factors that affect food supply

The supply of food depends on the space available to grow crops and rear animals, and may vary over time. For example, one valley may be fertile and support more crops at one point in time but may eventually become infertile. Many physical and human factors can help either improve or reduce food supply.

Physical factors

Climate
The weather conditions prevailing in an area over a long period of time make up its climate, which in turn controls temperature and rainfall, and affects the suitability of growing crops.

Water supply
Farmers need water for crops and livestock. If water is not readily available, crops and animals may die, decreasing the food supply.

Pests and diseases
An attack by pests can wipe out entire crops in one go, decreasing the food supply. For example, a plant pest caused the Irish potato famine in 1845, in which one million people died.

Human factors

Conflict
A war can reduce the availability of food in an area by destroying crops, killing livestock, and leaving fewer people to tend to the land. The shortage of food can also cause further conflict.

Poverty
The UN recognizes poverty as one of the key causes of food insecurity across the world. The poor spend a large share of their income on food and are, therefore, the most affected when food prices increase.

Technology
People have been using technologies to improve the food supply for years. The more efficient the methods of farming, the lower the cost of food production and the greater the overall yield.

Ways to improve food security

Global hunger is a real international concern and there is a need to improve food security. Ending hunger is the second of the 17 Sustainable Development Goals of the UN, which estimates that global agricultural production must increase by 50 percent to meet it.

◁ **Produce genetically modified (GM) crops**
When crops are modified to make them pest-resistant and able to survive harsh conditions with limited water, they can grow on previously unsuitable land.

◁ **Use improved methods of irrigation**
Small-scale, affordable, and sustainable methods of irrigation can help address water scarcity and increase the production of food.

◁ **Mechanize agriculture**
The use of machinery in farming improves the yield of a piece of land. This can protect farmers from changes in crop prices and alleviate poverty.

◁ **Reduce intake of meat**
The meat industry uses a large amount of water, which is crucial for growing crops. Eating less meat can make more water available for agriculture.

◁ **Reduce food waste**
High-income countries tend to throw away a huge amount of food. Reducing this food waste can make more food available to a larger number of people.

Water security

RELIABLE AND SECURE ACCESS TO CLEAN WATER IS ESSENTIAL FOR EVERYONE AROUND THE WORLD.

SEE ALSO
❮ 80–81 The hydrological cycle
❮ 140–141 Health
❮ 166–167 Human impact
❮ 184–185 Sustainability
❮ 188–189 Big geographical issues

Plants and animals depend on water for their survival. In addition to needing clean water for drinking and washing, humans need it for agriculture and industry.

Limited fresh water

The Earth's hydrological cycle is a closed system, which means that water is a finite resource. Water is not created or destroyed but is continually recycled. Despite 70 percent of the Earth's surface being covered in water, only a tiny amount of it is fresh water that can be easily accessed by people; the rest is salt water.

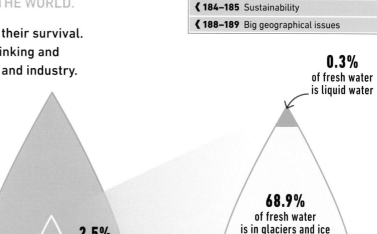

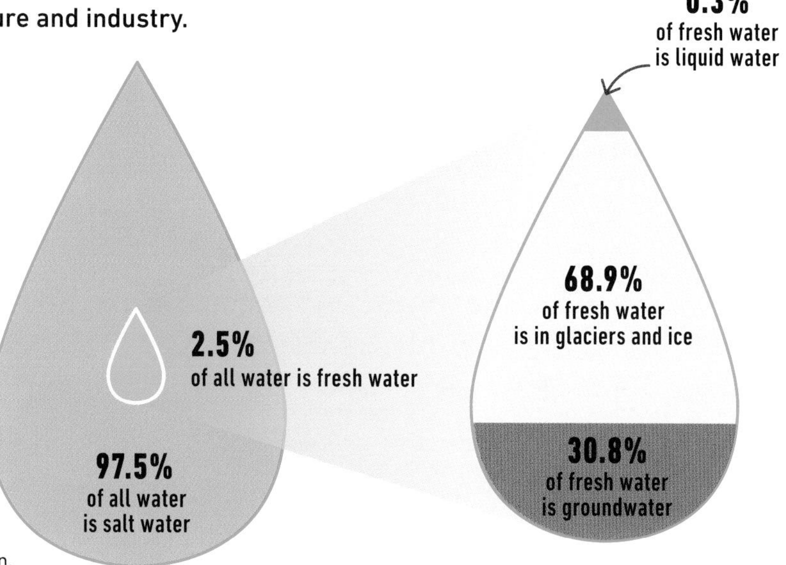
0.3%
of fresh water is liquid water
68.9%
of fresh water is in glaciers and ice
2.5%
of all water is fresh water
30.8%
of fresh water is groundwater
97.5%
of all water is salt water

The Earth's resources ▷ The vast majority of the Earth's water is salt water, which cannot be used for drinking without desalination.

Access to water

The global distribution of water is uneven—accessing fresh water in certain parts of the world is easier than in others. Areas with high rainfall and extensive river systems have the best access. Underground supplies called aquifers can be drilled in areas with scarce resources.

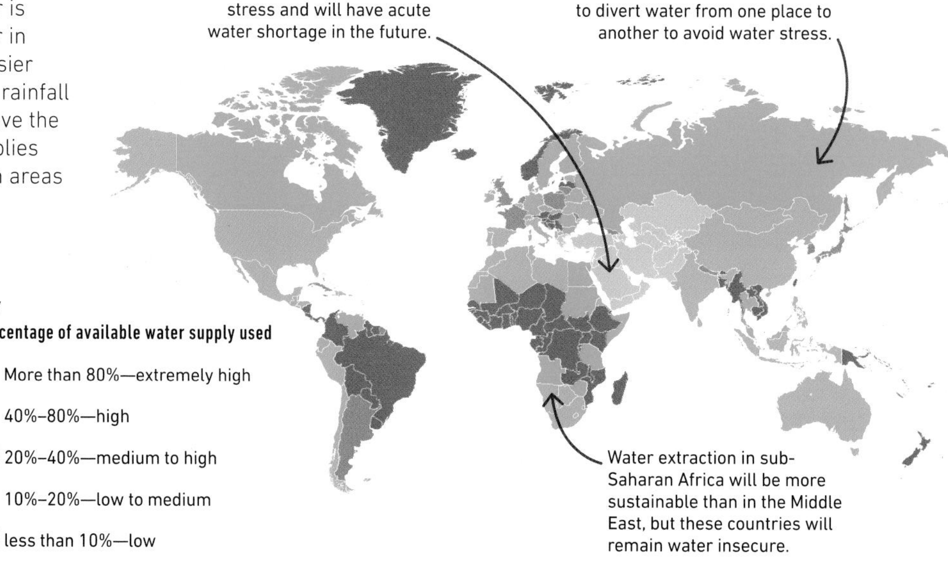
Countries in the Middle East are facing extreme water stress and will have acute water shortage in the future.
Large countries with abundant water supplies, such as Russia, will be able to divert water from one place to another to avoid water stress.
KEY
Percentage of available water supply used
More than 80%—extremely high
40%–80%—high
20%–40%—medium to high
10%–20%—low to medium
less than 10%—low
Water extraction in sub-Saharan Africa will be more sustainable than in the Middle East, but these countries will remain water insecure.

Projected water stress by 2040 ▷ In the coming years, the countries where water usage exceeds the natural supplies of the region will face the most critical levels of water stress.

Factors affecting the availability of water

A combination of physical and human factors affects the availability of water. Climate and geological features are physical factors, but human factors are also significant. Human actions often exhaust resources beyond the natural system's ability to recover effectively.

Physical factors

Climate
The general rainfall and temperature of an area affect the availability of water. Areas with high rainfall and lower temperatures have sufficient water.

Geology
In areas where the bedrock lets water seep through, aquifers can store water. Where water cannot seep through, it collects on the surface of the ground and evaporates away.

Human factors

Population growth
More water is extracted to sustain a rapidly growing population. If the demand for water exceeds the supply, sources may be depleted.

Pollution
Water can become unusable due to human activity. Sources of water close to mining and textile industries are particularly polluted.

Infrastructure
The construction of a dam or a viaduct can improve the availability of water in an area, while reducing its supply elsewhere—both at the same time.

Poverty
Poorer countries have limited infrastructure to pump, transport, divert, store, treat, and deliver clean, usable fresh water to their people.

Impact of water insecurity

Water insecurity is defined as the lack of a reliable source of water, of appropriate quality and quantity to meet the needs of the local population and environment. Poorer countries that cannot afford technological solutions often face water insecurity. If an area experiences a prolonged period of water insecurity, it may lead to significant social, political, and environmental consequences.

Unclean water ▷
When there is no alternative, people may resort to drinking contaminated water. Half a million people die every year by consuming unclean water.

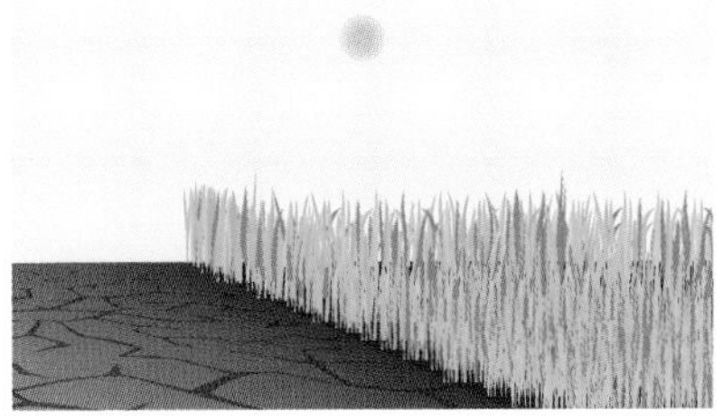

△ **Limited food production**
Crops can fail if there is an insufficient supply of water. This can then lead to localized food insecurity, malnutrition, and even famine.

△ **Conflict due to damming rivers**
A dam can improve the water security of the country that built it, but it may affect the availability and quality of water in countries downstream, leading to disputes.

△ **Decline in industrial output**
Water is essential to most industries, especially those relating to food, chemicals, and paper. The lack of it decreases production, affecting the economy.

Conflict and resolution

WHEN COUNTRIES OR GROUPS OF PEOPLE DISAGREE WITH EACH OTHER, IT CAN LEAD TO A CONFLICT OR EVEN WAR.

SEE ALSO
❮ 124–125 Migration
❮ 138–139 The spread of cultures
❮ 182–183 Sources of energy
❮ 184–185 Sustainability

Conflicts can occur at a local, national, or global level and range in intensity from minor disputes to major wars. Conflicts can arise over water or land rights, access to resources, or over borders. They can have serious, long-lasting consequences.

Reasons for conflict

Conflicts have taken place throughout history. Their causes can be complex and diverse but are often connected to beliefs, culture, or the control of land. In order for a conflict to be resolved, the reasons behind it need to be properly understood.

Economic
The control of important economic resources, such as oil, precious minerals, or water may lead to conflict.

Political
These conflicts may arise from disagreement about trade or over the control of land or territorial waters.

Cultural
Some conflicts occur because of differences in ethnicity, culture, or religion between groups.

Boundary
These conflicts arise because of disagreements about where a border should lie between countries.

Ideological
Conflicts can take place between groups with differing ideas or ideals about politics or religion.

Ways to reduce and resolve conflict

Millions of people across the world are affected by conflict each year. The United Nations recognizes that it is one of the major barriers to achieving its Sustainable Development Goal of ending world poverty. Organizations that try to resolve conflict make use of different methods.

Laws and regulations

An effective method to resolve conflict is through the use of laws that all sides must follow. These can be enforced by governments or international bodies.

Mediation and reconciliation

The two sides in a conflict may not fully understand each other's position. A neutral organization or person can help with this.

Involvement of international bodies

Organizations such as the United Nations or NATO can send soldiers or observers to a conflict zone. They might remain after the conflict is over to keep the peace.

Cease-fire

A cease-fire is an agreement made by the sides involved in a conflict to stop fighting either briefly or permanently. It usually takes place following mediation, or involvement of an international body.

Effects of conflict

A conflict can lead to a great many problems in the area where it takes place: people may be wounded or killed, and homes, factories, and infrastructure damaged or destroyed. The conflict, however, may affect other areas or groups of people, too.

Loss of life
Conflicts may result in the deaths of many soldiers. Civilians who play no direct part in the conflict may also be killed, either deliberately or accidentally.

Forced displacement of people
When homes are destroyed in a conflict, people are forced to flee. At times, people may have to leave because they belong to a particular ethnic or religious group.

Loss of infrastructure
Infrastructure, such as bridges, schools, and roads, is often destroyed. This seriously affects a country's ability to recover once the conflict is over.

Conflict can spread to new areas
A conflict may escalate, involving more people in other regions. The arrival of people fleeing the conflict may result in tensions in the new areas they move to.

People leave looking for security elsewhere
A conflict may create many refugees, as people seek a safe place to live. The countries they flee to may not be able to host them properly.

Sanctions
The country or individuals held responsible for a conflict can be punished by the international community. Sanctions stop a country from buying or selling certain things.

Cooperation

In 1945, at the end of World War II, the international community came together to think of ways to prevent a similar war happening in the future. This led to the creation of the United Nations (UN), its aim being to uphold human rights, promote global cooperation, and maintain peace around the world.

Structure of the United Nations

General Assembly
(Recommends)
193 member countries

Security Council
(Decides)
Has five permanent members and 10 nonpermanent members.

Secretariat
(Implements)
Responsible for the day-to-day work of the UN, it is headed by the Secretary General.

Economic and Social Council
Gives recommendations on economic, social, and environmental issues and reviews UN policies.

International Court of Justice
In charge of settling legal disputes and questions submitted to it by member nations or other UN organs.

UN organs and specialized agencies
These include, the United Nations High Commissioner for Refugees (UNHCR), the United Nations International Children's Emergency Fund (UNICEF), the World Health Organization (WHO), and the United Nations Educational, Scientific and Cultural Organization (UNESCO).

International Criminal Court
Working in cooperation with the UN, it holds trials of individuals charged with crimes concerning the international community.

3

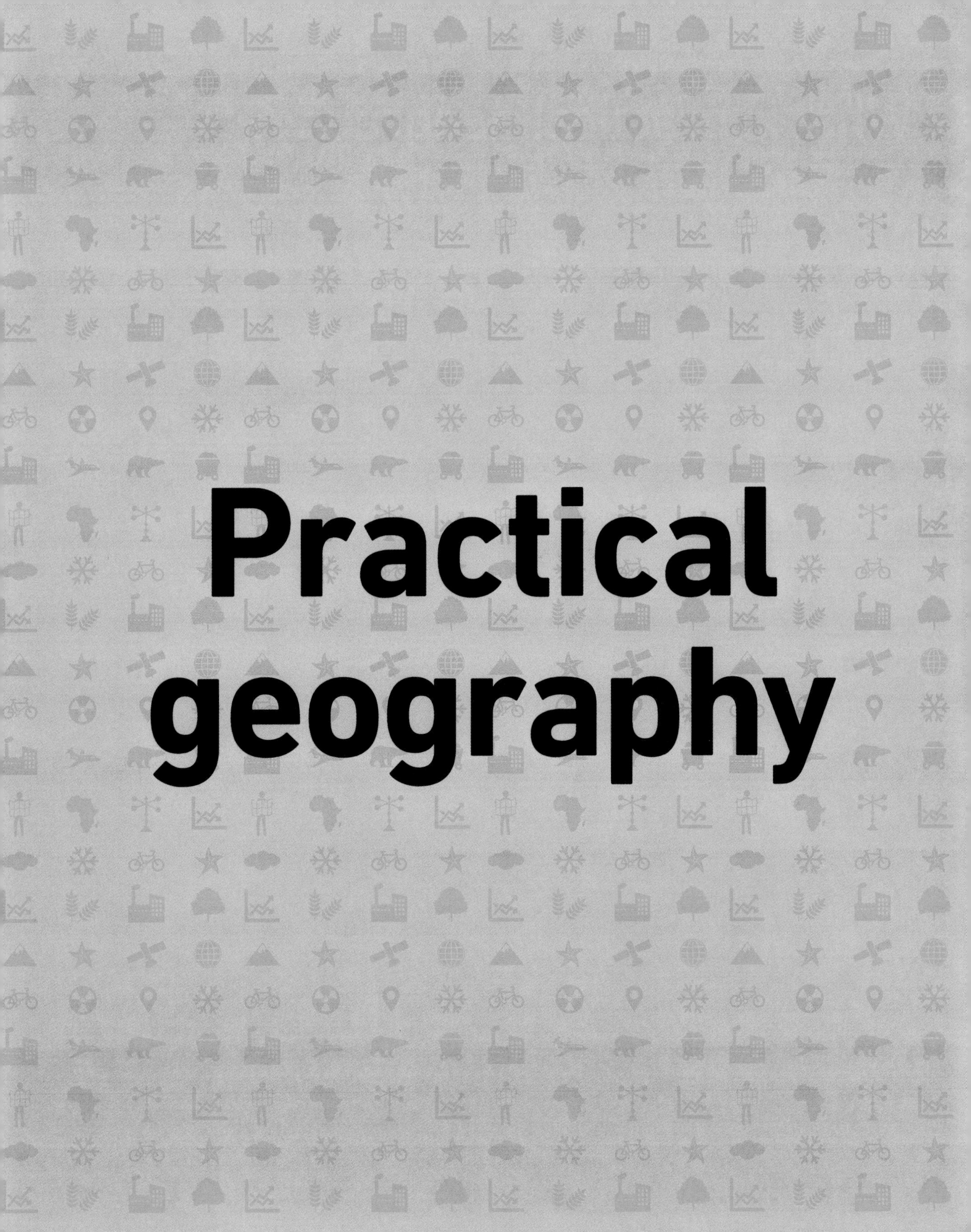
Practical
geography

What is practical geography?

GEOGRAPHERS RECORD, MEASURE, AND INTERPRET THE WORLD AROUND THEM IN ORDER TO UNDERSTAND IT.

The world is much too big and diverse for geographers to study from experiments in laboratories. Instead, geographers must go out into the real world using a range of practical skills, such as understanding location, reading maps, and collecting data.

Location and place

Since ancient times, geographers have tried to map the known world. Gradually, and especially since the age of exploration in the 15th century, they have been able to measure distance accurately. Using measuring devices such as latitude and longitude, they have built the basic building' blocks of geography. Geographers understand how places fit together on a global scale, with Earth divided into land and sea and countries and regions.

Longitude

Latitude

△ **Lines of longitude**
Lines of longitude enable every place on Earth's position east or west to be located with precision.

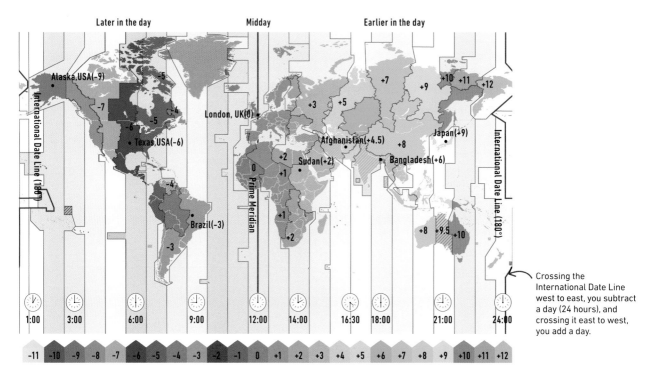

Later in the day
Midday
Earlier in the day
Alaska,USA(-9)
-7
-6
-5
-4
Texas,USA(-6)
London, UK(0)
+3
+5
+7
+9
+10
+11
+12
Afghanistan(+4.5)
+8
Japan(+9)
Bangladesh(+6)
+2
Sudan(+2)
0
+1
-4
Brazil(-3)
-3
+1
+2
+8
+9.5
+10
International Date Line (180°)
Prime Meridian
International Date Line (180°)
1:00
3:00
6:00
9:00
12:00
14:00
16:30
18:00
21:00
24:00
-11 -10 -9 -8 -7 -6 -5 -4 -3 -2 -1 0 +1 +2 +3 +4 +5 +6 +7 +8 +9 +10 +11 +12
Crossing the International Date Line west to east, you subtract a day (24 hours), and crossing it east to west, you add a day.

△ **Time zones**
Maps can provide different types of information, such as the Earth's time zones. Because the Earth rotates, the sun rises at different times in different places.

Map skills

Maps not only show geographers where things are or how to get from one place to another; they are also an effective way of presenting data. Maps come in different forms to show different kinds of information. A book of maps is called an atlas. Being able to read a map (and even make your own), as well as use a map and compass, are core skills for geographers.

A compass needle points to magnetic north.

△ **Using a compass**
A compass points to magnetic north and allows you to figure out your direction on a map.

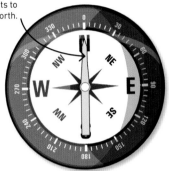

The grid lines help you to find your exact location.

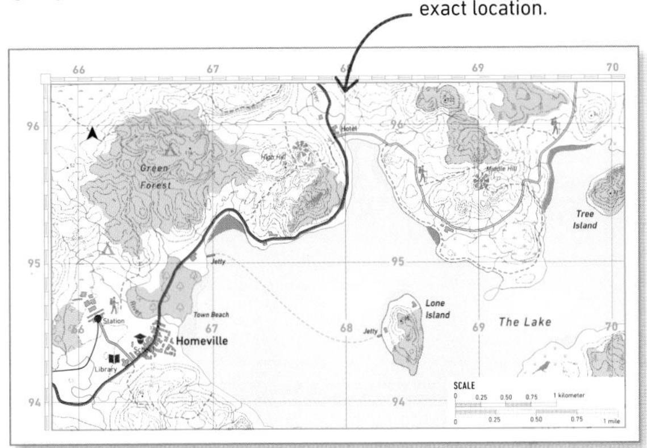

66
67
69
70
96
95
94
Hotel
Green Forest
Tree Island
Jetty
Town Beach
Station
Homeville
Library
Lone Island
The Lake
SCALE

△ **Using maps**
Maps provide information about roads, rivers, infrastructure, topography, and other features. The symbols on a map can be understood using a key.

Collecting data

Geographers observe the world around them and collect data to help them interpret and understand it. Often, they do this through "fieldwork", during which they visit a particular place to gather data about the people, built environments, or natural features there. Data can be qualitative (in images or words), quantitative (in numbers or statistics), or a mixture of both.

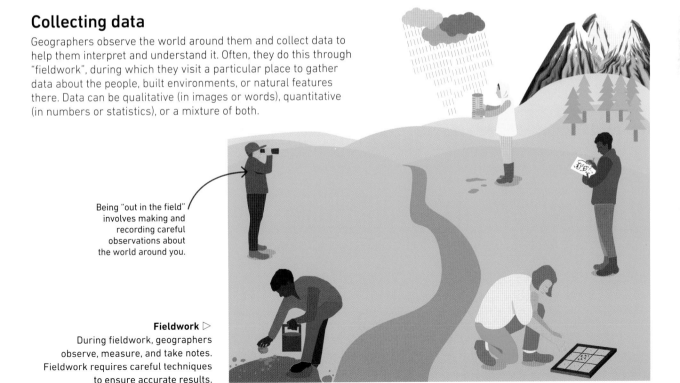

Being "out in the field" involves making and recording careful observations about the world around you.

Fieldwork ▷
During fieldwork, geographers observe, measure, and take notes. Fieldwork requires careful techniques to ensure accurate results.

Continents and oceans

THE EARTH'S SURFACE IS DIVIDED INTO SEVEN LAND MASSES, OR CONTINENTS, AND FIVE OCEANS.

SEE ALSO
❮ 24–25 Moving plates and boundaries
❮ 26–27 Shifting continents
❮ 38–39 Physical map of the world
❮ 112–113 Oceans and seas map

Together, the seven continents cover about 57 million sq miles (148 million sq km) of land. But the expanse of the oceans is well over double this – they cover 139 million sq miles (361 million sq km).

Earth's continents and oceans

The seven continents are the world's main geopolitical regions. They are also geologically important, being made mostly of ancient continental rocks that sit on top of the heavier rocks of the Earth's crust. Islands off the coast of continents are often included in the continent. So the British Isles are part of Europe, and Indonesia is part of Asia.

The **deepest part** of the **ocean** is the *Mariana Trench*, which is about 7 miles (11 km) deep.

△ **Asia**
The largest continent includes 48 countries and is home to about two-thirds of the world's population.

△ **Africa**
The second-largest continent contains 54 countries.

△ **North America**
Dominated by the US and Canada, this continent has 23 countries.

△ **South America**
This has 14 countries and a chain of mountains on its west side.

△ **Antarctica**
This land mass has no countries and is covered almost entirely with ice.

△ **Europe**
This shares some countries, such as Turkey and Russia, with Asia.

△ **Australasia**
The smallest continent includes Australia and New Zealand.

Subcontinents

A subcontinent can be either a large landmass that is smaller than a continent, such as Greenland, or part of an even larger continent, such as the Indian subcontinent. Even when attached to a continent, it may have its own distinct wildlife or climate.

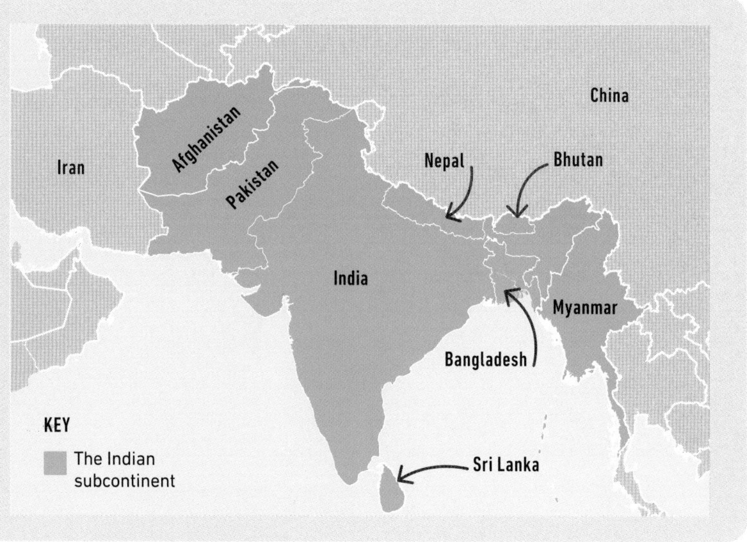
China
Iran
Afghanistan
Pakistan
Nepal
Bhutan
India
Myanmar
Bangladesh
Sri Lanka
KEY
The Indian subcontinent

The Indian subcontinent ▷
The Indian subcontinent contains eight countries. It is bordered in the north and west by the Himalayan and Hindu Kush mountains. Until 55 million years ago, it was an entirely separate continent.

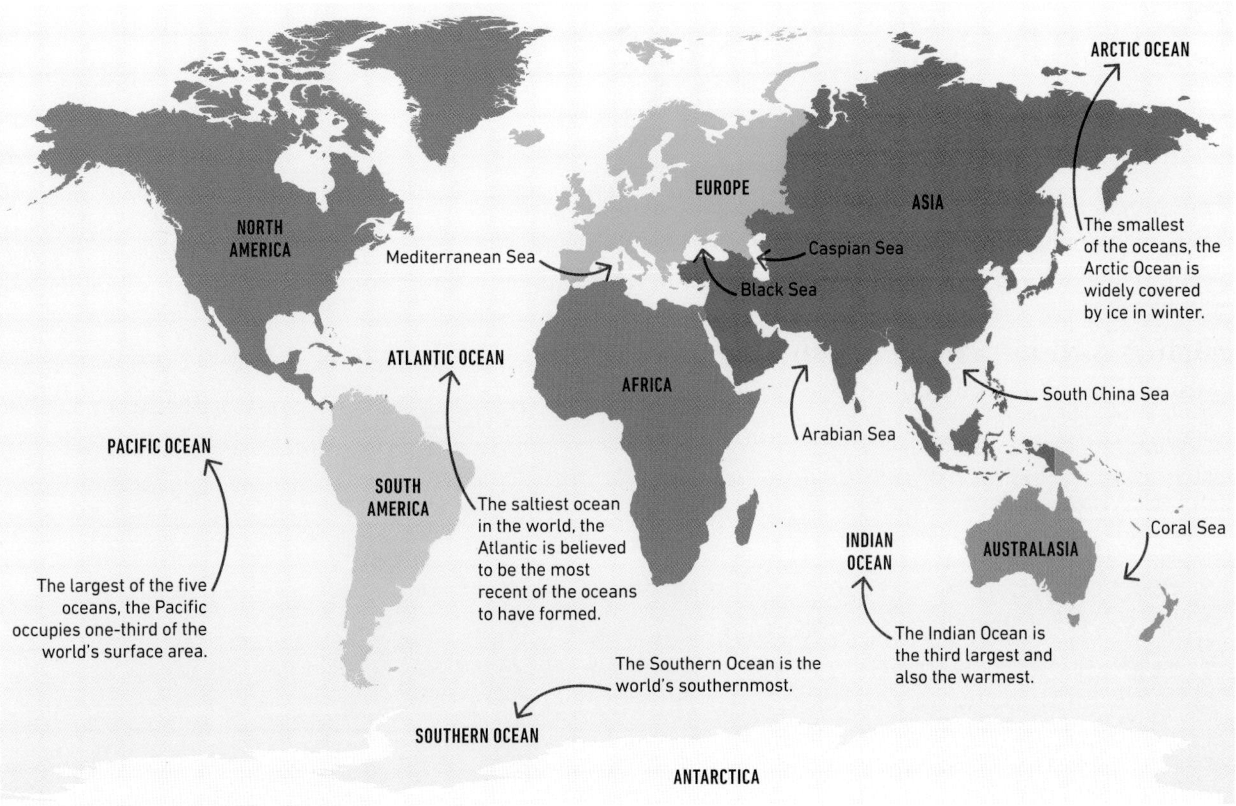

ARCTIC OCEAN
NORTH AMERICA
EUROPE
ASIA
Mediterranean Sea
Caspian Sea
Black Sea
The smallest of the oceans, the Arctic Ocean is widely covered by ice in winter.
ATLANTIC OCEAN
AFRICA
South China Sea
Arabian Sea
PACIFIC OCEAN
SOUTH AMERICA
The saltiest ocean in the world, the Atlantic is believed to be the most recent of the oceans to have formed.
INDIAN OCEAN
AUSTRALASIA
Coral Sea
The largest of the five oceans, the Pacific occupies one-third of the world's surface area.
The Indian Ocean is the third largest and also the warmest.
The Southern Ocean is the world's southernmost.
SOUTHERN OCEAN
ANTARCTICA

△ **Position of the continents and oceans**
Africa, Europe, and Asia together form Afro-Eurasia—the largest connected landmass on the Earth. All the oceans join to form a single world ocean.

Oceans

The Earth has five great oceans, filled with salty water. The largest oceans—the Pacific, Atlantic, and Indian—all merge into the other two, the Antarctic and Arctic, at each end of the world. The Pacific is by far the largest, with twice the area of the Atlantic Ocean. The Indian Ocean is a little smaller than the Atlantic.

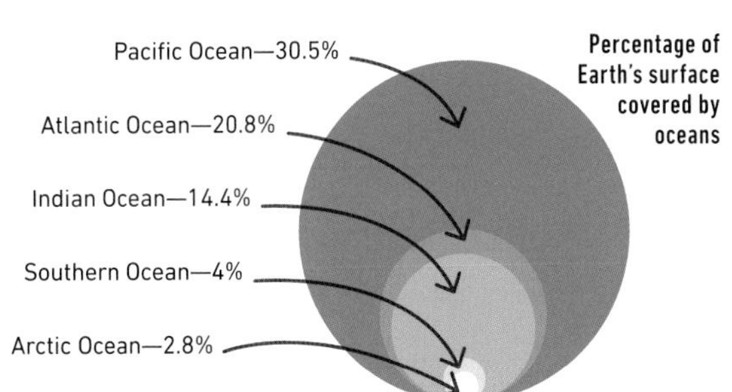
Pacific Ocean—30.5%
Atlantic Ocean—20.8%
Indian Ocean—14.4%
Southern Ocean—4%
Arctic Ocean—2.8%
Percentage of Earth's surface covered by oceans

The edge of a continent

Countries with coasts claim much if not all of the shelf as their own territory. In Exclusive Economic Zones, the area up to 124 miles (200 km) from the coast, a country has fishing and oil exploration rights but cannot block ships. Territorial waters, up to 14 miles (22 km) out, are entirely owned by the country.

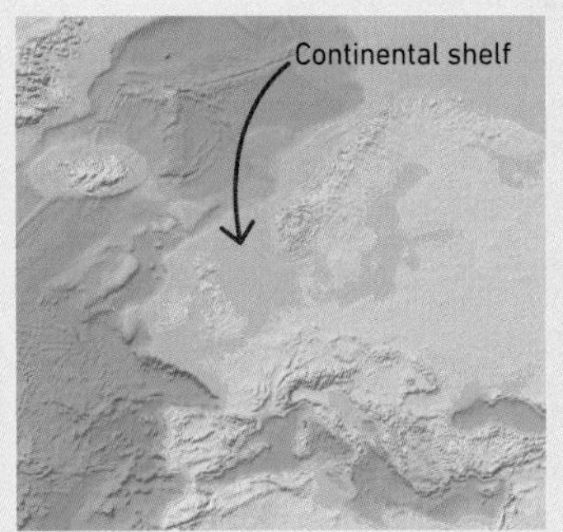
Continental shelf

△ **Continental shelf**
Geologically, continents do not stop at the end of the land. The continental shelf is the shallow margin of sea around the coast and is part of the continent.

Countries and nations

THE WORLD IS DIVIDED INTO COUNTRIES AND NATIONS OF VARIOUS SIZES.

SEE ALSO
❮ 196–197 Conflict and resolution
Capitals and major cities 208–209 ❯
Geopolitics 216–217 ❯

A country is an area of land with borders and its own laws, while a nation describes people with a shared culture or language. The words "country" and "nation" are often used interchangeably.

What is a country?

Countries, also known as states, are territories (regions of land) defined and recognized by international law. Their right to govern themselves is officially recognized, and it is illegal for other countries to interfere inside their boundaries. Every country has its own government and its own set of laws. The boundaries between countries are called international borders, and you may need a passport to cross them.

In **1914**, there were only **57 sovereign countries**. Many of today's countries were then part of global empires.

4 main components of a country

Territory
Every country has its own territory (land area), with clearly defined borders. Territory is the clearest and strongest feature of any country.

Population
A country needs to have a population that lives in it. The size of populations varies widely.

Sovereignty
The right of a country to govern itself without outside interference is called sovereignty. All United Nations member states have sovereignty.

Government
Every country is run by its own government, which creates and administers its laws, though it also needs to follow international laws.

Number of countries

The number of countries in the world is not internationally agreed on because the sovereignty of some countries is disputed. The United Nations (UN), the world's largest international organization, has 193 member states, or countries, but it partially recognizes 13 more. Other organizations, such as international sports bodies, allow some nonsovereign countries to take part in their competitions.

193
Member states of the UN

206
UN member states plus states partially recognized by the UN

206
Olympic nations

211
Recognized by FIFA (soccer's governing body)

What is a nation?

A nation is a group of people who share the same culture, history, traditions, and, usually, language. If the majority of a country's population belong to the same nation, it can be called a "nation-state." However, it is unusual for the entire population of a country to have the same national identity, and most countries are "multicultural" to a greater or lesser degree. It is also important to recognize that not all nations have their own sovereignty.

Changing borders

Sometimes nations are split across more than one country, often as a result of war, and seek to join together again. At other times, a smaller national group within a nation-state may be unhappy being governed by the country's government and wish to become an independent nation-state in its own right.

Germany's division even affected its capital, Berlin, which was split into East and West Berlin, divided by a wall.

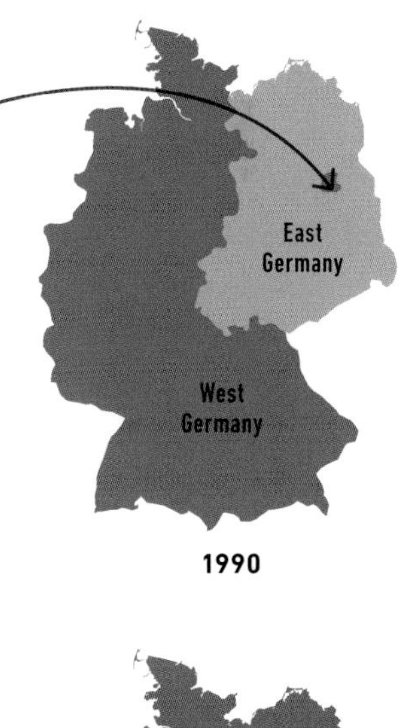
East Germany
West Germany
1990

Yugoslavia

1991

Slovenia
Croatia
Bosnia and Herzegovina
Serbia
Montenegro
Kosovo (disputed)
North Macedonia

TODAY

Serbia was the dominant nation in Yugoslavia. The concentration of Serb power caused increasing friction with the other Yugoslav nations.

Germany

TODAY

△ **The breakup of Yugoslavia**
Yugoslavia was a country originally created after World War I, uniting a group of separate nations. After a series of conflicts that began in the early 1990s, it split apart into today's separate countries.

△ **German reunification**
Between 1949 and 1990, Germany was split into two countries divided by their political ideas: capitalism and communism. The two halves were reunited after the fall of communism in Eastern Europe.

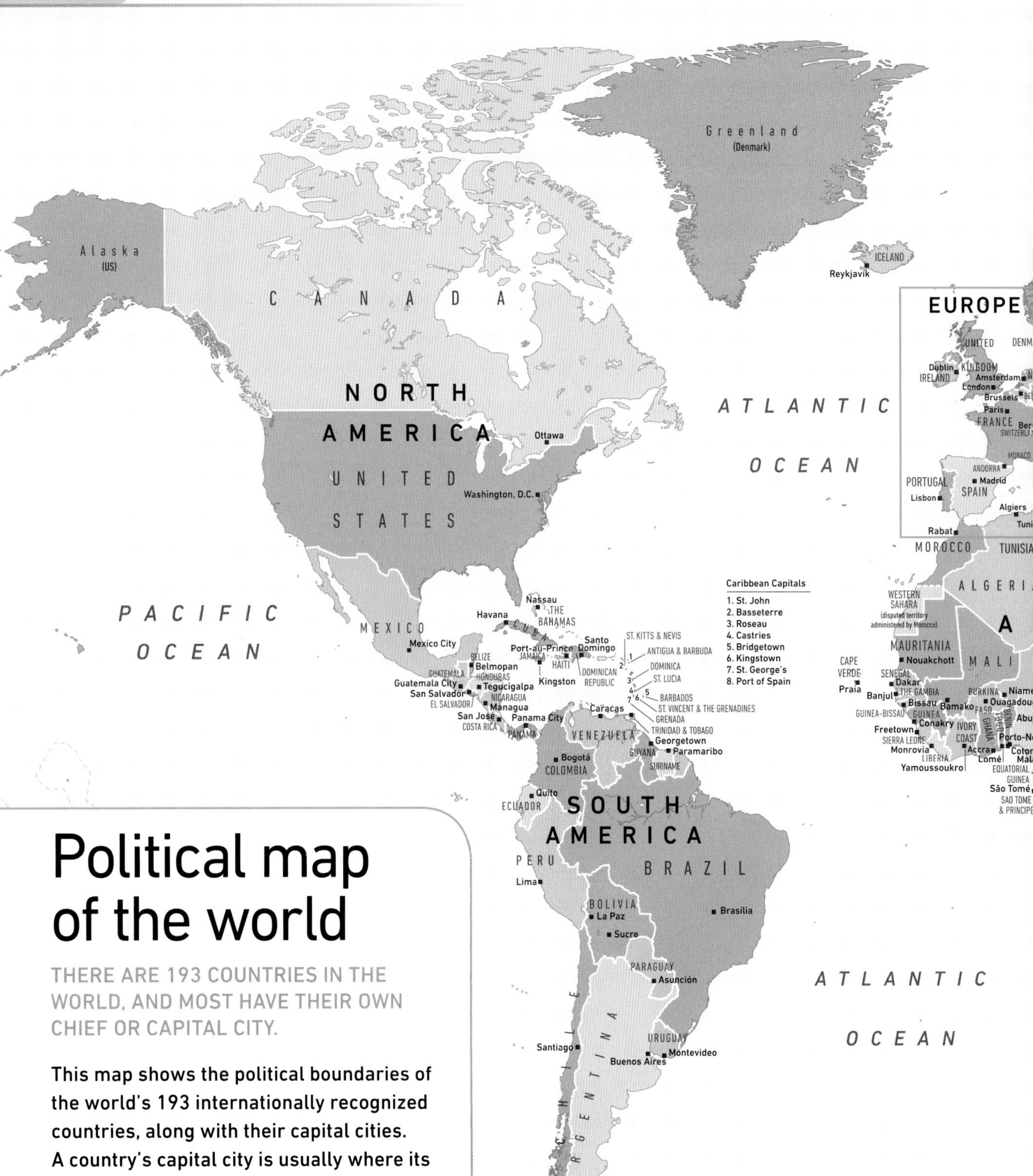
Greenland
(Denmark)
ICELAND
Reykjavík
Alaska
(US)
CANADA
NORTH AMERICA
Ottawa
UNITED STATES
Washington, D.C.
ATLANTIC OCEAN
PACIFIC OCEAN
EUROPE
UNITED KINGDOM
IRELAND
Dublin
London
Amsterdam
Brussels
Paris
FRANCE
SWITZERLAND
MONACO
ANDORRA
PORTUGAL
Lisbon
Madrid
SPAIN
Algiers
Rabat
MOROCCO
TUNISIA
ALGERIA
WESTERN SAHARA
(disputed territory administered by Morocco)
MAURITANIA
Nouakchott
MALI
CAPE VERDE
Praia
SENEGAL
Dakar
Banjul
THE GAMBIA
Bissau
GUINEA-BISSAU
GUINEA
Conakry
Bamako
BURKINA FASO
Ouagadougou
Freetown
SIERRA LEONE
Monrovia
LIBERIA
IVORY COAST
Yamoussoukro
GHANA
Accra
TOGO
Lomé
BENIN
Porto-Novo
EQUATORIAL GUINEA
São Tomé
SAO TOME & PRINCIPE
Caribbean Capitals
1. St. John
2. Basseterre
3. Roseau
4. Castries
5. Bridgetown
6. Kingstown
7. St. George's
8. Port of Spain
MEXICO
Mexico City
Havana
CUBA
Nassau
THE BAHAMAS
Santo Domingo
Port-au-Prince
JAMAICA
HAITI
DOMINICAN REPUBLIC
Kingston
ST. KITTS & NEVIS
ANTIGUA & BARBUDA
DOMINICA
ST. LUCIA
BARBADOS
ST. VINCENT & THE GRENADINES
GRENADA
TRINIDAD & TOBAGO
BELIZE
Belmopan
GUATEMALA
Guatemala City
HONDURAS
Tegucigalpa
San Salvador
EL SALVADOR
NICARAGUA
Managua
San José
COSTA RICA
Panama City
PANAMA
Caracas
VENEZUELA
Georgetown
GUYANA
Paramaribo
SURINAME
Bogotá
COLOMBIA
Quito
ECUADOR
SOUTH AMERICA
PERU
Lima
BRAZIL
Brasília
BOLIVIA
La Paz
Sucre
PARAGUAY
Asunción
CHILE
ARGENTINA
Santiago
URUGUAY
Montevideo
Buenos Aires

Political map of the world

THERE ARE 193 COUNTRIES IN THE WORLD, AND MOST HAVE THEIR OWN CHIEF OR CAPITAL CITY.

This map shows the political boundaries of the world's 193 internationally recognized countries, along with their capital cities. A country's capital city is usually where its government is based.

Oslo
Stockholm
Helsinki
Tallinn
ESTONIA
Riga
LATVIA
LITHUANIA
Vilnius
Moscow
UNITED KINGDOM
DENMARK
Copenhagen
Dublin
IRELAND
Amsterdam
The Hague
NETH.
Berlin
GERMANY
POLAND
Warsaw
Minsk
BELARUS
London
Brussels
BELGIUM
LUX.
Prague
CZECH REP. (CZECHIA)
Kyiv
UKRAINE
SLOVAKIA
Paris
FRANCE
LIECH.
Vaduz
Vienna
Bratislava
AUSTRIA
Budapest
MOLDOVA
Chisinau
Bern
SWITZERLAND
SLVN.
Ljubljana
HUNGARY
ROMANIA
Zagreb
Belgrade
SAN MARINO
CROATIA
Bucharest
MONACO
Sarajevo
SERBIA
BOS. & HERZ.
KOS.
Pristina
Sofia
BULGARIA
ANDORRA
Rome
Podgorica
Skopje
MAC.
MONTENEGRO
Madrid
VATICAN CITY
Tirana
ALBANIA
Ankara
PORTUGAL
SPAIN
ITALY
Lisbon
Athens
TURKEY
Algiers
Tunis
GREECE
MALTA
Valletta
Svalbard (Norway)
ARCTIC OCEAN
NORWAY
SWEDEN
FINLAND
RUSSIA
Astana
KAZAKHSTAN
Ulaanbaatar
MONGOLIA
GEORGIA
Tbilisi
UZBEKISTAN
Tashkent
Bishkek
KYRGYZSTAN
ASIA
ARMENIA
AZERBAIJAN
Yerevan
Baku
TURKMENISTAN
TAJIKISTAN
Dushanbe
Ashgabat
Beijing
Pyongyang
NORTH KOREA
JAPAN
Seoul
SOUTH KOREA
Sejong City
Tokyo
CYPRUS
Nicosia
SYRIA
Damascus
Tehran
LEBANON
Beirut
IRAN
Kabul
Islamabad
AFGHANISTAN
CHINA
Tripoli
ISRAEL
Amman
Baghdad
Jerusalem
JORDAN
IRAQ
Kuwait
KUWAIT
Cairo
LIBYA
EGYPT
PAKISTAN
NEPAL
BHUTAN
New Delhi
Kathmandu
Thimphu
BAHRAIN
Manama
QATAR
Doha
Abu Dhabi
Riyadh
U.A.E.
Muscat
SAUDI ARABIA
OMAN
BANGLADESH
Dhaka
INDIA
MYANMAR (BURMA)
Nay Pyi Taw
Hanoi
LAOS
VIETNAM
Taipei
TAIWAN
PACIFIC OCEAN
AFRICA
NIGER
CHAD
SUDAN
Khartoum
ERITREA
Asmara
Sanaa
YEMEN
Viangchan
THAILAND
Bangkok
CAMBODIA
Phnom Penh
Manila
PHILIPPINES
N'Djamena
DJIBOUTI
Djibouti
Addis Ababa
ETHIOPIA
CENTRAL AFRICAN REPUBLIC
Bangui
SOUTH SUDAN
Juba
SOMALIA
Colombo
SRI LANKA
Sri Jayewardenapura Kotte
Maale
MALDIVES
Ngerulmud
PALAU
MICRONESIA
Palikir
MARSHALL ISLANDS
Majuro Atoll
Bandar Seri Begawan
BRUNEI
Kuala Lumpur
Putrajaya
MALAYSIA
Singapore
SINGAPORE
UGANDA
Kampala
KENYA
Nairobi
Mogadishu
DEM. REP. CONGO
CONGO
Brazzaville
Kinshasa
Kigali
RWANDA
Bujumbura
BURUNDI
Dodoma
TANZANIA
Victoria
SEYCHELLES
INDIAN OCEAN
INDONESIA
Jakarta
Dili
EAST TIMOR
PAPUA NEW GUINEA
Port Moresby
Honiara
SOLOMON ISLANDS
NAURU
KIRIBATI
Tarawa Atoll
TUVALU
Funafuti Atoll
SAMOA
Apia
CABINDA (Angola)
Luanda
ANGOLA
COMOROS
Moroni
MALAWI
Lilongwe
ZAMBIA
Lusaka
Harare
ZIMBABWE
MOZAMBIQUE
MADAGASCAR
Antananarivo
MAURITIUS
Port Louis
VANUATU
Port-Vila
Suva
FIJI
TONGA
Nuku'alofa
NAMIBIA
Windhoek
BOTSWANA
Gaborone
Pretoria
Maputo
Mbabane
Lobamba
ESWATINI (formerly SWAZILAND)
Bloemfontein
LESOTHO
Maseru
SOUTH AFRICA
Cape Town
AUSTRALIA
AUSTRALASIA
Canberra
NEW ZEALAND
Wellington
SCALE
0
1,000
2,000
km
0
1,000
2,000
miles

Capitals and major cities

SEE ALSO
❮ 130–131 Cities
❮ 132–133 Megacities
❮ 136–137 Urbanization
❮ 204–205 Countries and nations

THE CITY WHERE A COUNTRY'S GOVERNMENT IS BASED IS KNOWN AS THE CAPITAL.

Usually, the capital is also the country's biggest city, like Paris in France and Tokyo in Japan, but not always. Washington, D.C., is the capital of the US, but New York is its biggest city.

Why some cities become capitals

Several factors influence why a city is chosen as the capital of a country. Most cities become capitals because of their accessibility, while some are capitals because of the presence of important institutions, such as the country's government and financial centers.

◁ **Transportation hub**
Many capitals develop where the country's major transportation routes meet, or at ports such as London, which historically was one of the country's major trading centers.

◁ **Secure location**
Some capitals, especially ancient capitals such as Athens, grew up on secure sites, such as defensible hilltops, from where the government could function effectively.

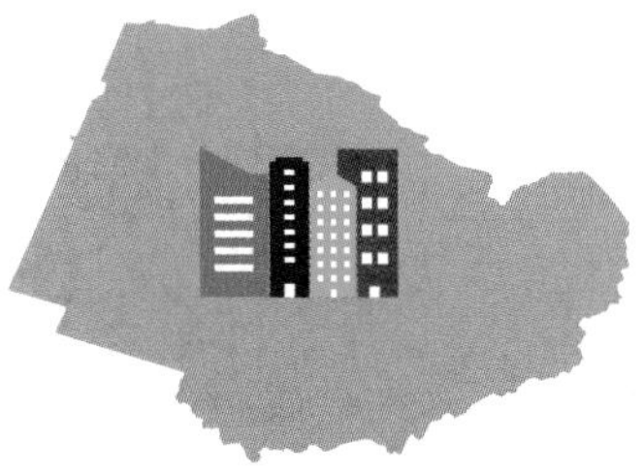

◁ **Central point**
Many capitals are located near the geographic or economic center of the country, such as Madrid in Spain, Warsaw in Poland, and Brussels in Belgium.

New capitals

Sometimes, a government can choose to make a different city the capital or build a new capital from scratch for political reasons. A new capital represents a fresh start and may need to be established in a neutral location to prevent the rest of the country from feeling left out.

Country	Capital city
Nigeria	**Abuja** Nigeria moved its capital in 1991 from Lagos on the coast to Abuja inland because of its politically neutral location.
Brazil	**Brasília** In 1960, Brazil built a new capital at Brasília, because Rio de Janeiro was overcrowded and far from much of the country.
Australia	**Canberra** In 1913, the dispute over which city should be the capital was resolved by building a new capital, Canberra.
Pakistan	**Islamabad** In the 1960s, Islamabad was built as the new capital since it is centrally located and less vulnerable to attack than Karachi.
India	**New Delhi** India moved its capital to New Delhi in 1931 because Calcutta, the old capital, was politically unstable.
US	**Washington, D.C.** A political compromise in 1790 led to Washington, D.C., being made the capital of the US.

Global cities

Cities such as New York, London, and Tokyo have an important economic impact on the rest of the world and are known as global cities or world cities. These are different from megacities, which are cities with a population of more than 10 million. Global cities have strong connections with each other and the region, as well as world-class infrastructure and facilities.

Alpha cities

Some organizations rank cities in order to study them and make comparisons. One system divides cities into alpha, beta, and gamma based on their economic activities and their connectivity with the rest of the world.

Alpha ++	London and New York are the only two cities in this top category, with maximum connectivity with the world.
Alpha +	Hong Kong, Beijing, Singapore, Shanghai, Sydney, Dubai, Paris, and Tokyo are the eight cities in this category.
Alpha	There are 23 alpha cities including Frankfurt, Mumbai, Mexico City, Madrid, and Istanbul.
Beta	While alpha cities link major regions, beta cities such as Seoul integrate smaller countries or regions into the world economy.
Gamma	Gamma cities such as Melbourne link local regions, such as provinces or states, with the world economy.

Common features ▷ Growing connections between global cities means that they have many features in common. A city can qualify as a global city only if it has most of these features.

Major transportation links

Major trade center

Global media hub

Major theaters and concert venues

Headquarters of global corporations

Universities and research centers

Concentration of services

International political connections

High-quality hospitals

Center for innovation

Primate cities

These cities are disproportionately larger, in terms of size, population, wealth, and influence, than any other city in a country. The figures below refer to the populations of each city's "urban agglomeration" (the city and its surrounding built-up areas), according to data collected by the UN.

KEY

Primate city

Second-largest city

Córdoba

Buenos Aires, Argentina
Argentina's primate city of Buenos Aires has a population of nearly 15 million—10 times as much as that of Córdoba, the second largest city.

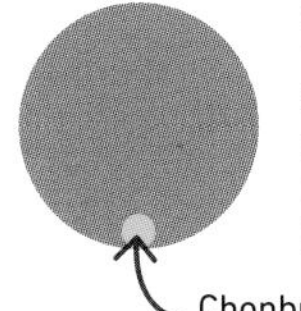
Chonburi

Bangkok, Thailand
Thailand's largest city, Bangkok, has a population of over 10 million and is more than 7 times larger than its second largest city, Chonburi.

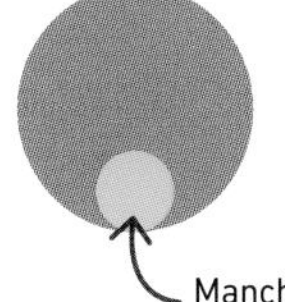
Manchester

London, UK
The UK's largest city, London, has a population of around 9 million, while Manchester, the second largest city, is home to around 2.7 million people.

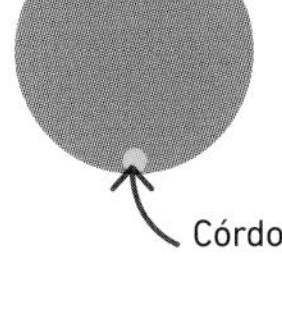
Guadalajara

Mexico City, Mexico
With over 20 million people, Mexico City is the largest city, compared to Guadalajara, the second largest city with a population of around 5 million.

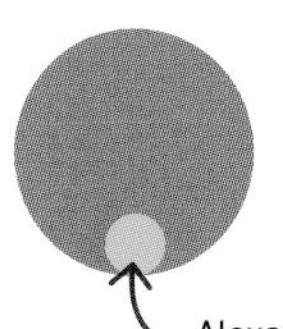
Alexandria

Cairo, Egypt
One of world's biggest primate cities, Cairo has a population of over 20 million, compared to around 5 million in Alexandria, Egypt's second largest city.

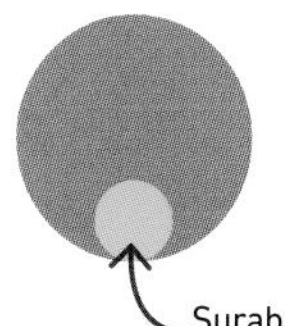
Surabaya

Jakarta, Indonesia
Home to 10.5 million people, Jakarta's population is around 3 times larger than that of Indonesia's second most populous city, Surabaya.

Hemispheres and latitude

SEE ALSO
❮ 76–77 Seasons
❮ 78–79 Climate zones
Longitude, time zones, and coordinates 212–213 ❯
Globes 222–225 ❯

LATITUDE IS THE MEASURE OF HOW FAR NORTH OR SOUTH YOU ARE FROM THE EQUATOR—FROM 0° TO 90° NORTH OR SOUTH.

The Earth is divided horizontally into two halves, or hemispheres, by an imaginary line called the equator. This line lies exactly halfway between the poles and is the widest point of the Earth.

The tropics and the polar bands

The Earth can be divided into five imaginary bands. Around the middle is the equator with the Tropic of Cancer to the north and the Tropic of Capricorn to the south. The Arctic Circle rings the North Pole and the Antarctic Circle the South Pole.

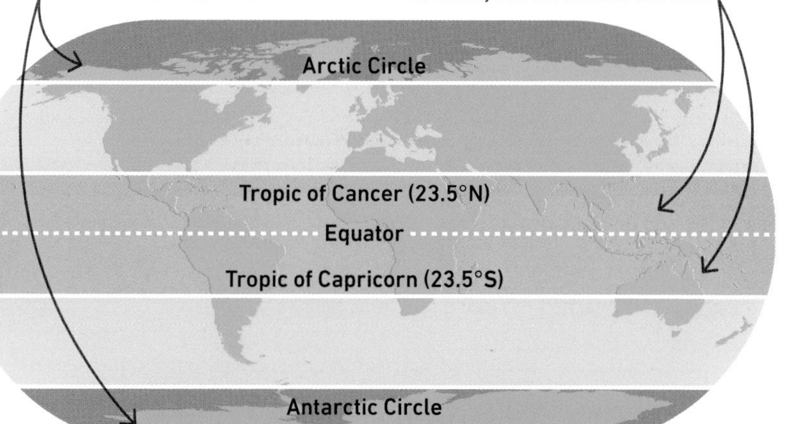
The polar bands cover the latitudes around the poles where the midnight sun can be seen in summer.
"The tropics" refers to the zone between the lines of Cancer and Capricorn, north and south of the equator, where the midday sun can be found overhead.
Arctic Circle
Tropic of Cancer (23.5°N)
Equator
Tropic of Capricorn (23.5°S)
Antarctic Circle

The Earth's bands ▷
These imaginary lines around the Earth mark the areas where the sun can be seen overhead at midday or at midnight. Their limits mark the farthest point where this is possible.

The hemispheres

The northern hemisphere is the half of the Earth north of the equator, which contains more than two-thirds of the Earth's land. The southern hemisphere is the half of the Earth below the equator, with less than one-third of the Earth's land. The Earth can also be divided vertically into eastern and western hemispheres.

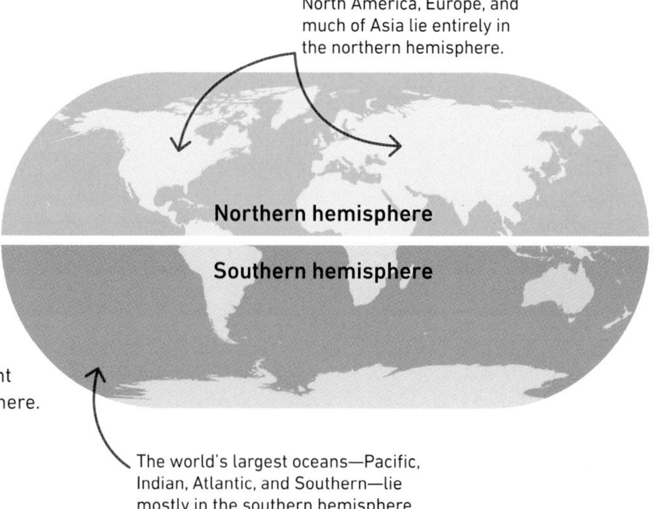
North America, Europe, and much of Asia lie entirely in the northern hemisphere.
Northern hemisphere
Southern hemisphere
The world's largest oceans—Pacific, Indian, Atlantic, and Southern—lie mostly in the southern hemisphere.

Divided planet ▷
Around 90 percent of the world's population lives in the northern hemisphere because it has more land. Only 10 percent lives in the more empty southern hemisphere.

Latitude

Lines of latitude are imaginary circles drawn around the Earth parallel to the equator, which is why they are also called parallels. Latitudes are expressed in degrees north or south of the equator. The equator itself is 0°, the North Pole is 90°N, and the South Pole is 90°S. The Tropic of Cancer is at 23.5°N and the Tropic of Capricorn is at 23.5°S.

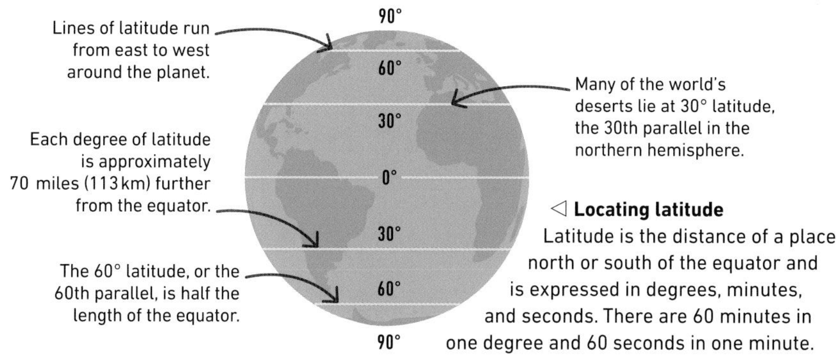
Lines of latitude run from east to west around the planet.
Each degree of latitude is approximately 70 miles (113 km) further from the equator.
The 60° latitude, or the 60th parallel, is half the length of the equator.
90°
60°
30°
0°
30°
60°
90°
Many of the world's deserts lie at 30° latitude, the 30th parallel in the northern hemisphere.

◁ **Locating latitude**
Latitude is the distance of a place north or south of the equator and is expressed in degrees, minutes, and seconds. There are 60 minutes in one degree and 60 seconds in one minute.

The sun seen from different latitudes in the northern hemisphere

The daily path of the sun through the sky changes with the latitude, climbing highest in the tropics and lowest in the polar regions. The path changes through the year, too. For six months, the highest point of the sun, or zenith, moves north across the equator, and for the next six months it moves south. This gives rise to the solstices, the shortest and longest days of the year. Equinoxes are the midpoints between the solstices when the day and night are both 12 hours long. The opposite season, path of the sun, and daylight hours are happening at the same time in the southern hemisphere.

The **sun's path** is **longest** on the **summer solstice** and **shortest** on the **winter solstice**.

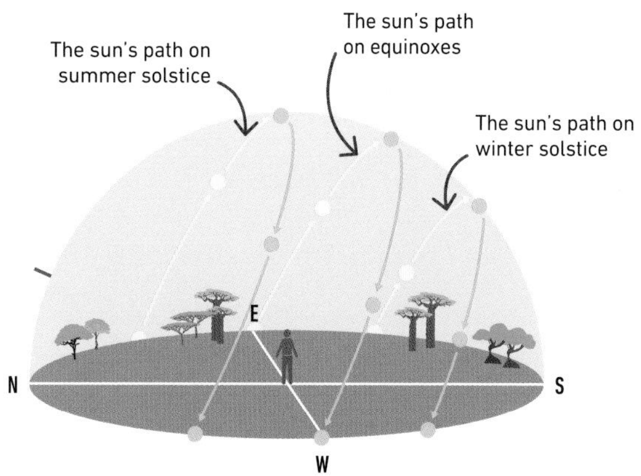
The sun's path on summer solstice
The sun's path on equinoxes
The sun's path on winter solstice
E
N
S
W

23.5° North
At the Tropic of Cancer (23.5° North), the sun is directly overhead at noon on the summer solstice. On the winter solstice, the sun is lower in the sky at noon and daylight is shorter.

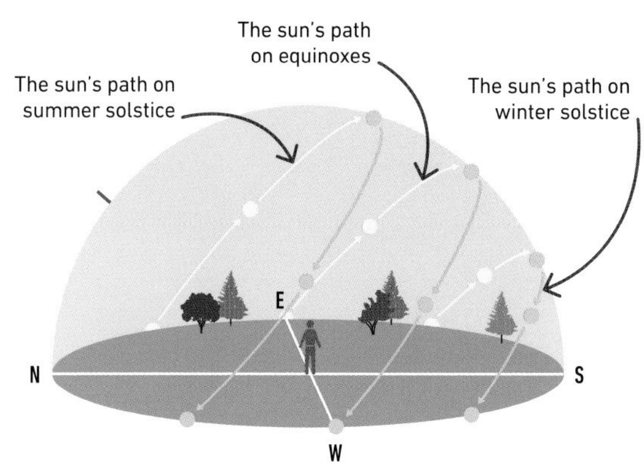
The sun's path on equinoxes
The sun's path on summer solstice
The sun's path on winter solstice
E
N
S
W

50° North
At 50° North, the path of the sun is at a lower angle across the sky, similar to winter at the Tropic of Cancer.

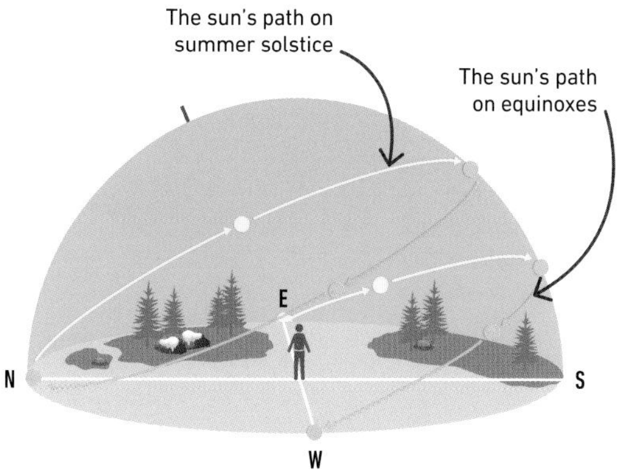
The sun's path on summer solstice
The sun's path on equinoxes
E
N
S
W

66.5° North
At the Arctic Circle (66.5° North), the sun is always low. It is just visible above the horizon at midnight on the summer solstice but is not visible for most of the winter.

Longitude, time zones, and coordinates

SEE ALSO
❮ 210–211 Hemispheres and latitude
Globes 222–225 ❯
Geographic Information Systems (GIS) 230–231 ❯

THE IMAGINARY LINES OF LONGITUDE RUN NORTH TO SOUTH ON THE EARTH AND LIE AT RIGHT ANGLES TO THE EQUATOR.

Lines of longitude are used to indicate how far east or west a place is. They divide the Earth into different time zones and, along with lines of latitude, help form a location's coordinates.

Longitude

Lines of longitude, or meridians, run between the North Pole and the South Pole, dividing the world like the segments of an orange—wide near the equator and narrow at the poles. The Prime Meridian is the longitude at 0° and runs through Greenwich in the UK. A longitude is expressed east or west of the Prime Meridian.

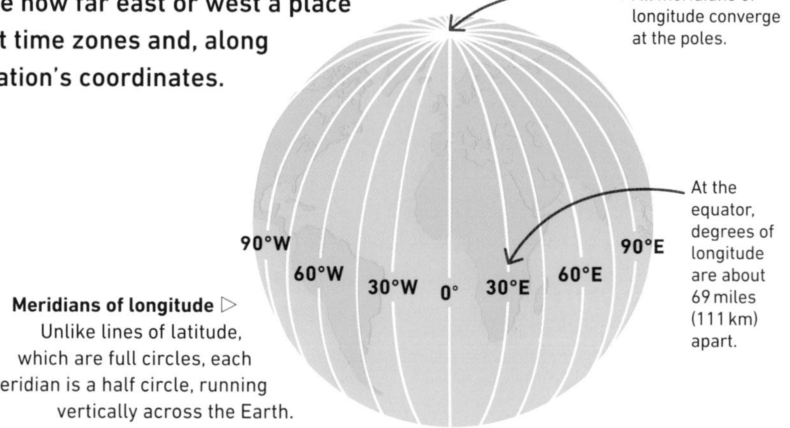
All meridians of longitude converge at the poles.
At the equator, degrees of longitude are about 69 miles (111 km) apart.
90°W
60°W
30°W
0°
30°E
60°E
90°E

Meridians of longitude ▷ Unlike lines of latitude, which are full circles, each meridian is a half circle, running vertically across the Earth.

Chronometers and longitude

Before satellite systems, navigators relied on clocks called chronometers to find longitude. Their position at sea could be determined by comparing the chronometer's clock—set at a fixed location—to the local time around the ship's location, which was established by measuring the angle of the sun.

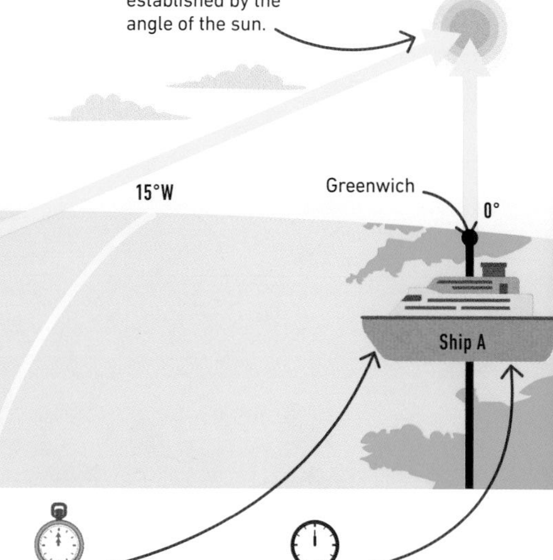
Local time is established by the angle of the sun.
Greenwich
0°
15°W
30°W
45°W
Ship B
Ship A
15° longitude = 1 hour

Time and position ▷ A chronometer helps work out the longitude by referring to the difference between the angle of the sun at a particular place and its angle at the Prime Meridian, at the same time of the day.

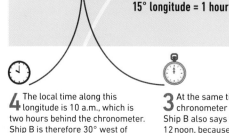

KEY
Local clock
Ship's chronometer

4 The local time along this longitude is 10 a.m., which is two hours behind the chronometer. Ship B is therefore 30° west of Greenwich, or the Prime Meridian.

3 At the same time, the chronometer aboard Ship B also says it is 12 noon, because it shows the fixed universal time.

2 The chronometer matches the clock that has been set at 12 noon at the fixed location, which is Greenwich.

1 Ship A is at 0° longitude. The sun is currently directly overhead along this longitude. It is 12 noon at Greenwich.

Time zones

Because of the Earth's rotation from west to east, the time of day changes around the world. When it is dawn in North America, it might be noon in Europe and sunset in China. To make things simpler, the world is divided into 24 time zones, one for each hour of the day.

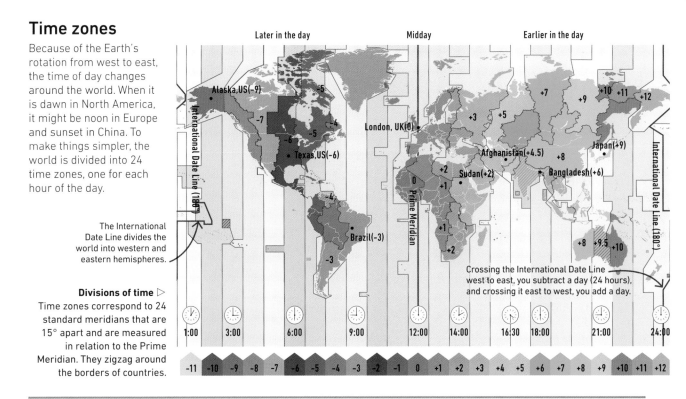

Later in the day
Midday
Earlier in the day
Alaska,US(-9)
Texas,US(-6)
Brazil(-3)
London, UK(0)
Sudan(+2)
Afghanistan(+4.5)
Bangladesh(+6)
Japan(+9)
International Date Line (180°)
Prime Meridian
Crossing the International Date Line west to east, you subtract a day (24 hours), and crossing it east to west, you add a day.
1:00 3:00 6:00 9:00 12:00 14:00 16:30 18:00 21:00 24:00
-11 -10 -9 -8 -7 -6 -5 -4 -3 -2 -1 0 +1 +2 +3 +4 +5 +6 +7 +8 +9 +10 +11 +12

The International Date Line divides the world into western and eastern hemispheres.

Divisions of time ▷
Time zones correspond to 24 standard meridians that are 15° apart and are measured in relation to the Prime Meridian. They zigzag around the borders of countries.

Locating places with coordinates

The position of places on the Earth's surface is identified using coordinates that combine lines of latitude and longitude. These work in the same way as the x and y axes on a graph. The "origin" is the point where the Prime Meridian crosses the equator.

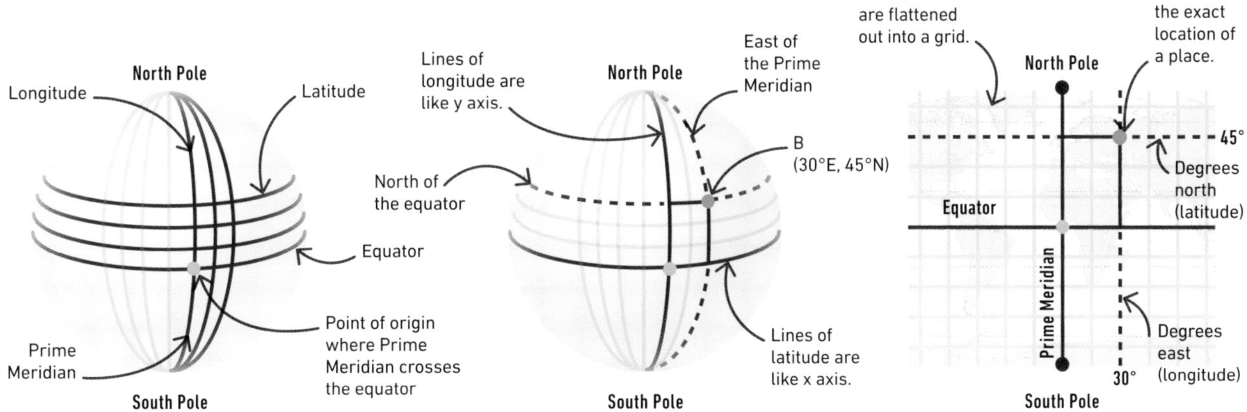
North Pole
Longitude
Latitude
Equator
Prime Meridian
Point of origin where Prime Meridian crosses the equator
South Pole
Lines of longitude are like y axis.
North Pole
East of the Prime Meridian
B (30°E, 45°N)
North of the equator
Lines of latitude are like x axis.
South Pole
Curved lines are flattened out into a grid.
"B" shows the exact location of a place.
North Pole
45°
Degrees north (latitude)
Equator
Prime Meridian
Degrees east (longitude)
30°
South Pole

△ **Finding coordinates**
Every point on Earth can be located precisely with its coordinates, by referring to the lines of longitude and latitude, which cross each other at right angles.

△ **Numerical location**
The coordinates of a place depend on the degrees north or south and the degrees east or west it is from the point of origin, just like on a graph with an x and y axis.

△ **Coordinates on a map**
On many flat maps, the lines of longitude and latitude make a grid of squares. The position of a place can then be identified in relation to these squares.

Distance and relative distance

SEE ALSO
❮ 138–139 The spread of cultures
❮ 154–155 Transportation and distribution
Topographic maps 226–227 ❯

DISTANCE IS A MEASURE OF SPACE BETWEEN TWO POINTS. THERE ARE DIFFERENT WAYS TO MEASURE IT.

The physical distance between two places can be measured in a variety of ways. Apart from units of length, such as miles or paces or blocks, there are other kinds of distance a geographer may consider.

Absolute distance

The most basic measure of distance—absolute distance—tells you how far apart two points are physically. It can be measured to provide a precise figure. The shortest distance between two points is a straight line, but the distance covered will vary according to the route taken.

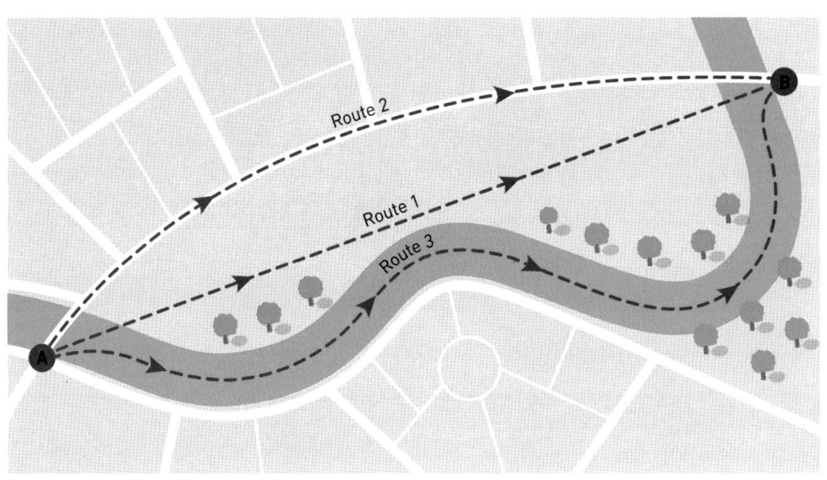
Route 2
Route 1
Route 3
A
B

Measuring absolute distances ▷ On a map, a straight line (Route 1) between A and B is the simplest distance to measure. The distances by road (Route 2) and along the river (Route 3) are longer and harder to measure.

Relative distance

Unlike absolute distance, relative distance measures how far apart places are socially, culturally, and economically. Two affluent town centers linked by a fast train may be closer to each other in many ways than to poor local neighborhoods that can be accessed only on foot. It is possible to measure, for instance, time distance, convenience distance, social distance, or even the economic distance between places.

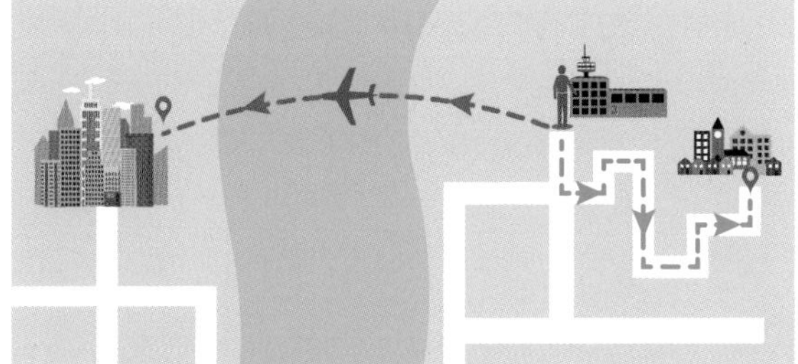

◁ Time distance The time it takes to travel between two points is the time distance. If someone lives very near the airport, the time distance to a foreign city might be less than that to a nearby town.

◁ Convenience distance This measure of distance considers how much time obstacles can add to a trip. The destination may be physically nearer, but if the route passes through a muddy stream and forests, its convenience distance is greater.

Effects of distance

There are many ways of thinking about the effects of distance. In addition to the physical distance, a trip may involve a time delay, effort, and discomfort. All these factors must be taken into account. Places that are easy to get to may not necessarily be closer; they may just involve an easier route with fewer obstacles (convenience distance). Transferability considers the ease or cost of moving between two places.

Communication links

Telecommunication, such as internet video links, can reduce the relative time distance between places on opposite sides of the world to almost nothing. Cell phones and messaging services have replaced letters, dramatically reducing relative time distance.

Hong Kong
New York City

Distance decay

Physical distance affects levels of interaction significantly. People and businesses interact a great deal with others that are local, less so with people and businesses in a town farther away, and very little with those in a far away town.

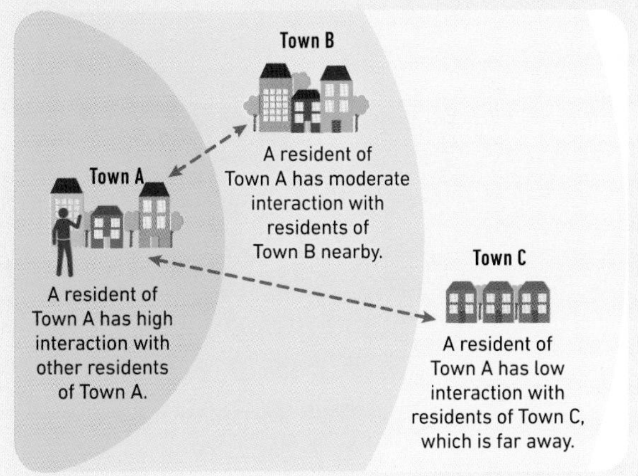
Town B
Town A
A resident of Town A has moderate interaction with residents of Town B nearby.
Town C
A resident of Town A has high interaction with other residents of Town A.
A resident of Town A has low interaction with residents of Town C, which is far away.

Friction of distance

Covering a distance requires time, energy, effort, and resources. The greater the distance, the more of each of these may be required. This is called the friction of distance. Businesses typically plan their operations to minimize the friction of distance.

△ **Long distance but low friction**
A long physical distance may have low friction. Even if the road is long and winding, in the absence of obstacles, the trip will be easy.

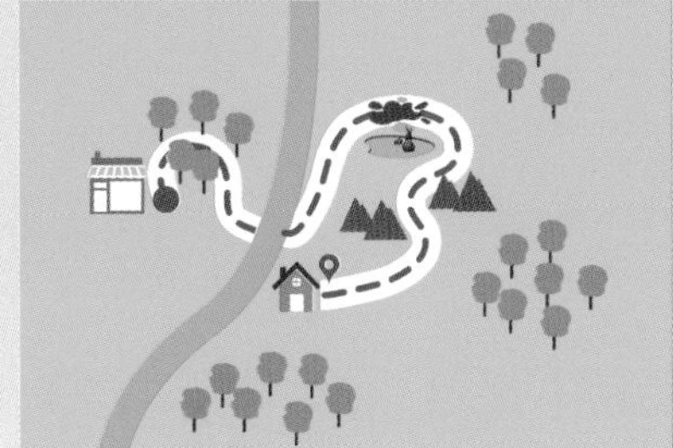

△ **Short distance but high friction**
A short physical distance may have high friction. If there are difficult features to cross, such as a mountain pass or a river, the trip may be very hard.

△ **Medium distance and friction**
Trips with medium distance and medium friction are the hardest to assess, and so it is important to weigh their costs against their rewards.

Geopolitics

SEE ALSO
❮ 162–163 Globalization
❮ 196–197 Conflict and resolution
❮ 204–205 Countries and nations
❮ 208–209 Capitals and major cities

THE STUDY OF HOW GEOGRAPHICAL FACTORS AFFECT INTERNATIONAL RELATIONS IS CALLED GEOPOLITICS.

Geopolitics focuses on how the world is divided into countries and nations as well as many other kinds of groupings such as trading blocs, economic zones, political unions, and military alliances.

Key regions of the world

Geopolitics divides the world into units based on locality, politics, economics, and beliefs, grouping the 193 countries in different ways. The largest grouping is by geopolitical realms, which are further divided into geopolitical regions.

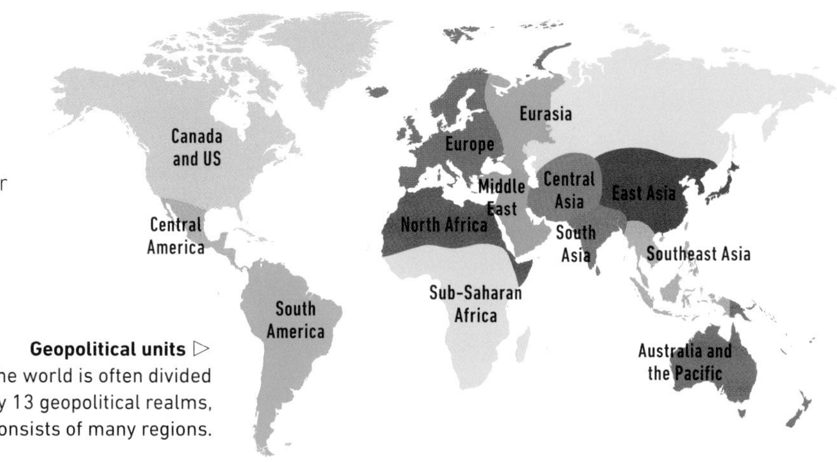
Canada and US
Central America
South America
Europe
Eurasia
Middle East
North Africa
Sub-Saharan Africa
Central Asia
South Asia
East Asia
Southeast Asia
Australia and the Pacific

Geopolitical units ▷ Today, the world is often divided into roughly 13 geopolitical realms, each of which consists of many regions.

Political alliances

Neighboring countries can form political alliances to provide each other with financial help or extra security against shared dangers. Alliances may be in the form of agreements to act together, or involve complex organizations or even joint governments.

The European Union (EU)
A political and economic union of 27 European nations, the EU has a joint parliament with representatives from all member nations, and joint laws.

Austria	Estonia	Italy	Portugal
Belgium	Finland	Latvia	Romania
Bulgaria	France	Lithuania	Slovakia
Croatia	Germany	Luxembourg	Slovenia
Cyprus	Greece	Malta	Spain
Czech Republic	Hungary	Netherlands	Sweden
Denmark	Ireland	Poland	

US
UK
Russia
China
France
Albania (2023)
Brazil (2023)
Gabon (2023)
Ghana (2023)
Japan (2024)
Malta (2024)
Mozambique (2024)
Ecuador (2024)
Switzerland (2024)
United Arab Emirates (2023)

△ The United Nations (UN) Security Council
Of the UN's 193 member nations, 15 serve on the UN Security Council, which tries to maintain international peace and security. The council has five so-called "permanent members."

KEY
- Permanent members
- Nonpermanent members, including the year in which membership expires

Military alliances

Agreements between countries to combine their armed forces are known as military alliances, meaning that they would not only fight together in a conflict but also use their combined strength to deter enemies from starting a war. Military alliances between aggressive countries can sometimes be threatening to others.

◁ **North Atlantic Treaty Organization** NATO is a military alliance between 30 North American and European countries, all of which have agreed to defend each other against attacks.

NATO members

Albania	Greece	Norway
Belgium	Hungary	Poland
Bulgaria	Iceland	Portugal
Canada	Italy	Romania
Croatia	Latvia	Slovakia
Czech Republic	Lithuania	Slovenia
Denmark	Luxembourg	Spain
Estonia	Montenegro	Türkiye
France	Netherlands	United Kingdom
Germany	North Macedonia	United States

Economic alliances

Countries may form economic alliances to make it easier for them to buy and sell things across borders. Trade blocs (groups of countries who trade together) ease trade by lowering tariffs (extra charges on goods crossing a border), for their members. Economic unions agree to share trade regulations and standards.

△ **NAFTA** The North American Free Trade Agreement between the US, Canada, and Mexico came into effect in 1994 and removes most tariffs on trade between them.

△ **ASEAN** With 10 members, the Association of Southeast Asian Nations is not just an economic alliance but a regional political organization.

G7

△ **G7** The Group of Seven is a forum that brings together seven of the world's richest nations—Canada, France, Germany, Italy, Japan, the UK, and the US.

WORLD TRADE
ORGANIZATION

△ **WTO** The World Trade Organization, whose members represent 98 percent of world trade, deals with the rules of trade between nations.

G20

△ **G20** The Group of Twenty is a forum for 19 nations and the EU, and the governors of their central banks. It accounts for most of the world's wealth.

OECD

△ **OECD** The Organisation for Economic Co-operation and Development is a forum of 38 rich countries, mostly in Europe and North America.

Former regions

The geopolitical map of the world changes constantly as governments rise and fall, and new partnerships are made while old ones break apart. For example, Macedonia became North Macedonia in January 2019 and might join the EU and NATO in the future, and the UK left the EU in 2020.

◁ **African colonies** At the end of the 19th century all African countries, except Ethiopia and Liberia (in yellow), were under European colonial rule. Today, most of these countries are independent.

△ **The Soviet Union** Also known as the Union of Soviet Socialist Republics (USSR), the Soviet Union included many countries, but was ruled by Russia as a single political bloc from 1922 to 1991.

△ **The Commonwealth of Nations** King Charles III currently heads this voluntary political association of 56 independent countries, nearly all of which were once colonies of the British Empire.

Types of maps

A MAP IS A GRAPHIC REPRESENTATION OF A PART OF EARTH ON A FLAT SURFACE. IT ILLUSTRATES SPECIFIC FEATURES OF THE LANDSCAPE.

SEE ALSO

How a map works	220–221 ❯
Topographic maps	226–227 ❯
Atlases	228–229 ❯
Quantitative data	234–235 ❯
Qualitative data	236–238 ❯
Geographical inquiry	244–245 ❯

General-purpose maps provide many types of information on one map. Most maps, however, are created to communicate specific information, such as variations in population density or the location of airports and railroad lines.

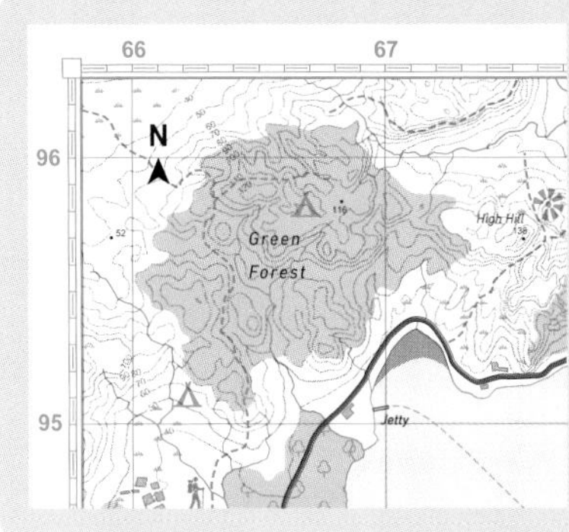

66
67
96
95
N
Green Forest
High Hill
Jetty

Topographic map
Often used for hiking or walking, topographic maps show features of the landscape in great detail, such as the shape and height of hills.

Common types of maps

A wide range of maps can be created to show different kinds of information. The most common types are political and physical. Political maps are especially useful because they show the locations and boundaries of different countries or states. Physical maps show the main features of the landscape, including mountains and valleys.

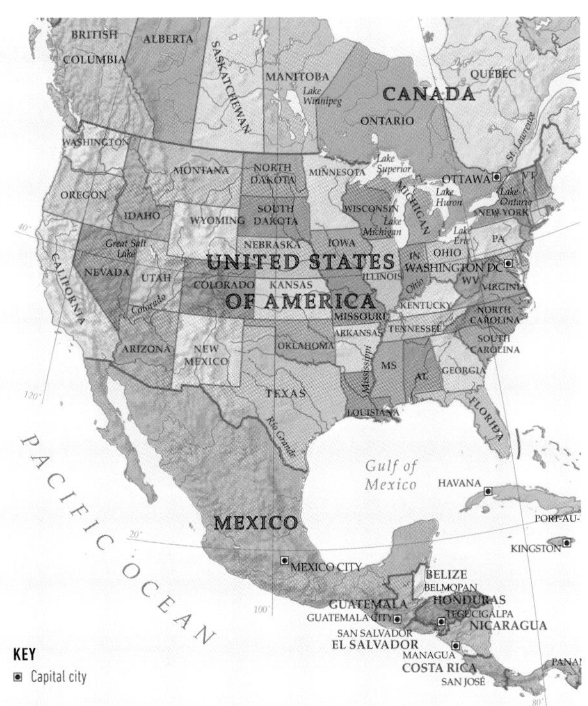

BRITISH COLUMBIA
ALBERTA
SASKATCHEWAN
MANITOBA
Lake Winnipeg
CANADA
QUÉBEC
ONTARIO
St. Lawrence
WASHINGTON
MONTANA
NORTH DAKOTA
MINNESOTA
Lake Superior
OTTAWA
VT
OREGON
IDAHO
WYOMING
SOUTH DAKOTA
WISCONSIN
MICHIGAN
Lake Huron
Lake Ontario
NEW YORK
Lake Michigan
Lake Erie
PA
Great Salt Lake
NEBRASKA
IOWA
IN
OHIO
UNITED STATES OF AMERICA
NEVADA
UTAH
COLORADO
KANSAS
ILLINOIS
WASHINGTON DC
WV
VIRGINIA
CALIFORNIA
Colorado
KENTUCKY
NORTH CAROLINA
MISSOURI
ARKANSAS
TENNESSEE
SOUTH CAROLINA
ARIZONA
NEW MEXICO
OKLAHOMA
Mississippi
MS
AL
GEORGIA
TEXAS
LOUISIANA
FLORIDA
Rio Grande
PACIFIC OCEAN
Gulf of Mexico
HAVANA
MEXICO
PORT-AU-
KINGSTON
MEXICO CITY
BELIZE
BELMOPAN
HONDURAS
GUATEMALA
GUATEMALA CITY
TEGUCIGALPA
NICARAGUA
SAN SALVADOR
EL SALVADOR
MANAGUA
COSTA RICA
SAN JOSÉ
PANAI
KEY
Capital city

△ **Political map**
These maps show the outlines of countries and states and the location of capital cities. This map shows the states of the US and the countries of North and Central America.

Canadian Shield
Laurentian Mountains
Rocky Mountains
Great Plains
Lake Winnipeg
Lake Manitoba
Mount Rainier
Mount St. Helens
Lake Superior
St. Lawrence
Great Lakes
Lake Huron
Lake Ontario
Great Basin
Missouri
Lake Michigan
Sierra Nevada
Great Salt Lake
Colorado
Lake Erie
Appalachian Mountains
Ohio
Death Valley
Arkansas
Mississippi
Brasstown Bald
Sonoran Desert
Baja California
Rio Grande
Sierra Madre Occidental
Sierra Madre Oriental
Mississippi Delta
Gulf of Mexico
Cuba
Pico de Orizaba
Yucatan Peninsula
Sierra Madre del Sur
Lake Nicaragua
ELEVATION
6,500–13,000 ft
3,200–6,500 ft
1,600–3,200 ft
800–1,600 ft
320–800 ft
0–320 ft

△ **Physical map**
Maps of this type show natural features of the land. This map shows the hills, valleys, rivers, and lakes in North America and the heights of the land above sea level.

Other types of maps

Some maps are created to show specific data, especially statistics. Statistical maps are especially useful to show how things such as population and resources are distributed. They often use a political or physical map as a base.

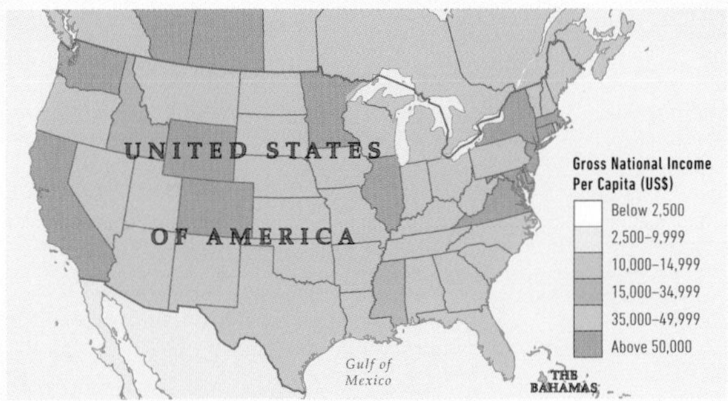
UNITED STATES
OF AMERICA
Gross National Income Per Capita (US$)
Below 2,500
2,500–9,999
10,000–14,999
15,000–34,999
35,000–49,999
Above 50,000
Gulf of Mexico
THE BAHAMAS

△ **Choropleth map**
Different coloring or shading is used to represent values for the subject of the map. This map shows how income varies across the US and Mexico.

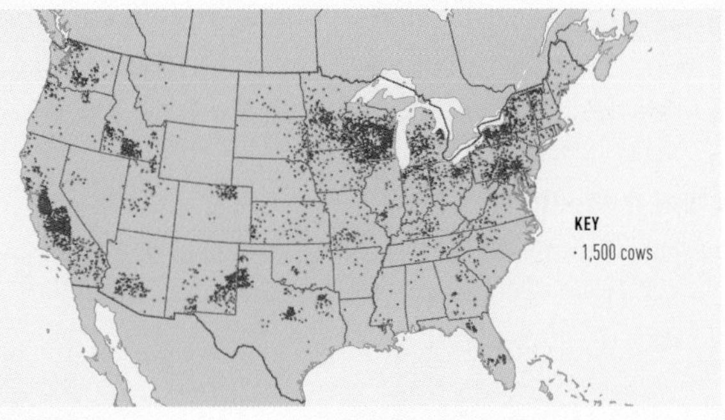
KEY
· 1,500 cows

△ **Dot map**
Dots are used to show data such as the distribution of population or other phenomena. Each dot represents a particular amount of something. In this map, one dot indicates 1,500 cows.

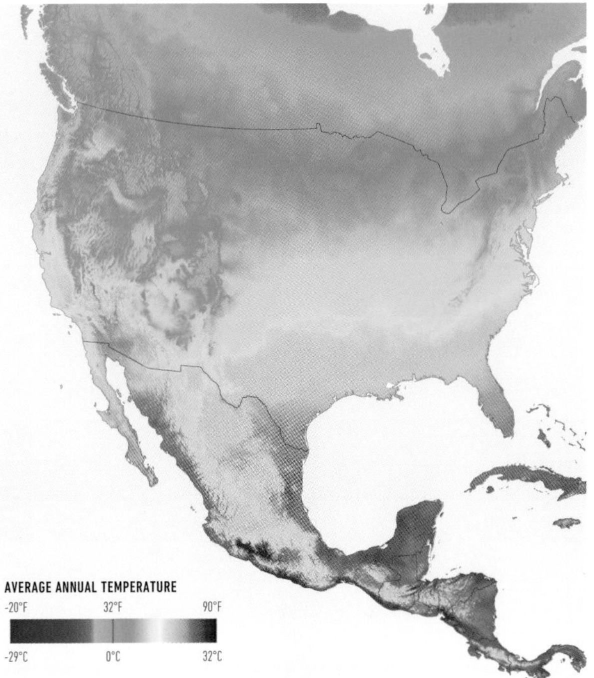
AVERAGE ANNUAL TEMPERATURE
-20°F
32°F
90°F
-29°C
0°C
32°C

△ **Temperature map**
Some maps show climate variations between places. A temperature map, such as the one above, shows how hot or cold it is in different parts of the same continent.

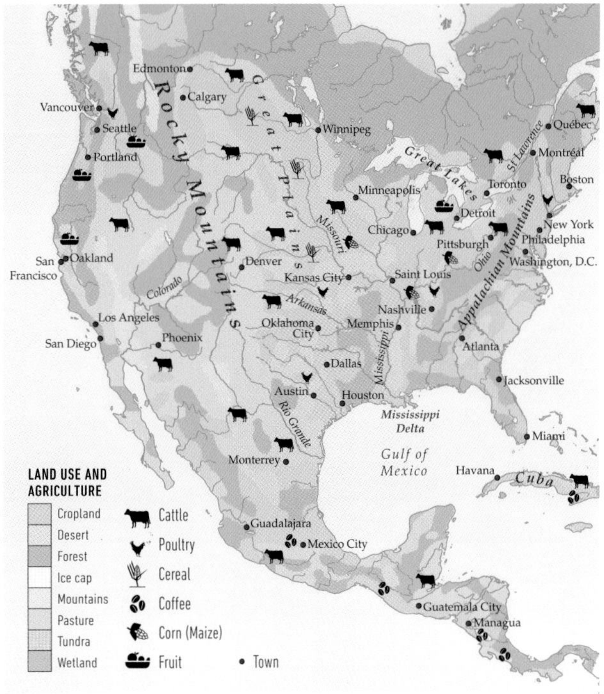
Edmonton
Calgary
Vancouver
Seattle
Portland
Winnipeg
Québec
Montréal
Boston
Toronto
Minneapolis
Detroit
New York
Philadelphia
Washington, D.C.
Chicago
Pittsburgh
San Francisco
Oakland
Denver
Kansas City
Saint Louis
Nashville
Los Angeles
San Diego
Phoenix
Oklahoma City
Memphis
Atlanta
Dallas
Jacksonville
Austin
Houston
Miami
Monterrey
Havana
Cuba
Guadalajara
Mexico City
Guatemala City
Managua
Rocky Mountains
Great Plains
Great Lakes
St Lawrence
Appalachian Mountains
Missouri
Colorado
Arkansas
Ohio
Mississippi
Rio Grande
Mississippi Delta
Gulf of Mexico
LAND USE AND AGRICULTURE
Cropland
Desert
Forest
Ice cap
Mountains
Pasture
Tundra
Wetland
Cattle
Poultry
Cereal
Coffee
Corn (Maize)
Fruit
• Town

△ **Land use map**
These maps represent the main use of the land in different places, usually employing different colors or symbols to illustrate the different uses.

How a map works

MAPS ARE A FLAT PLAN OF THE LANDSCAPE USING A SET OF SYMBOLS TO SHOW ITS FEATURES.

SEE ALSO
❮ 218–219 Types of maps
Globes 222–225 ❯
Topographic maps 226–227 ❯
Atlases 228–229 ❯

Maps are key tools for geographers. They communicate information about places and can be used for all kinds of purposes. The person drawing the map can choose what to show and what to leave out.

What a map shows

Maps vary according to their purpose, but most of them show features such as distances and directions using a scale. Some maps, such as metro maps, show only specific things such as railroad lines and stations.

Compass point

Direction is important on maps, so most include a compass showing how to align it with the real world. The compass shows the north or true north on a map. This is especially useful if you are using a map and a compass to find your way, also known as orienteering.

N
W E
S

Scale bar

The ratio of distances on the map compared to the real-world distances is the scale. Small-scale maps show large areas on a small map, whereas large-scale maps show small areas on a big map.

Linear scale
The line is marked off in units, which represent the real distance on the map.

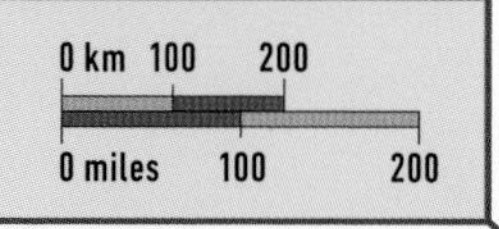
0 km 100 200
0 miles 100 200

Statement of scale
This states what a length on the map represents.

1 mm on the map represents 1 km on the ground

Representative fraction
This is the scale of the map shown as a fraction.

1:1,000,000, meaning one unit on the map would be equal to 1,000,000 units on the ground

▽ **Reading a map**
A small-scale map shows a large area, such as the whole continent of Australia, allowing the person reading the map to get a complete overview of where places are in relation to each other.

MAP OF AUSTRALIA

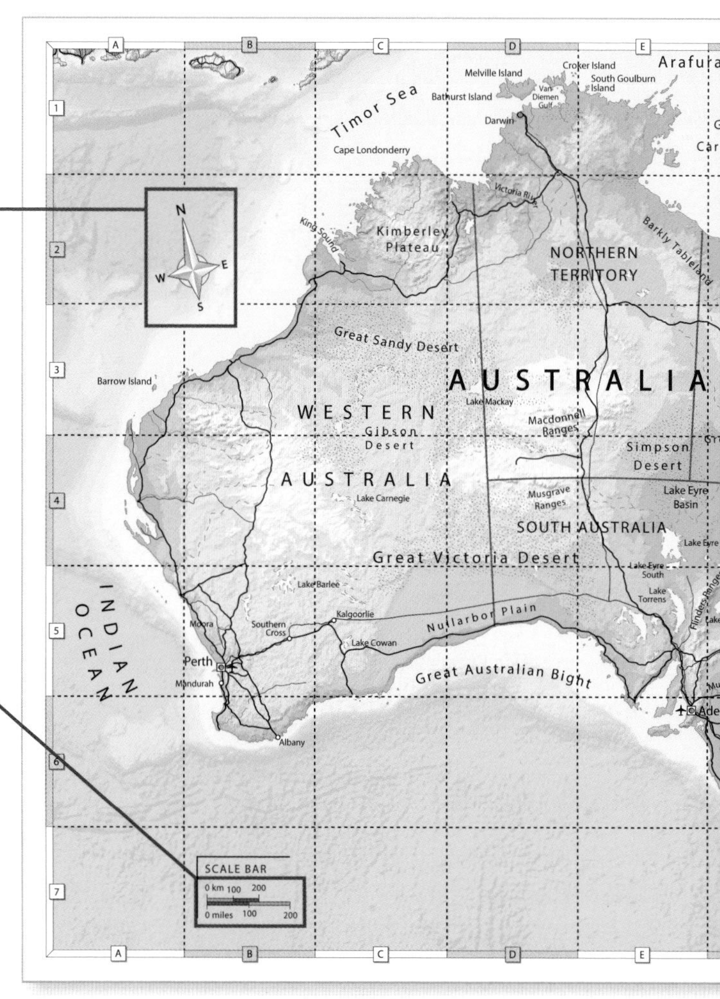
A
B
C
D
E
1
2
3
4
5
6
7
Timor Sea
Melville Island
Bathurst Island
Croker Island
South Goulburn Island
Van Diemen Gulf
Darwin
Arafura
Cape Londonderry
King Sound
Kimberley Plateau
NORTHERN TERRITORY
Barkly Tableland
N
W
E
S
Great Sandy Desert
Barrow Island
AUSTRALIA
WESTERN AUSTRALIA
Lake Mackay
Gibson Desert
Macdonnell Ranges
Simpson Desert
Lake Carnegie
Musgrave Ranges
Lake Eyre Basin
SOUTH AUSTRALIA
Great Victoria Desert
Lake Barlee
Lake Torrens
INDIAN OCEAN
Moora
Southern Cross
Kalgoorlie
Nullarbor Plain
Lake Cowan
Perth
Mandurah
Great Australian Bight
Albany
SCALE BAR
0 km 100 200
0 miles 100 200

Grid

Most maps are overlaid with a grid—a pattern of vertical and horizontal lines forming squares. Each square in the grid is given a letter and a number. This enables the map reader to locate a place on the map quickly and precisely from the numbers and letters of its square, known as its grid reference.

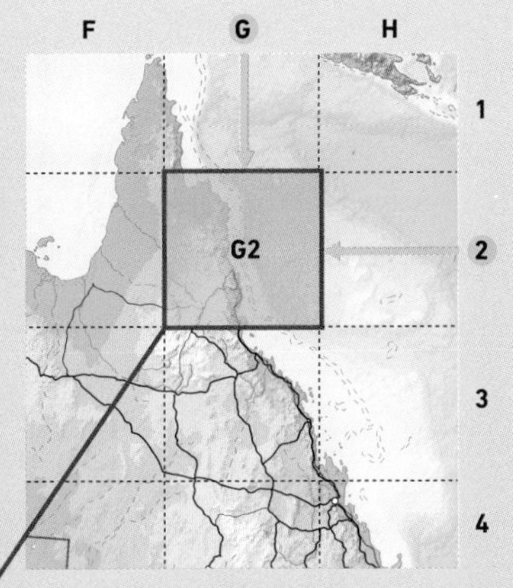
F
G
H
1
2
3
4
G2

Different scales

The choice of scale is very important to the map user. A very small-scale map shows a broad overview of a large region but has few details. In some cases, a large-scale map might be needed to show the detail required for making legal and planning decisions at a local level.

ATLANTIC OCEAN
British Isles
North Sea
Ireland
IRELAND
DUBLIN
UNITED KINGDOM
Britain
LONDON
Celtic Sea
Celtic Shelf
English Channel
BELGIUM
FRANCE

◁ **Broad view**

This small-scale map shows the layout of the entire British Isles. Very little detail can be seen at this scale, and only the names of countries and the largest cities can be seen.

SCALE 1:42,000,000

Grimsby
Skegness
ENGLAND
Nottingham
Norwich
Great Yarmouth
Leicester
Peterborough
Lowestoft
Birmingham
Coventry
Northampton
Cambridge
Ipswich
Milton Keynes
Luton
Colchester
Oxford
Harlow
Southend-on-Sea
Swindon
LONDON
Margate
Reading
Croydon
Canterbury
Maidstone
Dover
Crawley
Southampton
Brighton
Hastings
Portsmouth
Hove
Eastbourne
Isle of Wight
Bournemouth
English Channel

◁ **Major transportation links**

At a larger scale, major roads in the southeast region of the UK can be seen. Many towns are named, and the map reader can see the difference in their size and status.

SCALE 1:11,000,000

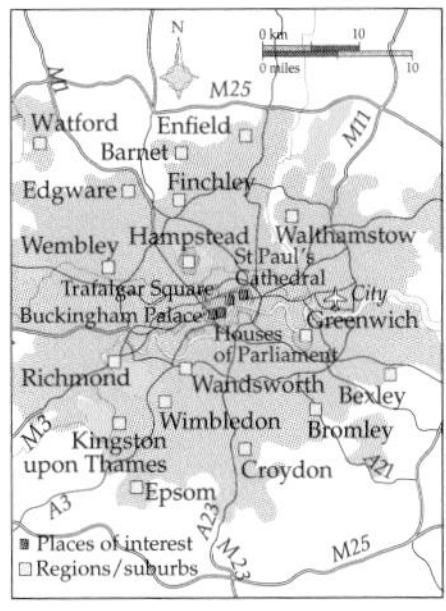
M25
Watford
Enfield
Barnet
Edgware
Finchley
Wembley
Hampstead
Walthamstow
St Paul's Cathedral
Trafalgar Square
Buckingham Palace
City
Greenwich
Houses of Parliament
Richmond
Wandsworth
Bexley
Wimbledon
Bromley
Kingston upon Thames
Croydon
Epsom
Places of interest
Regions/suburbs

◁ **Regional routes**

This map is at a much larger scale. One can see the major roads that lead out of London, along with the names of many suburbs and places of interest.

SCALE 1:1,800,000

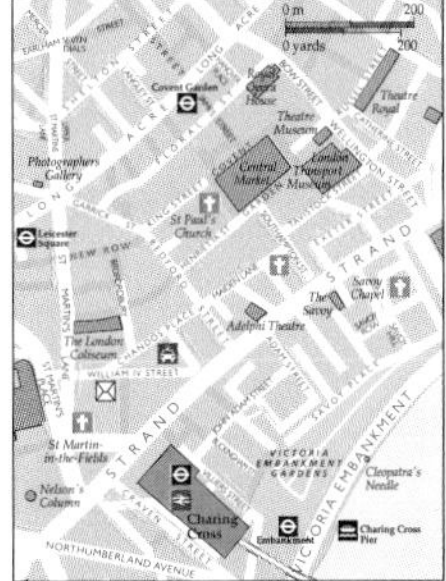

◁ **Street map**

To navigate around a city such as London, one needs a street map—a map that is large enough in scale to show all the individual streets, complete with street names and key landmarks.

SCALE 1:25,000

Key or legend

Most maps use a variety of symbols to show the different features on it, such as major cities, airports, mountains, and so on. Maps also have a key, or "legend," which explains the meaning of each symbol clearly.

- State border
- Major road
- Railroad
- River
- Seasonal river
- Capital of country
- Province capital
- Major city
- Other city
- Airport
- Lake

Globes

THE SURFACE OF THE EARTH CAN BE REPRESENTED ON A SPHERICAL GLOBE OR ON THE FLAT SURFACE OF A MAP.

SEE ALSO
❮ 210–211 Hemispheres and latitude
❮ 212–213 Longitude, time zones, and coordinates
Atlases 228–229 ❯

The Earth is a sphere, so a globe is the most accurate way to represent it. It is difficult to represent the Earth as a map on a flat surface, because we always distort it in some way.

What globes show

Globes are typically mounted on a rod, or axis, so that they can be spun around and viewed from all sides. The axis of a globe runs between the poles and is tilted at 23.5 degrees, matching the Earth's axis. This makes the globe spin just like the Earth.

Satellite imagery ▷ Photographs taken by satellites can be used to show the Earth, together with its weather systems, as a globe.

Representing the Earth ▷ Globes show larger details of the Earth's surface, such as continents, rivers, and mountain ranges. They may also show countries, cities, and shipping routes.

Making a globe

Most globes are made by printing everything to be shown on 12 or more flat, shield-shaped pieces of paper called gores. These segments are then stuck carefully onto the globe so that they match up precisely. Relief globes are made in a mold and depict hills and troughs using a raised surface.

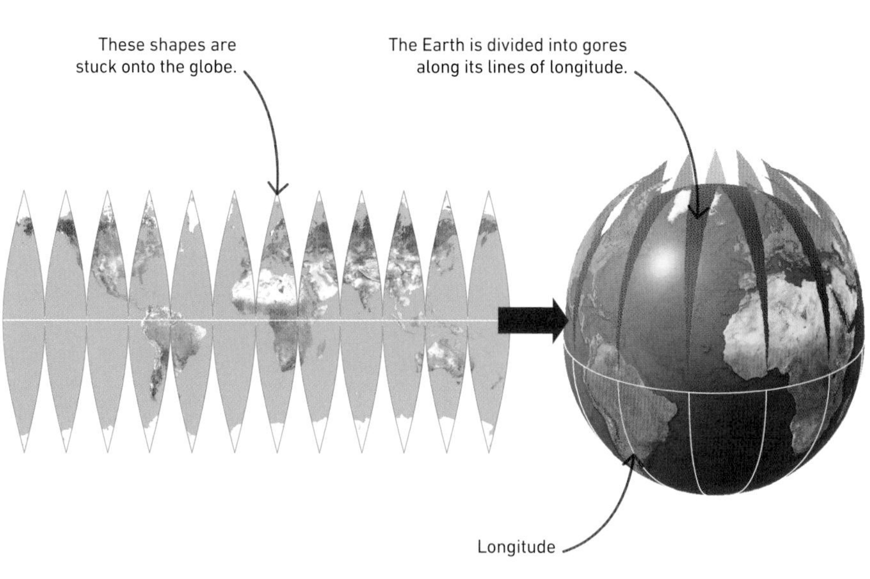
These shapes are stuck onto the globe.
The Earth is divided into gores along its lines of longitude.
Longitude

Using gores ▷ Gores are widest at the equator and narrower toward the poles so that they can wrap around the globe perfectly.

Latitude

Lines of latitude, or parallels, are imaginary circles drawn around the Earth, parallel to the equator. They show how far north or south a point is from the equator as degrees north or south. The equator is at 0°, and the north and south poles are 90°N and 90°S, respectively.

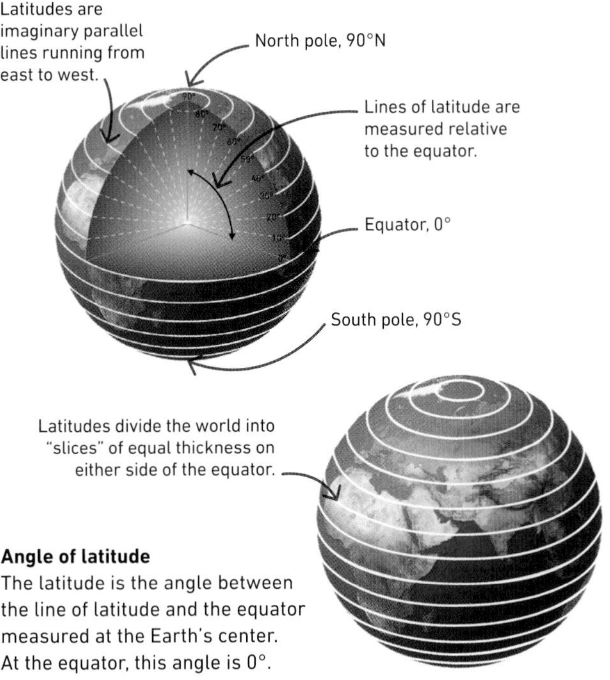
Latitudes are imaginary parallel lines running from east to west.
North pole, 90°N
Lines of latitude are measured relative to the equator.
Equator, 0°
South pole, 90°S
Latitudes divide the world into "slices" of equal thickness on either side of the equator.

Angle of latitude
The latitude is the angle between the line of latitude and the equator measured at the Earth's center. At the equator, this angle is 0°.

Longitude

Lines of longitude, or meridians, are imaginary lines drawn around the Earth between the North and South Poles. The longitude of a place shows how far east or west it is from the Prime Meridian as degrees east or west. The Prime Meridian is the reference point at 0°.

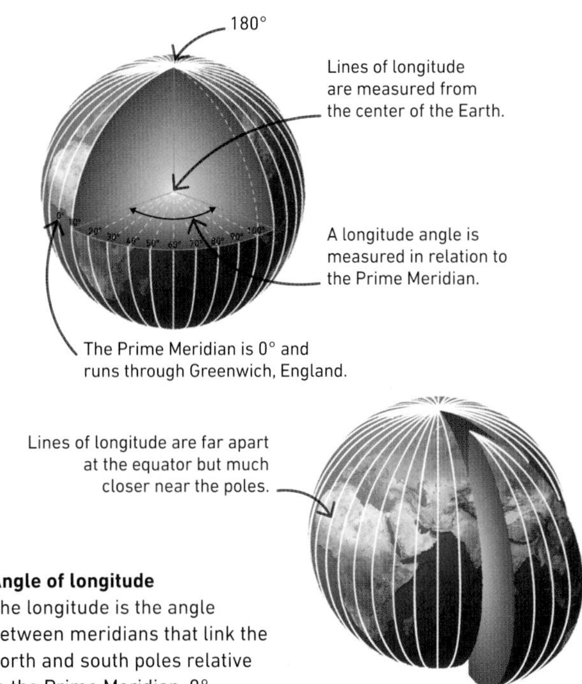
180°
Lines of longitude are measured from the center of the Earth.
A longitude angle is measured in relation to the Prime Meridian.
The Prime Meridian is 0° and runs through Greenwich, England.
Lines of longitude are far apart at the equator but much closer near the poles.

Angle of longitude
The longitude is the angle between meridians that link the north and south poles relative to the Prime Meridian, 0°.

Calculating distance

In the past, sailors steered their boats on a "rhumb" line. A rhumb is a steady compass setting and appears on Mercator maps as a straight line. Although straight on the map, it is not the shortest distance between two places, because the Earth's surface is curved. The shortest distance is a "great circle." This is any circle drawn around the world's center, and appears as a curved line on the map.

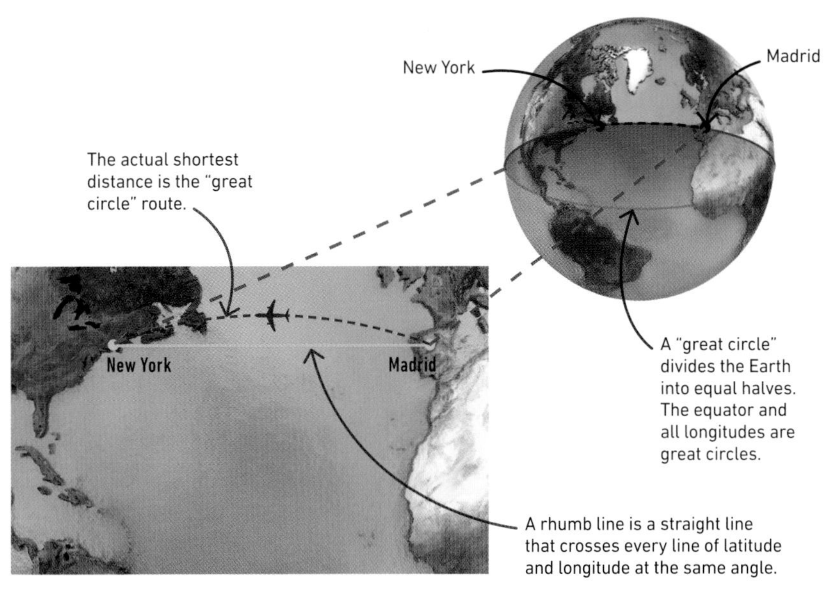
New York
Madrid
The actual shortest distance is the "great circle" route.
New York
Madrid
A "great circle" divides the Earth into equal halves. The equator and all longitudes are great circles.
A rhumb line is a straight line that crosses every line of latitude and longitude at the same angle.

The shortest distance ▷
A straight rhumb line may look like the shortest distance between two places, but the curve of a "great circle" is actually the shortest distance.

Map projections

Transforming the curved surface of a globe onto a flat map is not easy. There are several ways of doing this, but each way of representing the world (or creating a "projection") contains some distortion. Some projections are better at showing distances, some at showing the area of continents, and others the overall position of continents.

Cylindrical projections

These maps are called "cylindrical" because an image of the surface of the globe is transferred onto a surrounding cylinder, or tube. This cylinder is then rolled out to give a flat map. These maps are very useful for showing the whole world, but they distort the size of countries that are near the poles.

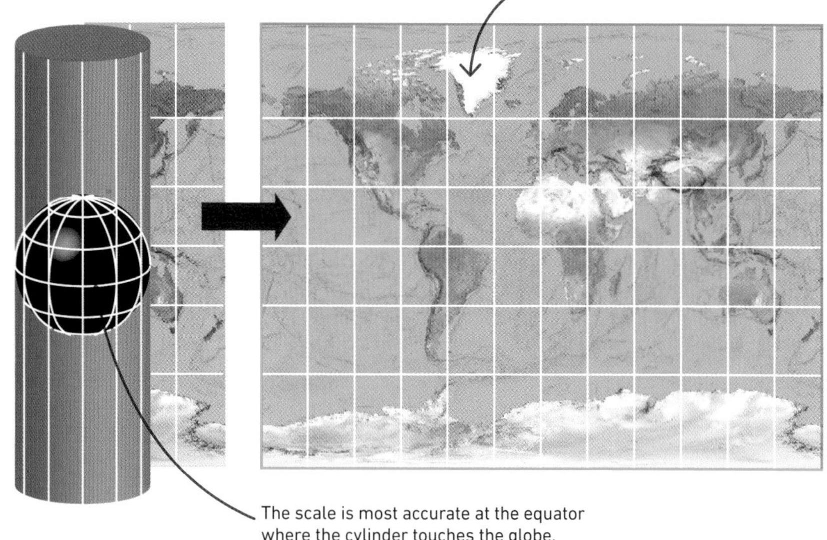
The points farthest away from the equator are the most distorted.
The scale is most accurate at the equator where the cylinder touches the globe.

Putting a map onto a cylinder ▷ This type of map is made by rolling a sheet of paper around the globe. Features are then projected, or plotted out, onto the cylinder using a mathematical formula.

Conic projections

These kind of maps are made by transferring an image of one-half of the globe onto a "cone," which rests on top of it. These projections show shapes almost as accurately as cylindrical projections and areas much better. They are often used for smaller areas of the globe or country maps.

Longitude lines

Latitude lines

This projection is most accurate at the point where the cone touches the globe.

The greatest distortion is at the point farthest away from the globe.

When a conic projection is cut from the tip to the bottom, it results in a flat, fan-shaped map.

A conic projection map is usually cut to give it a more usable shape.

◁△ Putting a map onto a cone The cone rests on top of the globe. The image of the section of globe it is resting on is then transferred onto the cone, which is cut to form a map.

Azimuthal projections

With this type of map, an image of the globe is transferred onto a flat, circular piece of paper that lies underneath it. The full map is usually round but may then be trimmed to a rectangular shape. This projection is very often used for viewing polar regions, but it can also be used to display individual continents.

The polar regions can be displayed without the distortion found in some other projections.

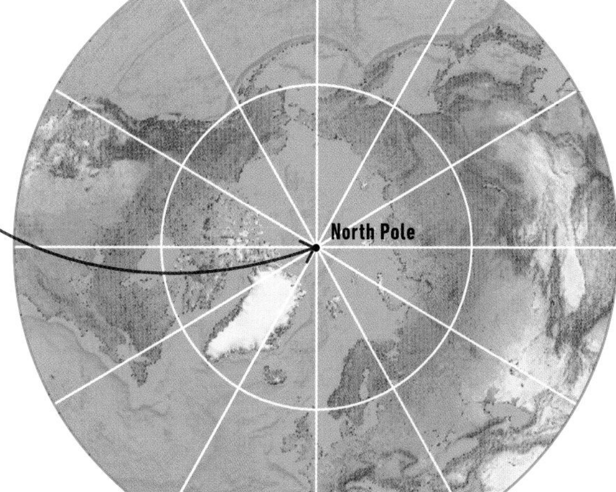
North Pole

Mapping around the poles

▽ **Distance from the center**
Every point on a map created using azimuthal projections is at a proportionally correct distance from its center.

The surface of the globe is displayed on a flat circle.

The scale is most accurate at the central point where the circle touches the globe.

Mercator and Peters projections

Invented in 1569 by Flemish mapmaker Gerardus Mercator, the Mercator projection keeps latitude and longitude lines at right angles to each other. It allows a straight course to be plotted on the map in any direction. The Peters projection, invented in 1967 by Arno Peters, is an equal area projection. Your choice of projection depends on what you want to use the map for.

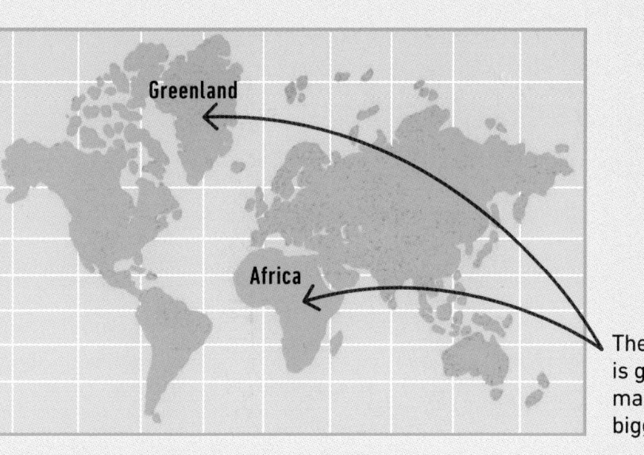
Greenland
Africa

The size of Greenland is greatly distorted, making it appear bigger than Africa.

△ **Mercator's projection**
The main disadvantage of the Mercator projection is that it makes those countries nearer the poles appear much larger than they really are.

The size of land masses is more accurate than in the Mercator projection, allowing for a fairer comparison.

These projections cannot be used for measuring distances when traveling along the lines of latitude.

△ **Peters projection**
Continents near the equator appear stretched in the Peters projection, but the relative sizes of countries are more accurate than in the Mercator projection.

Topographic maps

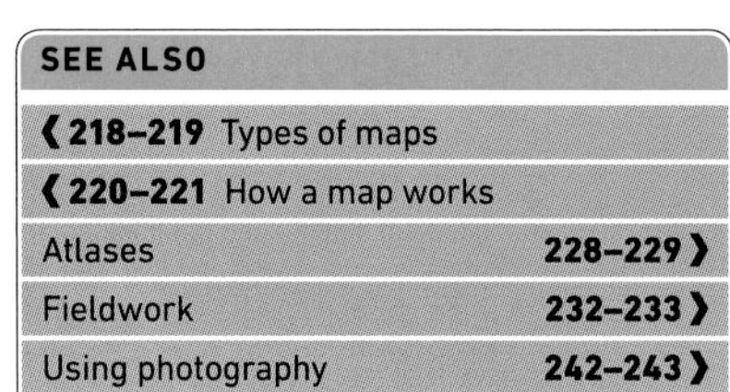
SEE ALSO
❮ 218–219 Types of maps
❮ 220–221 How a map works
Atlases 228–229 ❯
Fieldwork 232–233 ❯
Using photography 242–243 ❯

THESE ARE MAPS THAT PROVIDE ACCURATE DETAILS ABOUT THE EARTH'S SURFACE, IN PARTICULAR, TERRAIN AND HEIGHT.

The topography of a landscape is its physical shape. Topographic maps focus on a small area at a large scale and are ideal to use when hiking.

Reading maps

Topographic maps use symbols to show the physical features of the land. A key, or legend, explains what the symbols mean. Learning to read maps helps walkers or hikers explore the outdoors better and navigate through unfamiliar terrain. A skilled map reader may decipher details such as the types of woodlands or other features of the terrain.

Map features

The symbols shown on a map are identified in the key. Maps may show features such as roads, railroads, and buildings, types of vegetation such as woodlands and marsh, and the height of hills.

KEY

Roads and paths

Main road

Secondary road

Public footpath

Railroads

Double-track railroad

Railroad station

Leisure

Campsite

Viewpoint

Nature reserve

Walking trail

General features

Building

Library

School

Ferry route

Vegetation

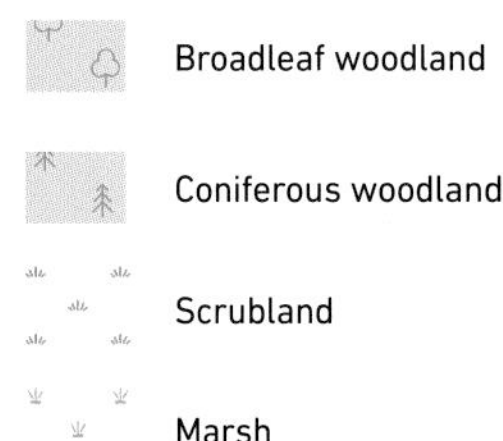
Broadleaf woodland
Coniferous woodland
Scrubland
Marsh

Terrain features

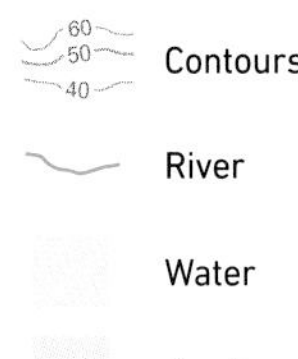
60
50
40
Contours
River
Water
Sand

Mud

A WALKING MAP

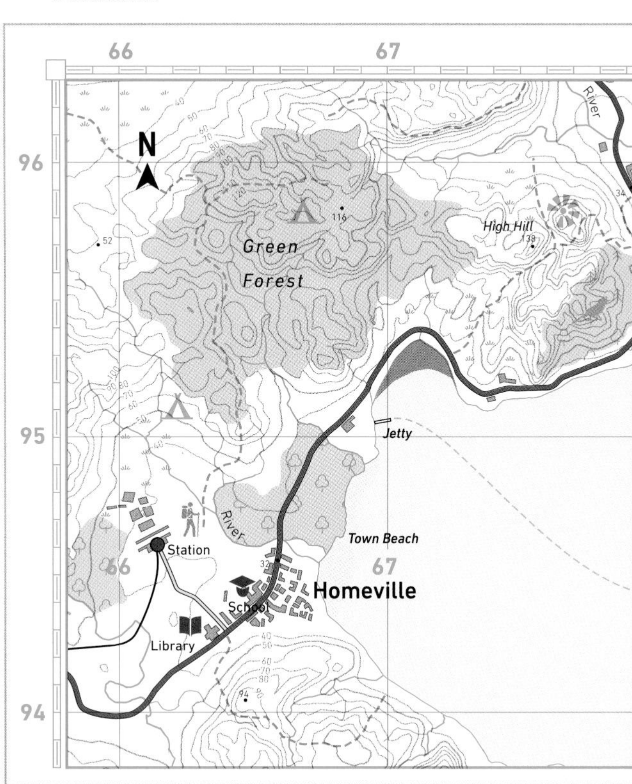
66
67
96
95
94
N
River
High Hill
Green
Forest
Jetty
Town Beach
Station
River
66
67
Homeville
School
Library

△ **Using a topographic map**
Understanding the landscape on a topographic map takes practice. Skilled map readers can see that the roads on this map hug the coast and run through valleys.

Finding north

The compass on a map often has two directions for north. True north is the North Pole. Magnetic north is where a magnetic compass points.

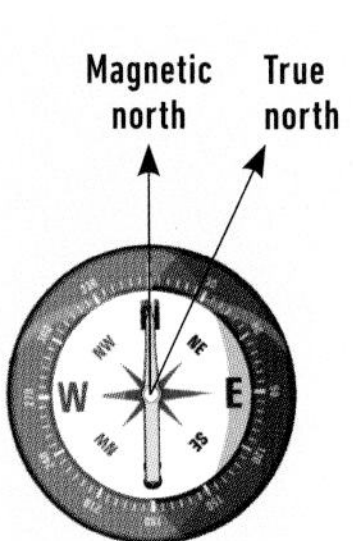
Magnetic north
True north
N
W
E

Grid references

A system of numbered squares called a grid is used to locate places accurately on a map. Eastings, or vertical lines, and northings, or horizontal lines, make up a square. Eastings increase in value from west to east and northings from south to north. When these lines cross, a grid reference is created.

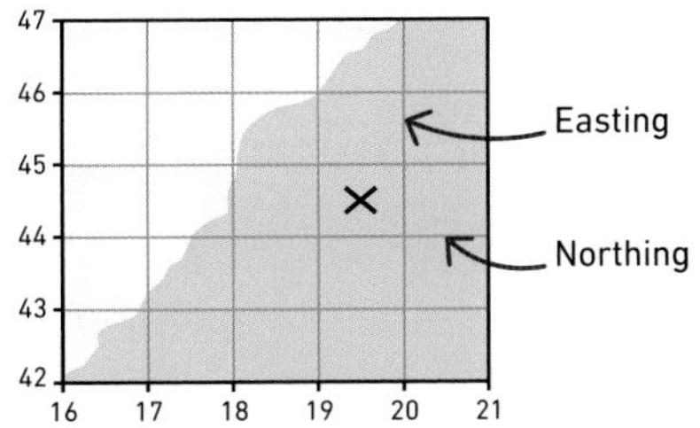
47
46
45
44
43
42
16
17
18
19
20
21
Easting
Northing

◁ **Locating places**
In the six-figure grid reference 195445, the first two figures are the easting, and the fourth and fifth indicate the northing. The third and sixth figures narrow it down to a more exact location inside the square.

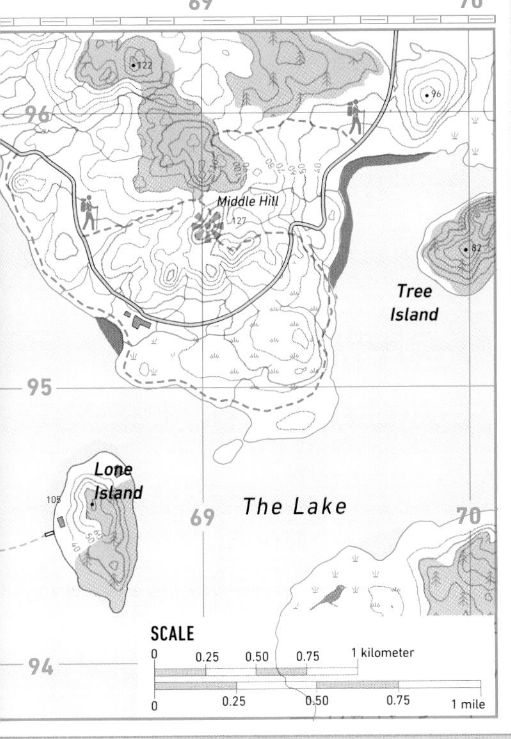
69
70
96
Middle Hill
127
Tree Island
95
Lone Island
105
69
The Lake
70
94
SCALE
0
0.25
0.50
0.75
1 kilometer
0
0.25
0.50
0.75
1 mile

Contour lines

Contours are lines that link places of equal height above sea level. They show how steep a terrain is and always form complete loops unless interrupted by a cliff. Lines that are closer together with increasing numbers represent a hill, and those with decreasing numbers show a crater, or valley. Contour interval is the difference in height between one contour and the next. The larger the map's scale, the smaller the interval.

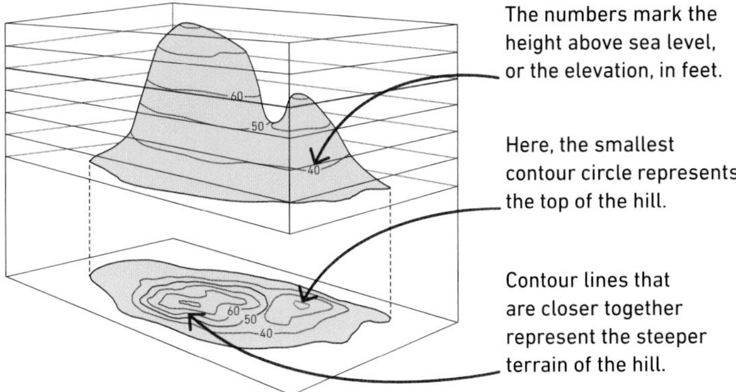
60
50
40
60
50
40
The numbers mark the height above sea level, or the elevation, in feet.
Here, the smallest contour circle represents the top of the hill.
Contour lines that are closer together represent the steeper terrain of the hill.

△ **How contours work**
Contours divide hills and valleys into horizontal layers, stacking one on top of the other. The contour line shows the outline of each layer.

Measuring distances

To figure out the shortest or the most direct route, it is important to be able to measure distances on a map. For example, 4 in on a map may represent 1 mile on the ground. A piece of string and a ruler are required.

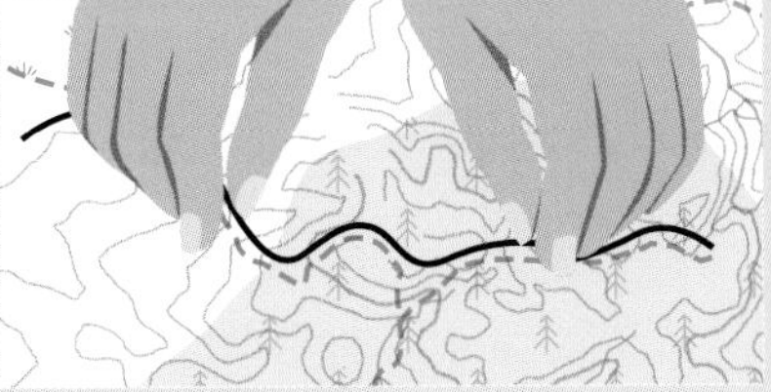

1 Using a string or solder wire
Lay a piece of string on the map, with one end on the starting place. Carefully shape it to follow the route on the map until you reach the destination.

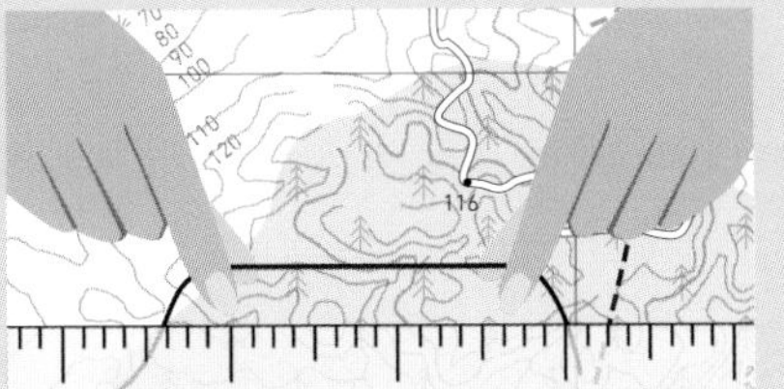

2 Using a ruler
Holding the string at the start and end points, measure the length on the ruler. Then convert the measurement using the scale bar.

Atlases

SEE ALSO
❮ 216–217 Geopolitics
❮ 218–219 Types of maps
❮ 220–221 How a map works
Geographic Information Systems (GIS) 230–231 ❯

THESE ARE COLLECTIONS OF TWO-DIMENSIONAL MAPS, OFTEN IN BOOK FORM, WHICH HAVE SHAPED PEOPLE'S IMAGINATIONS.

The maps in an atlas can cover the entire world, its main physical and political divisions, or a small region. Traditionally, atlases were printed on paper, but today many are in electronic form.

Early maps

The oldest known maps are clay tablets from Babylon dating back to 2,600 years ago. However, it was not until the end of the 1400s, when European sailors began to explore the world, that the first detailed world maps were made. The first atlas was created in 1570 by mapmaker Abraham Ortelius.

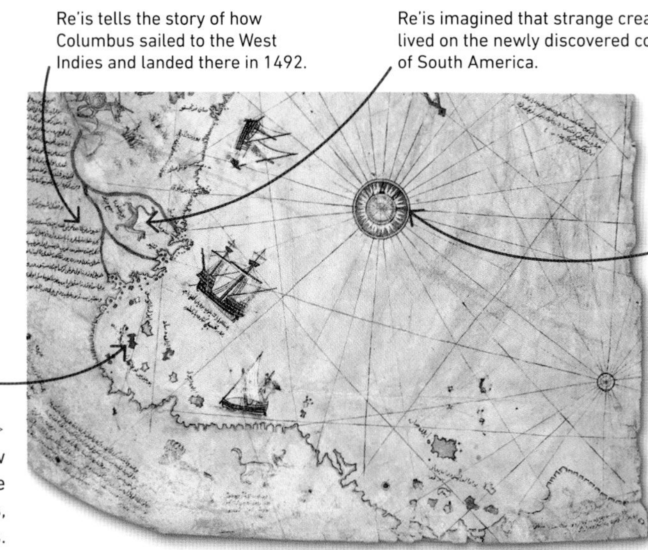

Re'is tells the story of how Columbus sailed to the West Indies and landed there in 1492.

Re'is imagined that strange creatures lived on the newly discovered continent of South America.

The map includes a compass rose with 32 lines showing winds and direction. The east-west line on the rose appears to demarcate the Tropic of Cancer.

The Falkland Islands are shown to be full of monsters.

Mapping expeditions ▷
Named after the Turkish admiral who drew it in 1513, the Piri Re'is map was one of the first to make use of seafarers' discoveries, but it had many gaps and inaccuracies.

Types of atlases

There are many types of atlases in use now. Some are small books that can fit in a pocket and have very simple maps. Others are giant books packed with several large and detailed maps. Atlases serve different purposes, too. Some of them provide an overview while others, such as road atlases, are very specific.

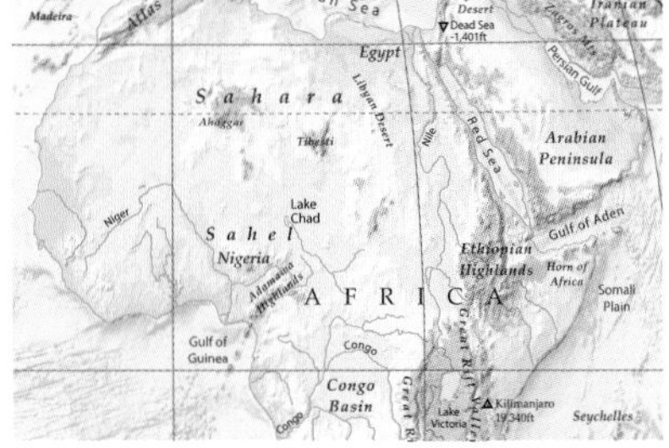
Egypt
Sahara
Arabian Peninsula
Red Sea
Gulf of Aden
Lake Chad
Sahel
Nigeria
AFRICA
Ethiopian Highlands
Somali Plain
Gulf of Guinea
Congo Basin
Seychelles

△ World atlas
World atlases have topographic maps that show physical features, such as mountains, as well as human features such as roads.

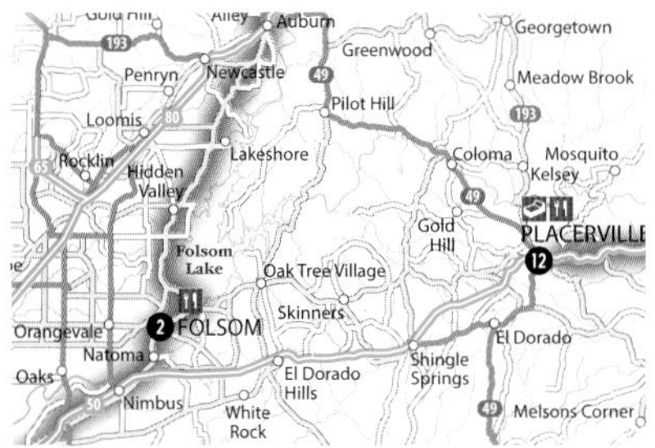
Auburn
Georgetown
Greenwood
Penryn
Newcastle
Meadow Brook
Pilot Hill
Loomis
Rocklin
Hidden Valley
Lakeshore
Coloma
Mosquito
Kelsey
Gold Hill
PLACERVILLE
Folsom Lake
Oak Tree Village
Skinners
Orangevale
FOLSOM
Natoma
El Dorado
Oaks
El Dorado Hills
Shingle Springs
Nimbus
White Rock
Melsons Corner

△ Road atlas
In the past, drivers used atlases showing roads to plan a route. Now they use their electronic satellite navigation maps instead.

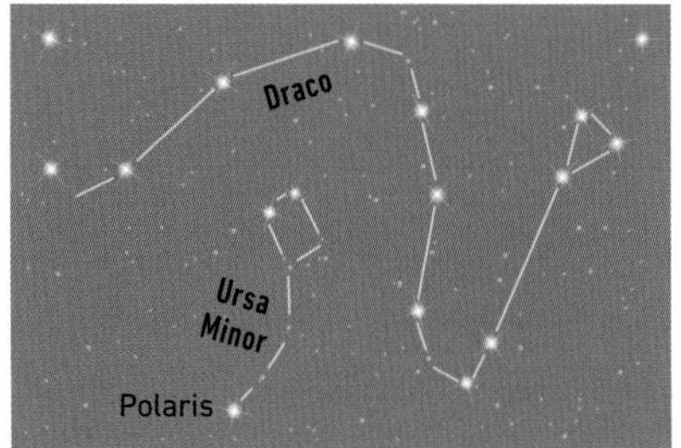
Draco
Ursa Minor
Polaris

△ Sky chart
The sky can also be mapped using an atlas. Sky atlases contain maps of the night sky showing stars and constellations.

Focus of maps

Maps in an atlas are rarely all at the same scale. An atlas might open with a map showing the entire world or a country at a small scale. Other maps might focus in on details in particular regions or even localities at larger scales. There may also be maps showing different political and physical features.

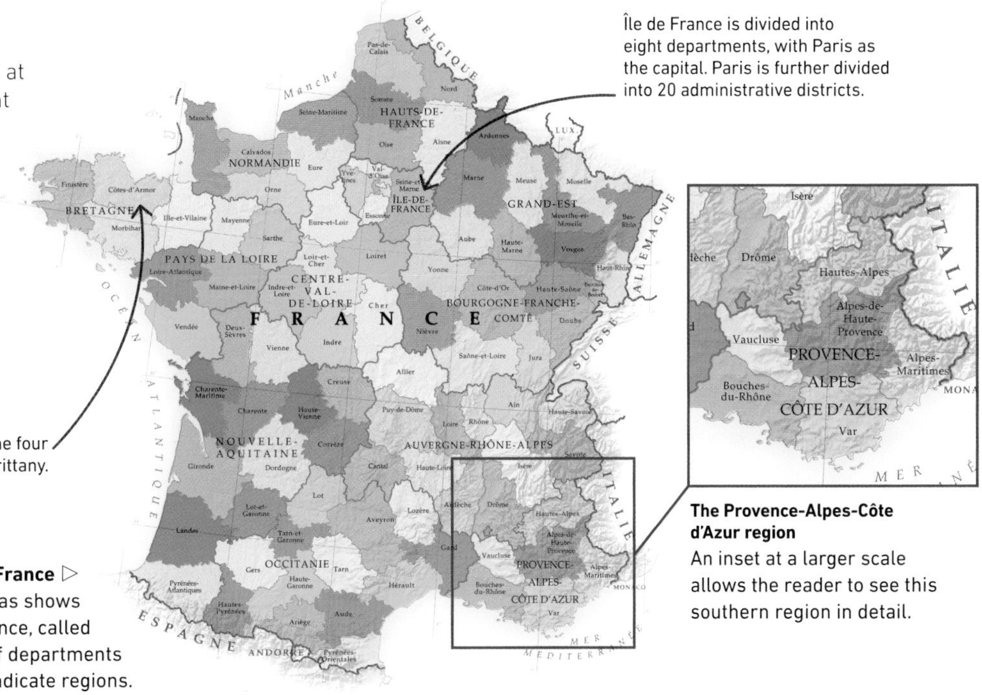

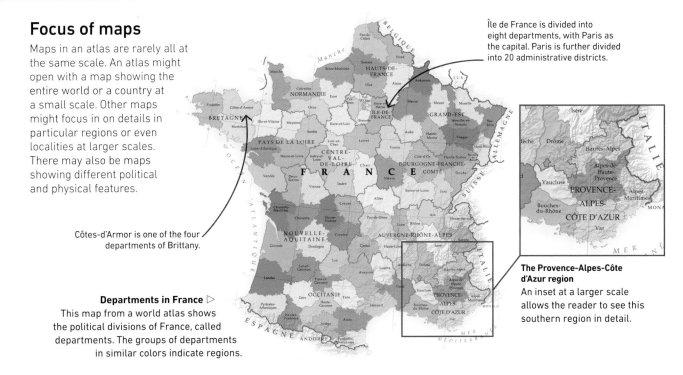

Île de France is divided into eight departments, with Paris as the capital. Paris is further divided into 20 administrative districts.

Côtes-d'Armor is one of the four departments of Brittany.

The Provence-Alpes-Côte d'Azur region
An inset at a larger scale allows the reader to see this southern region in detail.

Departments in France ▷
This map from a world atlas shows the political divisions of France, called departments. The groups of departments in similar colors indicate regions.

Using satellites

Satellites in space have transformed both how maps are made and how they are used. In the past, maps were made by taking measurements and surveys on the ground. Now they can be made and updated quickly using measurements and images from satellites, which are circling the Earth continuously.

> The largest **consumers** of **map data** in the future will be **mobile**.
> **Brian McClendon**, Vice President of Engineering, Google Maps, June 6, 2012

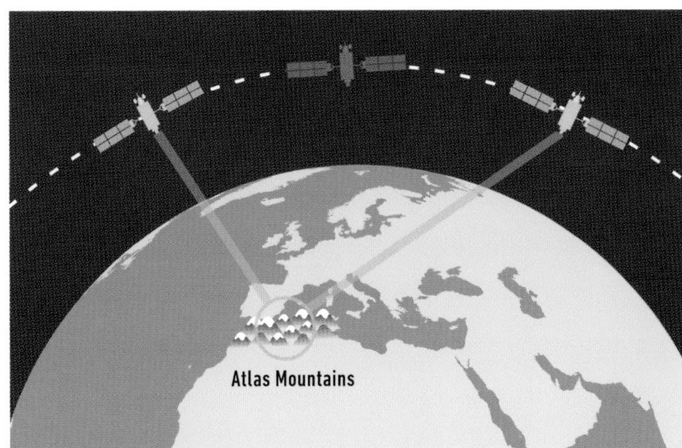
Atlas Mountains

△ **Mapping the Earth**
Linked measurements from satellites such as Radarsat-1 and TerraSAR-X help map the location and height of the physical features on the Earth's surface.

△ **View of Melbourne, Australia, from space**
Computer programs digitally combine satellite images, aerial photography, and Geographic Information System (GIS) data to produce 3-D images of the Earth.

Geographic Information Systems (GIS)

GIS USE SATELLITE LOCATIONS TO RECORD, STORE, AND PRESENT GEOGRAPHIC DATA ELECTRONICALLY.

SEE ALSO
❮ 214–215 Distance and relative distance
Fieldwork 232–233 ❯
Geographical inquiry 244–245 ❯

In GIS, every bit of data is tagged electronically with its precise location on Earth. This means a complete picture of data for every location can be stored on computers. This data can be accessed in many ways.

How GPS works

The Global Positioning System (GPS) uses a network of about 30 satellites circling the Earth in space. The GPS on a phone or other device is usually within radio reach of three or four satellites, each sending out a signal continuously, telling the phone where it is and when the signal was sent. The time taken for the signal to arrive indicates the distance of the phone from the satellite, and this information gives its exact location.

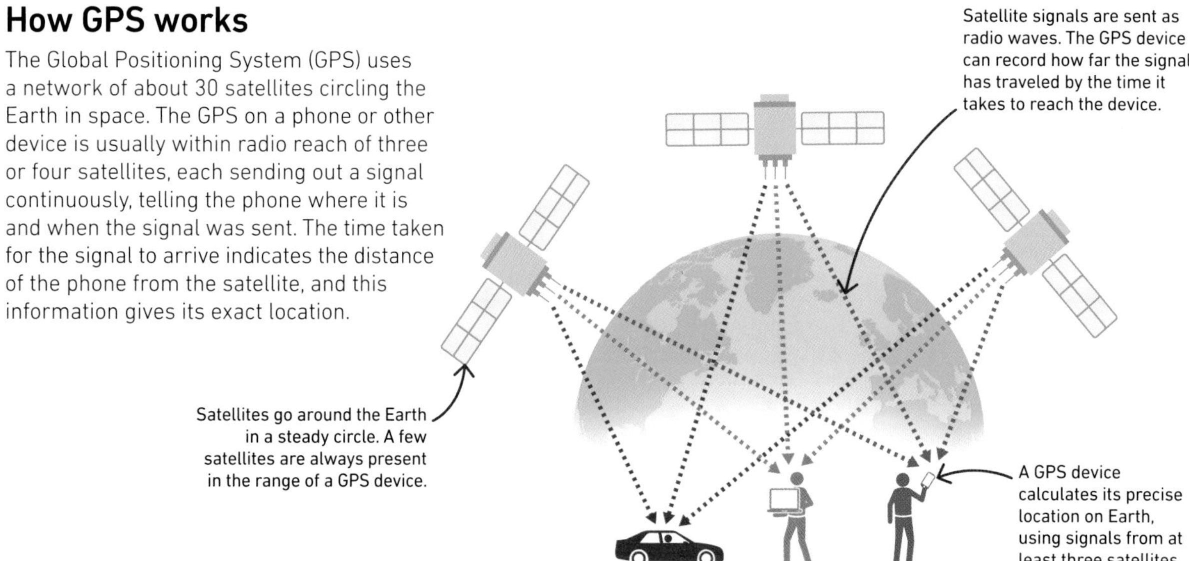
Satellite signals are sent as radio waves. The GPS device can record how far the signal has traveled by the time it takes to reach the device.
Satellites go around the Earth in a steady circle. A few satellites are always present in the range of a GPS device.
A GPS device calculates its precise location on Earth, using signals from at least three satellites.

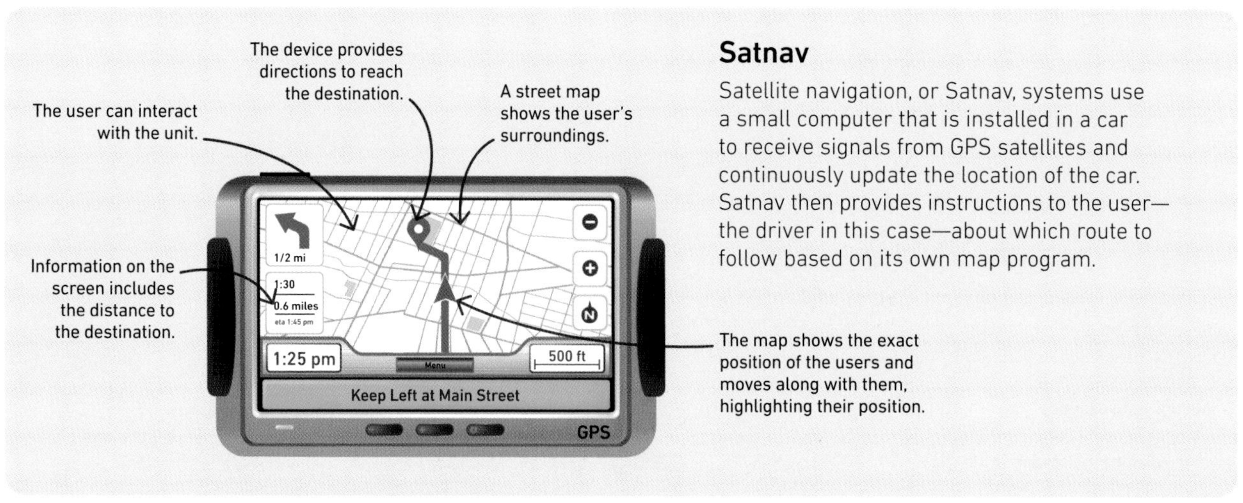
The device provides directions to reach the destination.
The user can interact with the unit.
A street map shows the user's surroundings.
Information on the screen includes the distance to the destination.
1/2 mi
1:30
0.6 miles
1:25 pm
Menu
500 ft
Keep Left at Main Street
GPS
The map shows the exact position of the users and moves along with them, highlighting their position.

Satnav

Satellite navigation, or Satnav, systems use a small computer that is installed in a car to receive signals from GPS satellites and continuously update the location of the car. Satnav then provides instructions to the user—the driver in this case—about which route to follow based on its own map program.

Building up the layers

GIS allows layers of computer maps to be created on screen to show different kinds of data for the same place. Data can be added to the maps and retrieved in different ways. Additional data may appear, for instance, when a place is clicked on in one layer.

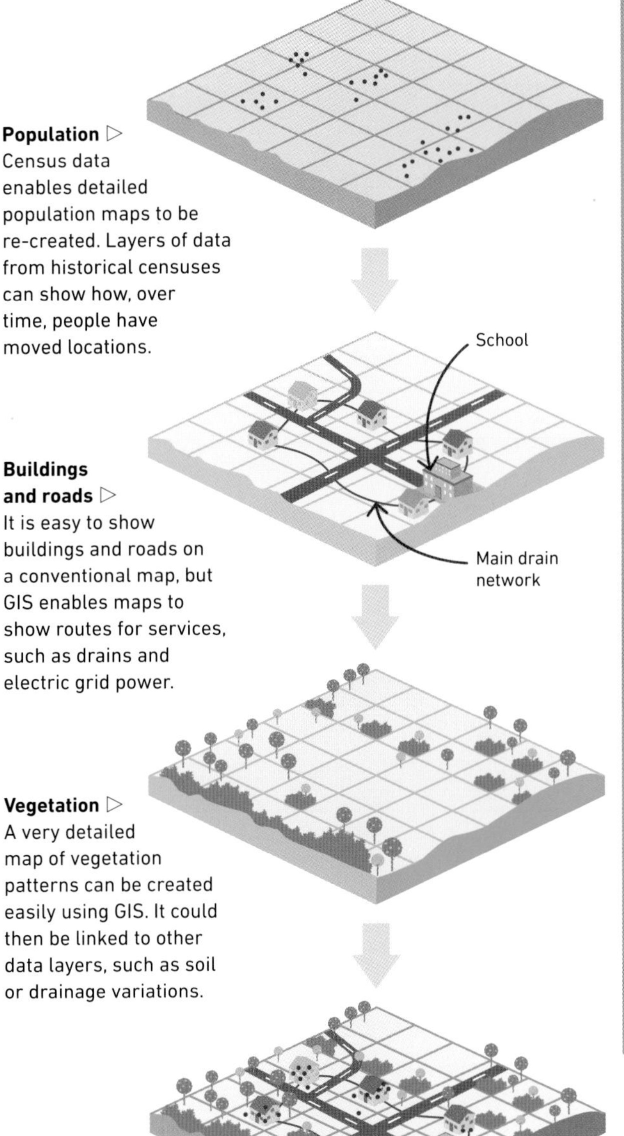
School
Main drain network

Population ▷
Census data enables detailed population maps to be re-created. Layers of data from historical censuses can show how, over time, people have moved locations.

Buildings and roads ▷
It is easy to show buildings and roads on a conventional map, but GIS enables maps to show routes for services, such as drains and electric grid power.

Vegetation ▷
A very detailed map of vegetation patterns can be created easily using GIS. It could then be linked to other data layers, such as soil or drainage variations.

Integrated data ▷
All the separate layers of data can be incorporated into the same computer program to integrate all the different data.

Finding links

GIS helps discover links between different kinds of data by integrating different layers. The integration of data sometimes helps confirm expected findings, such as certain plants being found mostly in damp ground, but can also reveal unexpected information.

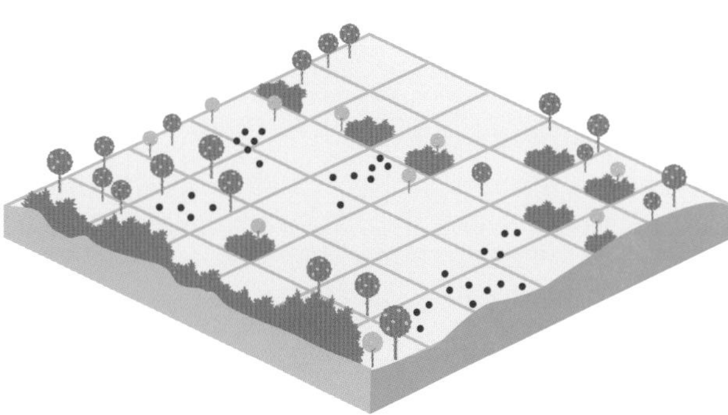

△ **Combining layers**
GIS confirms locational links quickly and easily. For example, integrating the vegetation and population layers shows in seconds that people mostly live away from woods.

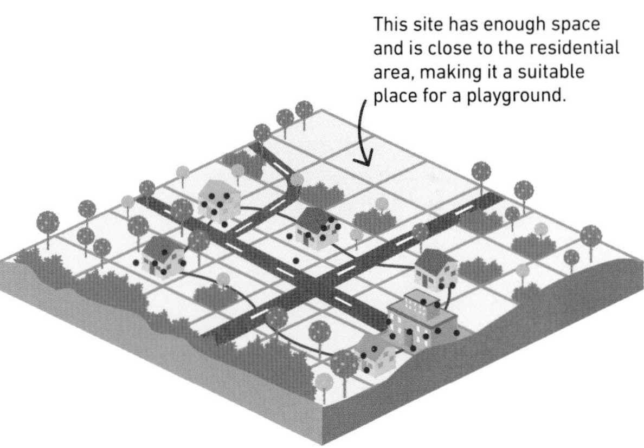
This site has enough space and is close to the residential area, making it a suitable place for a playground.

△ **Finding the right spot**
GIS helps plan locations for infrastructure projects when there are multiple factors to consider. For example, when creating a playground, GIS suggests an open location close to homes.

GIS is a technology that helps **reveal geographic patterns** and differences.

Fieldwork

SEE ALSO
Quantitative data 234–235 ❯
Qualitative data 236–238 ❯
Using photography 242–243 ❯
Geographical inquiry 244–245 ❯

FIELDWORK IS THE ACT OF MAKING DIRECT OBSERVATIONS IN THE REAL WORLD.

Working in the field often requires collecting data—by measuring phenomena such as rainfall or soil types, sketching or photographing landscape features, or interviewing people.

How to do fieldwork

Some fieldwork requires nothing more than going outside and making observations. This is called discovery fieldwork. However, most involves creating a structured plan, starting with clear aims, and following the stages below. Structured fieldwork is at the heart of geographical research.

1 Aims
It is important to pin down what you hope to achieve and keep the project at a realistic scale. It is unlikely that you will discover a new theory of mountain formation, but you might find out why rocks on a local hill are a certain shape.

2 Hypothesis
To shape a research project, you often begin with an idea about how or why something happens. Then, you look for evidence to support or contradict your idea. This is called hypothesis testing.

3 Primary and secondary data
There are two kinds of data you might choose to use. Primary data is data that you collect yourself, such as measurements or photographs. Secondary data is data that someone else has collected and made available.

4 Planning and risk assessment
Every step in fieldwork should be planned in advance in as much detail as possible. It is also crucial to carry out a risk assessment to avoid potentially dangerous situations, such as gathering data beside a rushing river.

What is fieldwork?

Fieldwork tends to be different for human and physical geography. For human geography, it often involves counting and observation to find patterns. For physical geography, it involves going out to collect samples and taking measurements.

Out in the field
Most fieldwork tends to be narrowly focused to give clear results, so it is likely to involve only one of the techniques shown here.

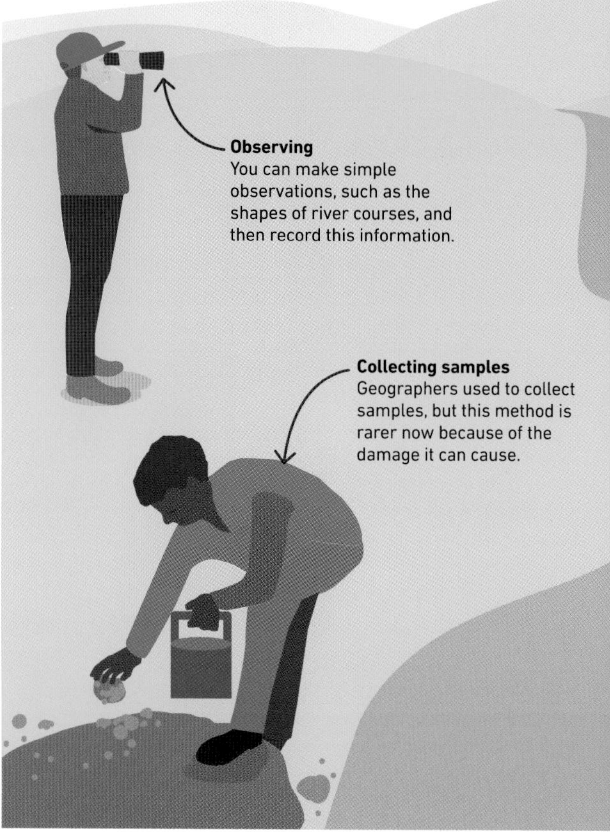
Observing
You can make simple observations, such as the shapes of river courses, and then record this information.
Collecting samples
Geographers used to collect samples, but this method is rarer now because of the damage it can cause.

Primary data is data you collect yourself, while **secondary data** has been collected by other researchers.

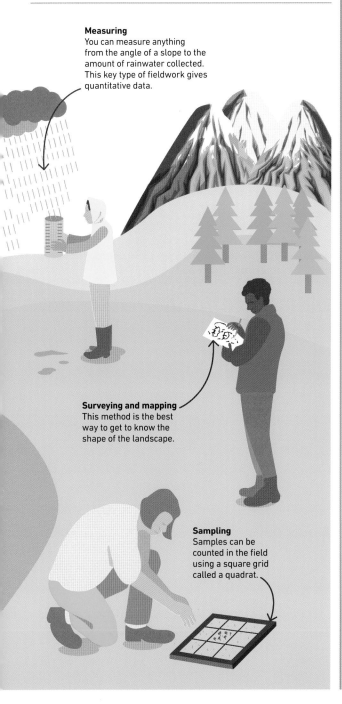

Measuring
You can measure anything from the angle of a slope to the amount of rainwater collected. This key type of fieldwork gives quantitative data.

Surveying and mapping
This method is the best way to get to know the shape of the landscape.

Sampling
Samples can be counted in the field using a square grid called a quadrat.

Fieldwork tool kit

Some items of equipment are needed for almost every kind of fieldwork. There are also special items you might need for particular research projects. When rock hunting, for example, you might need a hammer, a chisel, and a toothbrush for cleaning samples.

Cell phone or GPS device
Today, smartphones can replace many other tools, such as cameras and paper maps. A GPS device receives signals from a satellite and allows you to track your movement and locate your position.

Tablet
Tablets can replace or be used in addition to maps and reference books. Due to their larger screens, they can be useful for accessing databases in the field but may require internet access to do so.

Notebook and pencil
Unlike phones and tablets, notebooks and pencils do not rely on electricity to work. They can be used in the field to make notes, count with tally charts, and draw sketches of the landscape and its features on the go.

Knapsack
Large, sturdy knapsacks can be used to carry equipment from place to place, leaving your hands free to read maps or make notes. They can also be used to carry samples from the field to the lab.

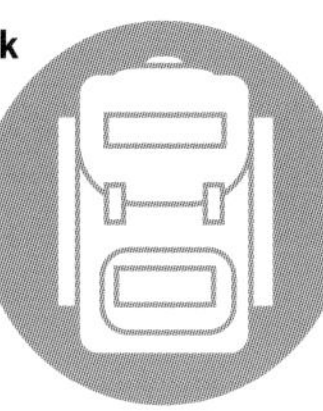

Good footwear
When in the countryside, you need practical footwear to protect your feet from mud, water, and plants. You will also be exposed to the weather and may need to wear a raincoat, sunscreen, or warm clothing.

Map and compass
A map and compass were once essential tools for navigating a landscape but are now often replaced by electronic devices. Paper maps are useful because they can be marked and written on.

Quantitative data

QUANTITATIVE DATA USES NUMBERS TO ANSWER QUESTIONS SUCH AS "HOW MANY?" OR "HOW OFTEN?".

A geographer may want to know how much it rains or how many people take the train each day. Since quantitative data is made up of numbers, it can be analyzed using math and put in statistical tables.

Two kinds of quantities

Quantitative data can be collected in two ways—counting or measuring. Data that you count might include the number of people who live in a village or animals that live in a nature reserve. Data you measure can be recorded on a scale and might include the height of your classmates or the amount of rain that falls over a week.

Counting or measuring ▷
The number of people with red or brown hair is counted. Weight, temperature, and length are measured.

Weight
11 lb

Temperature
90°F

Length
10 in

Number
3 red hair; 3 brown hair

Sampling

It is impossible to count every drop of rain that falls or every grain of sand on a beach, so geographers collect small amounts of data called samples. Samples give a snapshot of the big picture. Geographers use three ways to select the sample carefully to ensure it is fair: the samples may be random, systematic, or stratified.

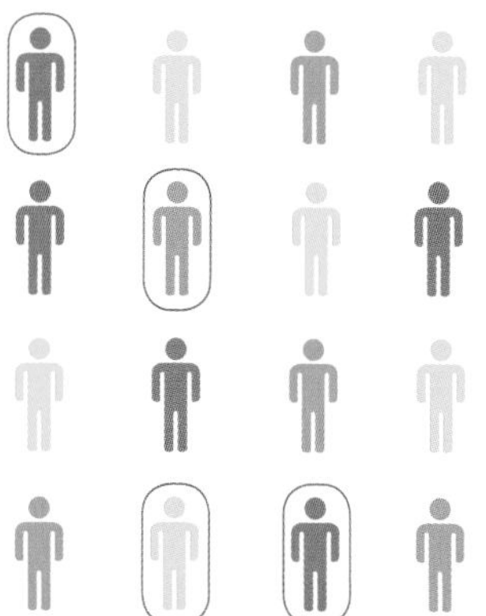

△ **Random sampling**
The geographer chooses a sample (of people to interview, for example) at random. This is the simplest form of sampling.

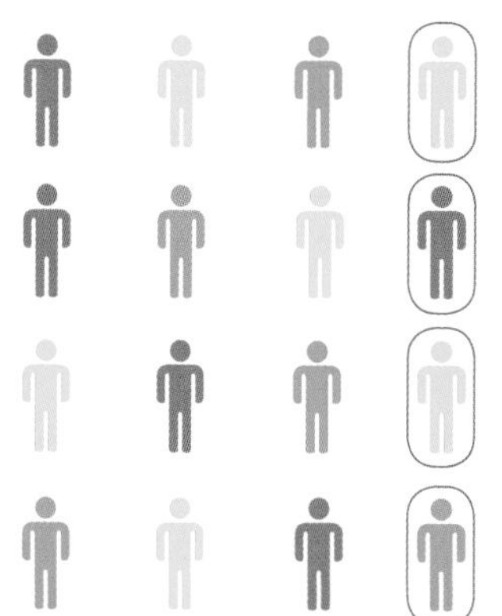

△ **Systematic sampling**
The geographer selects candidates at regular intervals. For example, every fourth person, or one stone every six feet.

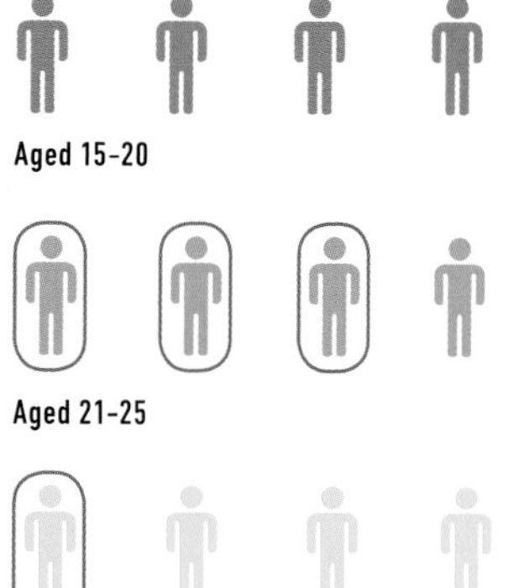
Aged 15-20
Aged 21-25
Aged 26-30

△ **Stratified sampling**
The geographer sorts the sample into groups (by age, for example) and selects a certain number from each.

Data types

Geographers work with different types of data. Nominal data puts things into groups (such as types of animals). Ordinal data puts things into order (of height, for example). Interval data allows geographers to note differences (in temperature, for instance). Ratio data allows geographers to compare numbers (of cars and buses in traffic, for example).

Nominal

Data may be nominal, meaning that it is grouped into categories. When collecting this kind of data, the geographer gives each group a label or category.

5 rabbits

3 ducks

1 fox

Types
Sorting animals into their different types gives us nominal data. Counting the number of each kind produces quantitative data.

Ordinal

When working with ordinal data, the order or ranking of the data is the most important information. The individual size of something is not relevant.

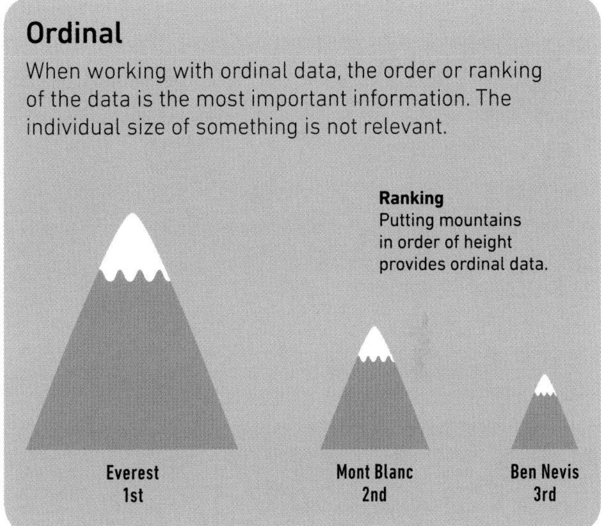
Everest
1st
Mont Blanc
2nd
Ben Nevis
3rd

Ranking
Putting mountains in order of height provides ordinal data.

Interval

This type of data is ordered, and the difference between each item is also recorded. For example, the temperature can be recorded each day and differences examined over the week.

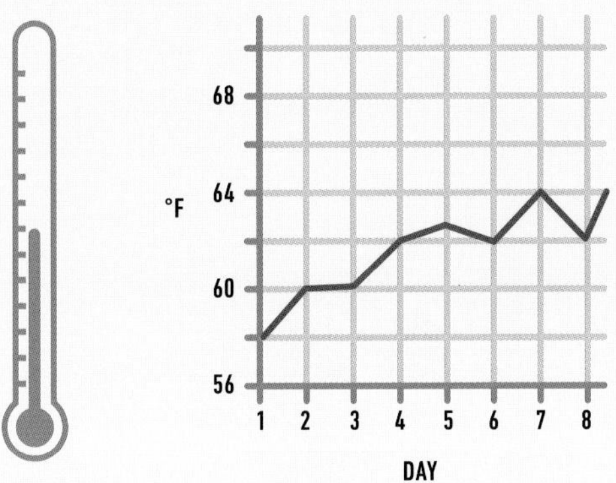
68
64
60
56
°F
1
2
3
4
5
6
7
8
DAY

Measuring
Data is recorded on an instrument such as a thermometer and any differences recorded.

Graph
Graphs are an effective way of displaying interval data. The squares on each axis display information about the interval.

Ratio data

Ratio data describes the relationship between two sets of numbers—say, the number of cars to buses in traffic. It is the most complete data, because it shows both the order and the interval.

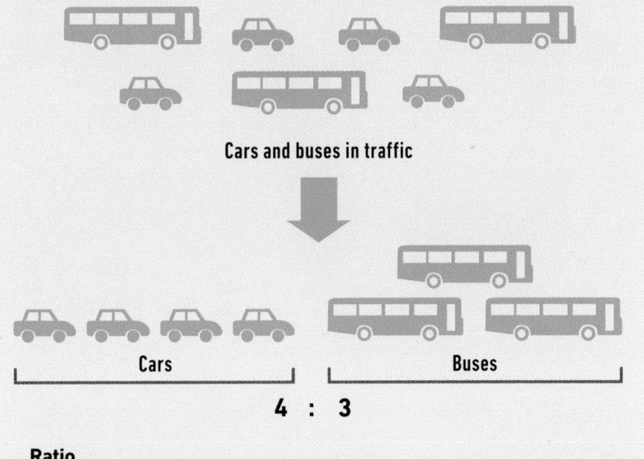
Cars and buses in traffic
Cars
Buses
4 : 3

Ratio
In this example, the traffic can be divided into cars and buses. The ratio shows there are four cars to three buses.

Qualitative data

SEE ALSO
❮ 232–233 Fieldwork
❮ 234–235 Quantitative data
Using photography 242–243 ❯
Geographical inquiry 244–245 ❯

QUALITATIVE DATA IS DATA THAT COMES IN THE FORM OF OBSERVATIONS, WORDS, AND IMAGES, RATHER THAN NUMBERS.

Qualitative data records or summarizes the quality of something, or describes what something is like. "The Sahara is a big, hot desert" is qualitative data. The Sahara's average temperature or rainfall is quantitative data. Both kinds of data are useful to geographers.

Forms of qualitative data

Qualitative data can take many forms. It can be a written description of landscape features (such as the shape of hills), photographs, sketches, or maps made in the field. Or it can be based on questionnaires that ask people about their habits and opinions—geographers are interested in what people *think* about places and landscapes.

"What do you think about...?"

"I think it's excellent."
"I can't stand it!"
"Too crowded."
"Great! I love it!"
"UGLY!"

Questionnaires

Questionnaires help reveal what people think, their characteristics, and their behavior. They can include "open questions" such as, "How do you feel about school?" or "closed questions" such as, "Do you feel safe at school—yes or no?" Offering a range of choices allows the questioner to turn qualitative data into quantitative data by recording the numbers of people who respond each way.

Most tourists find the hotels too expensive. Do you... ?
A strongly agree
B agree
C neither agree nor disagree
D disagree
E strongly disagree

How satisfied are you with local transportation?
How happy were you with the accommodations?

Feedback ▷
Interviewees may be invited to note how strongly they feel against a range of options, called a Likert scale.

Interviews

Questionnaires can be used with larger groups, but interviews often take place with a few individuals and provide more in-depth answers. Interviews need to be carefully planned, and answers properly recorded. The interviews can be "structured," in which each interviewee is asked the same questions in the same order, or "unstructured," in which questions come up spontaneously.

Ethics

The geographer must treat the interviewee with respect and always explain what data is being collected, why, and how it is going to be used. They should offer the interviewee anonymity and privacy whenever necessary.

Visual qualitative data

Geographers can collect visual evidence using sketches, maps, photos, and video. Comparing pictures from different places and times can reveal hidden connections and prove how things have changed.

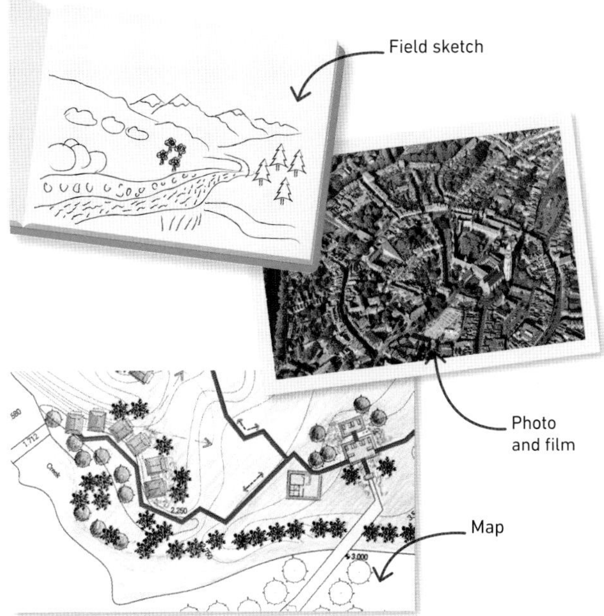
Field sketch
Photo and film
Map

Observation

Geographers may collect particular kinds of data, especially quantitative data, to prove a theory. However, sometimes it is valuable to simply observe, then analyze what you've seen afterward. This way geographers may learn something new. Observation might reveal that the course of a river is changing. This can lead to research that shows how and why this is happening. Observational data is often recorded as notes, sketches, or audio/video recordings.

Naturalistic
The simplest form of observation is just to watch and record, without any involvement. This could be an observation of how people interact in a street, for instance.

Participant
Your observations are made as an active participant, watching how your fellow travelers behave on a trip, for example. However, you could influence what happens.

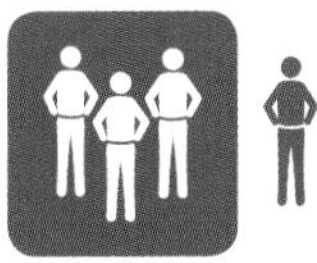

Controlled
In a controlled activity, participants' behavior is restricted—for example, where they go or how long they have to complete an activity while the observation takes place.

Sketches

Hand-drawn sketches, both pictures and maps, work in three ways. They can be a quick and effective way of making a note of the landscape and other features when out in the field. They can also be good visual aids for communicating information to others. When they are annotated, they are a useful summary of how the key features of the landscape are related.

Field sketches

Sketches made out in the field may be used to identify landforms such as bays, beaches, stacks, and headlands on the coast. They can also be used to record land use and other features of the human environment.

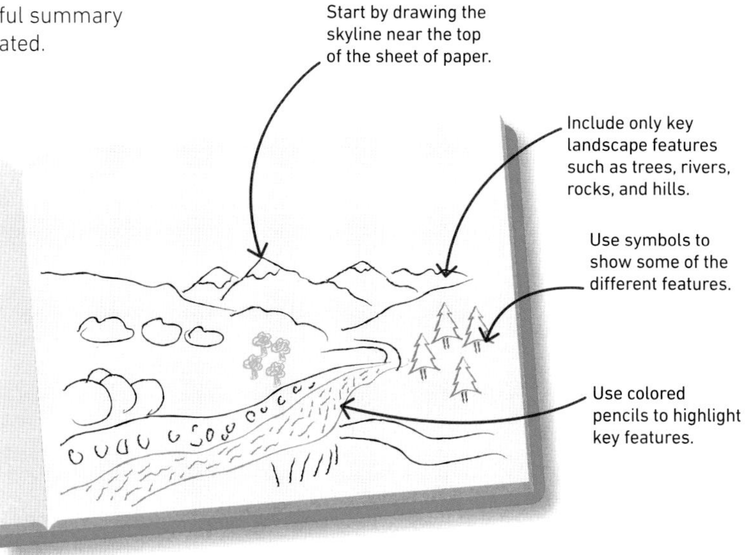
Start by drawing the skyline near the top of the sheet of paper.
Include only key landscape features such as trees, rivers, rocks, and hills.
Use symbols to show some of the different features.
Use colored pencils to highlight key features.

Making a sketch ▷
Use pencil so that you can make changes easily. It may help to fold your paper into quarters, flatten it, and then use the fold lines to help you place features.

Sketch maps

The basic principle of drawing a sketch map is to imagine you are drawing a bird's-eye view of the landscape below. So you have to remember well where things are in relation to each other—or go and find out for real by walking around. You might want to show the key features in the landscape around your campsite, for instance.

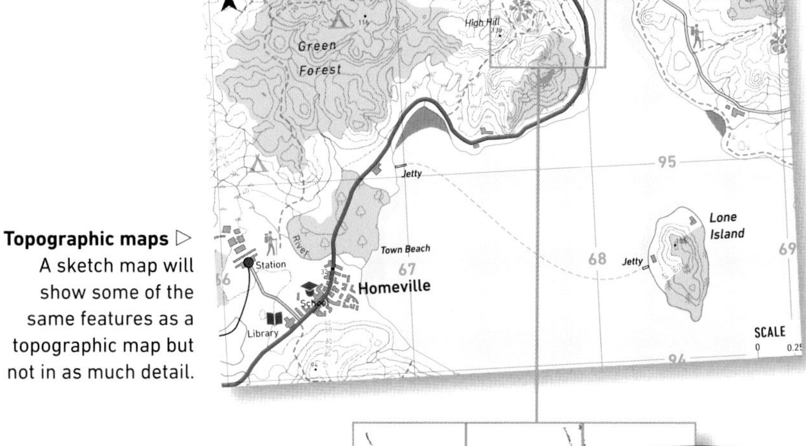
Hotel
High Hill
Green
Forest
95
Jetty
Lone
Island
River
Town Beach
67
Homeville
Station
School
Library
68
Jetty
69
SCALE

Topographic maps ▷
A sketch map will show some of the same features as a topographic map but not in as much detail.

Placing features
Check that each feature is in the right place in relation to the things around it.

Keep it simple
The features can be sketched roughly. It doesn't need to be a work of art!

Add a grid
Drawing a grid on your paper will help you get things in the right place.

Graphicacy

GRAPHICACY IS A WAY OF COMMUNICATING VISUALLY USING MAPS, CHARTS, AND GRAPHS.

SEE ALSO
❮ 234–235 Quantitative data
❮ 236–238 Qualitative data
Geographic inquiry 244–245 ❯

Data, often in numerical form, are just the starting point for the geographer. To begin to understand them, the numbers need to be presented and analyzed. Graphs are very effective tools for this.

Bar charts and histograms

Data can be shown in bar charts and histograms with columns of varying heights. The graph used depends on the kind of data. Bar charts show varying amounts of different but equal categories, while histograms show varying amounts of different and unequal categories.

Age (year)	Number of downloads in a month
10–15	12
16–18	15
19–25	28
26–29	12
>30	0

◁ **Table of data for histogram**
The age ranges here are not equal, so a histogram is best to show the frequency for each age range.

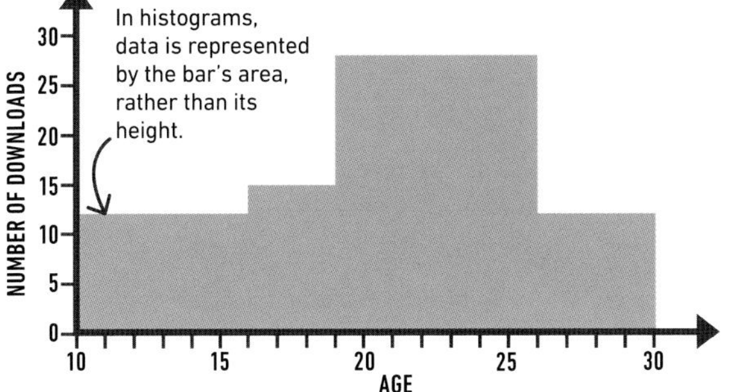
In histograms, data is represented by the bar's area, rather than its height.
NUMBER OF DOWNLOADS
0
5
10
15
20
25
30
10
15
20
25
30
AGE

△ **Histogram**
In a histogram, the columns touch. Here the heights show the varying numbers of downloads, but the bars are of different widths because the age ranges are not all of equal size.

Hobby	Boys	Girls
Reading	10	15
Sport	25	20
Computer games	20	10
Music	10	9
Collecting	5	10

◁ **Table of data for bar chart**
The data can first be recorded as a table. Figures in the table can then be used to create a bar chart.

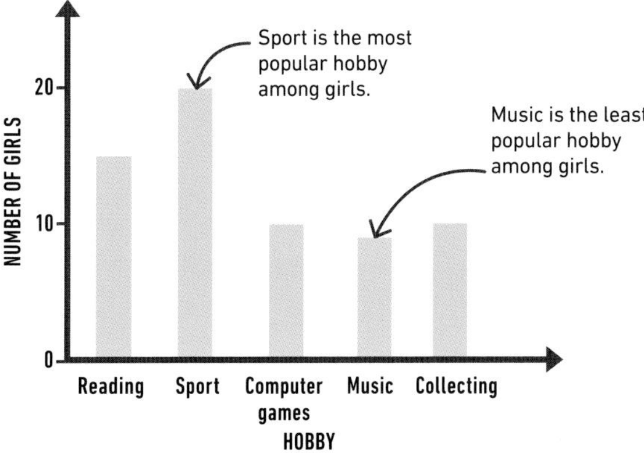
Sport is the most popular hobby among girls.
Music is the least popular hobby among girls.
NUMBER OF GIRLS
0
10
20
Reading
Sport
Computer games
Music
Collecting
HOBBY

△ **Bar chart**
In a bar chart, the columns are spaced apart, and column heights show the amount in each category. The order of the bars does not matter. This shows the number of girls who enjoy each hobby.

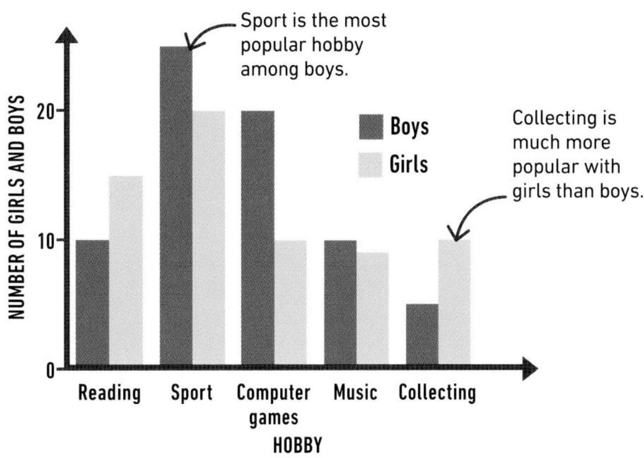
Sport is the most popular hobby among boys.
Boys
Girls
Collecting is much more popular with girls than boys.
NUMBER OF GIRLS AND BOYS
0
10
20
Reading
Sport
Computer games
Music
Collecting
HOBBY

△ **Compound (or multiple) bar chart**
Compound bar charts are two or more charts combined, with different-colored columns for the groups in each category. This chart shows how many girls and boys enjoy each hobby.

Line graph

A line graph is often used to show a trend over a number of days, months, or years, such as the number of hours of sunshine over a week. The graph has two scales (or axes) at right angles to each other. The data or changing values are plotted across the graph as a number of points at appropriate heights. These points are then linked by straight lines to show the trend.

Day	Monday	Tuesday	Wednesday	Thursday	Friday	Saturday	Sunday
Sunshine (hours)	12	9	10	4	5	8	11

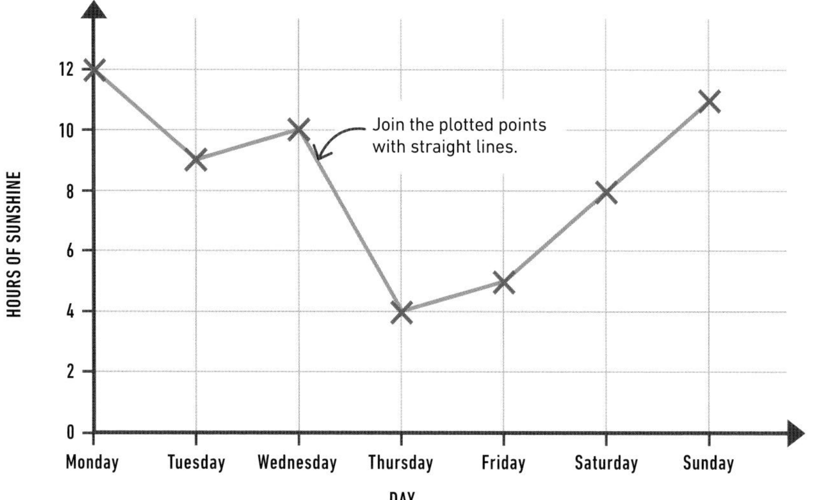
HOURS OF SUNSHINE
0
2
4
6
8
10
12
Join the plotted points with straight lines.
Monday
Tuesday
Wednesday
Thursday
Friday
Saturday
Sunday
DAY

Plotting the points ▷ The vertical axis shows the number of hours of sunshine and the horizontal axis lists the days of the week. A line is drawn with a ruler to link all the points and show how the hours vary.

Scattergraph

A scattergraph shows the relationship between two variables—values that vary—such as height or weight. The two are not necessarily connected, but by looking at a scattergraph, you may see that there is a link between them. This link is called correlation. It is important to remember that a correlation does not always mean one variable causes the other. Positive correlation means that one variable increases as the other does, while in negative correlation, one variable increases as the other decreases.

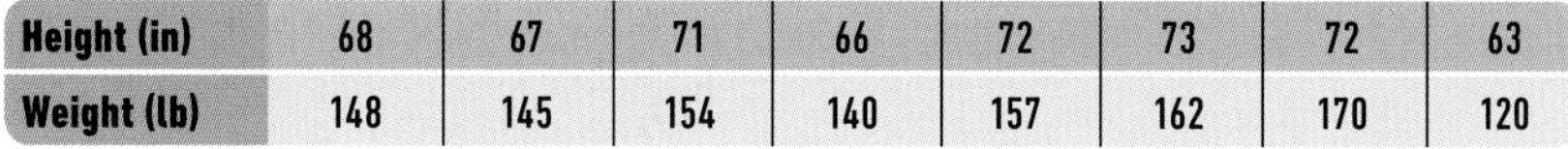

Height (in)	68	67	71	66	72	73	72	63
Weight (lb)	148	145	154	140	157	162	170	120

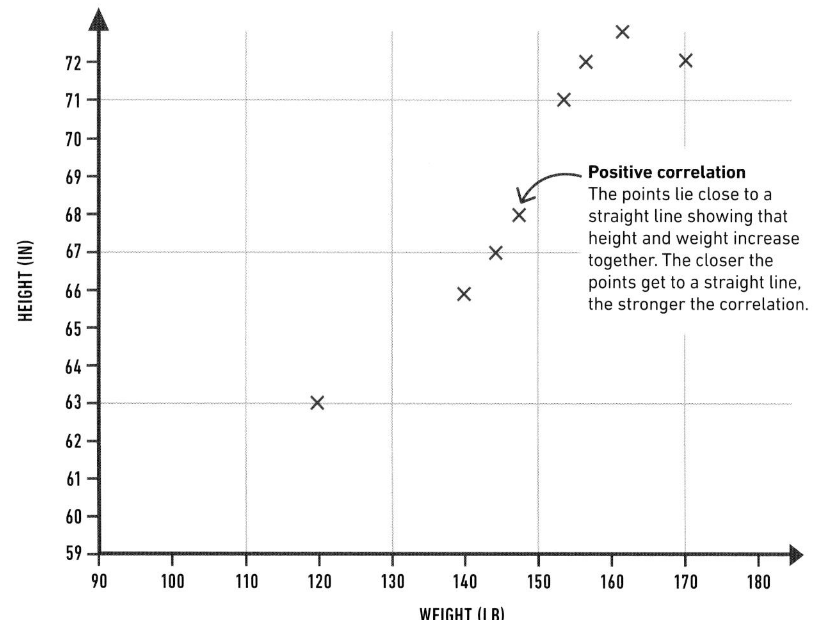
HEIGHT (IN)
59
60
61
62
63
64
65
66
67
68
69
70
71
72
Positive correlation
The points lie close to a straight line showing that height and weight increase together. The closer the points get to a straight line, the stronger the correlation.
90
100
110
120
130
140
150
160
170
180
WEIGHT (LB)

Height and weight ▷ A scattergraph can be used to plot different people's height and weight. The points are not in a straight line, but it is clear from the graph that weight increases with height.

Pie chart

A pie chart or pie graph is a circle divided into wedges like the slices of a pie. The size of the wedges shows the relative value of each category clearly, making it easy to compare them. Pie charts can be labeled, as they are here. Labels can point to the slices, or there might be a color-coded key. These are useful if the slices are too small to write on.

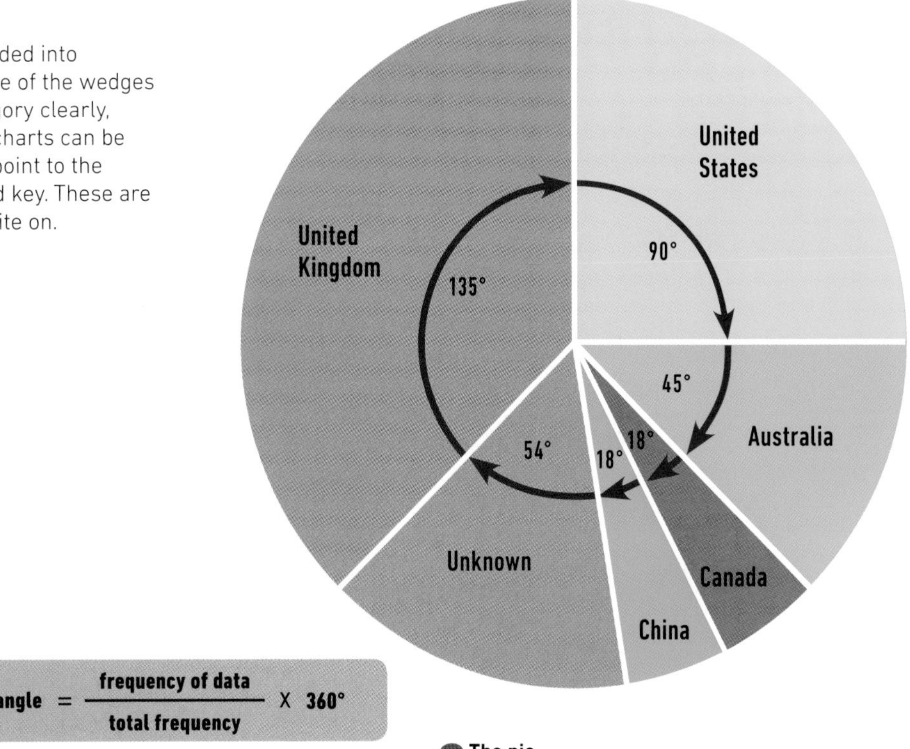
United States
90°
United Kingdom
135°
45°
Australia
54°
18°
18°
Unknown
Canada
China

Country of origin	Frequency of data
United Kingdom	375
United States	250
Australia	125
Canada	50
China	50
Unknown	150
TOTAL FREQUENCY	1,000

$$\text{angle} = \frac{\text{frequency of data}}{\text{total frequency}} \times 360°$$

1 The data
This table shows the number of hits on a website, split into the countries they came from.

2 The slice value
To find the angle for each slice of the pie chart, find the total for all groups and use it in this formula.

3 The pie
After drawing each slice on the circle, the pie chart can be labeled and color coded as necessary. As the angles add up to 360°, all of the slices fit into the circle exactly.

Proportional circles

Proportional circles are circles on a map that vary in size. The size of each circle shows the relative size of something where the circle is placed. Proportional circles are a good way of showing how measurements vary from place to place. They are often used to show varying crop yields in different locations, rainfall in different regions, or human factors such as the population of different towns. A city with a big population will have a large circle, while a village will have a small circle.

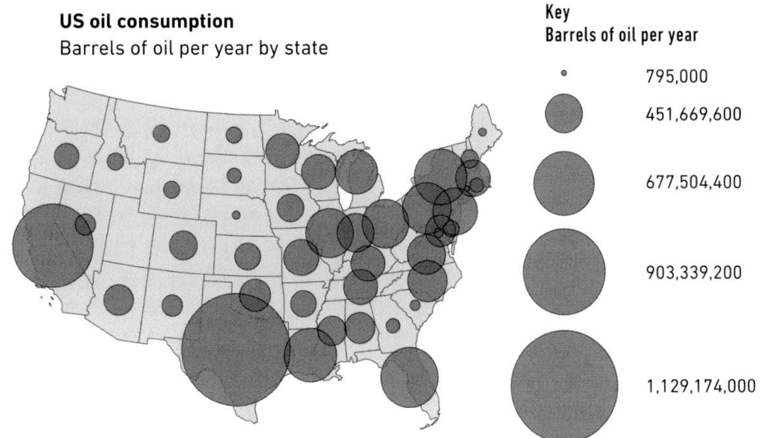
US oil consumption
Barrels of oil per year by state
Key
Barrels of oil per year
795,000
451,669,600
677,504,400
903,339,200
1,129,174,000

△ **Oil consumption**
This map shows the consumption of oil in different US states using five different-sized circles. The value of the circles is indicated in the key. It shows that the northeast, California, and Texas have the highest consumption of oil.

Using photography

PHOTOGRAPHS AND ELECTRONIC IMAGES ARE AN IMPORTANT WAY OF COLLECTING AND RECORDING DATA.

SEE ALSO
❮ 218–219 Types of maps
❮ 232–233 Fieldwork
❮ 236–238 Qualitative data
Geographical inquiry 244–245 ❯

Images can often reveal patterns and structures that are unclear or invisible any other way. This could be changes in the landscape over time or features the human eye cannot see, such as heat.

Interpreting photos

Geographers can use photographs in many ways. Researchers look for features in the landscape and photograph these in different locations, for instance. They might compare photos of the stages of different volcanic eruptions or the shapes of river valleys. Geographers recognize that photographs can also be used to deceive. What the photographer chooses not to show (to crop out) can give a misleading view. Interpreting photographs has to be done carefully.

◁ **Fog patterns**
Traffic pollution creates fog in cities, and photographs allow scientists to make direct visual comparisons of the effects in different cities.

Peak shapes ▷
Glaciation creates mountains with sharp peaks. Photographs let geographers study the shapes and compare the angles of the slopes.

Comparing photos

A photograph records an exact moment in time, so geographers can compare two or more different photos to study how things change. Looking at historic photographs can reveal long-term changes that might otherwise be missed. Geographers comparing photographs of a river might see how its course has changed over time. Photographs might be taken at different times of year to reveal seasonal changes or at different times of day. Time-lapse sequences combine shots taken at regular intervals to reveal changes that happen too slowly to be observed in real time.

Iceberg Glacier in 1939

Iceberg Glacier in 2008

△ **Shrinking glacier**
Before and after photographs provide definitive proof that many of the world's glaciers have shrunk over the last half century, mainly due to climate change.

Times Square during the day

Times Square during the night

△ **Changing city**
A time-lapse sequence of a particular location might reveal changes in how that space is used throughout the day.

Photography as data

Photographs from the air or from satellites in space have become very important for geographers, providing an instant overview of vast areas of the Earth's surface. There are now many different ways of creating images of the Earth, not just photographs. Images can be made with sound and radio pulses and also with scanners that record radiation that our eyes can't see.

△ **Aerial photography**
Vast areas can be mapped quickly with aerial photographs, but a normal photograph will provide only a flat picture. Stereoscopy mimics the way our eyes work by taking two pictures at the same time from slightly different viewpoints and overlapping them to give a 3-D view, which reveals the different heights of objects and buildings on the ground.

This satellite image shows mainland Italy, Sicily, Sardinia, and Corsica as seen from space.

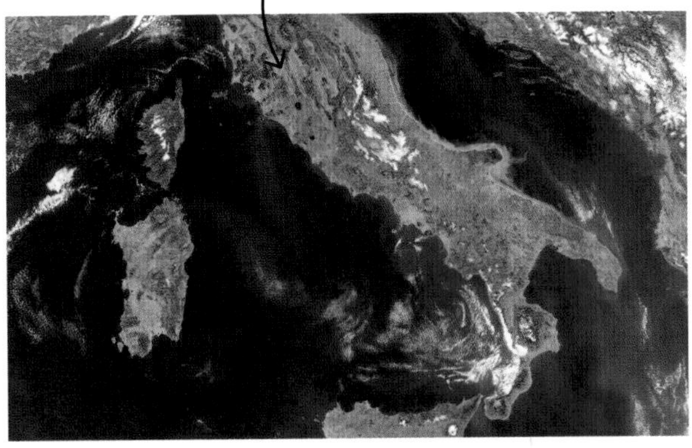

△ **Satellite**
Satellites orbiting the Earth can give an instant view of large areas of its surface. There are many kinds of satellites, but for geographers, the most important are weather satellites and Landsat, which circles the Earth 14 times a day recording the landscape in detail.

Infrared imaging here shows lava flows around Mount Etna in black.

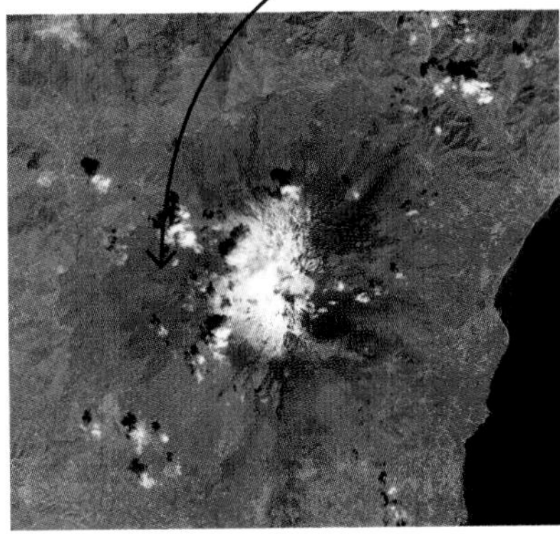

△ **False color**
Some types of light are invisible to the human eye, but they can be picked up by special equipment. What they record can then be shown in "false" colors. One of the most common kinds of false color images records and visualizes infrared radiation.

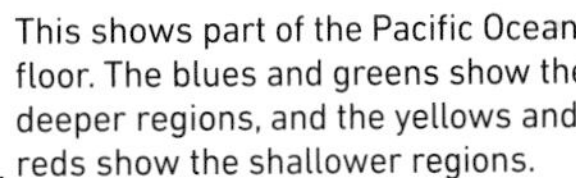

This shows part of the Pacific Ocean floor. The blues and greens show the deeper regions, and the yellows and reds show the shallower regions.

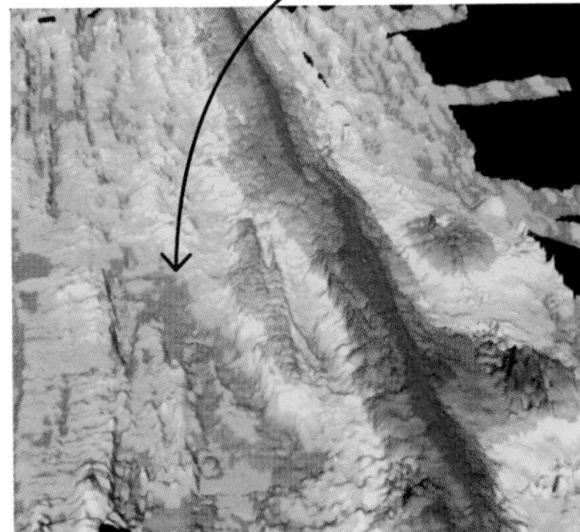

△ **Sonar**
Sonar uses sound to make images. Pulses of sound are sent out, and the image is made by recording the pattern made by the pulses as they bounce back off different surfaces. Sonar has been the main way scientists have mapped the ocean bed.

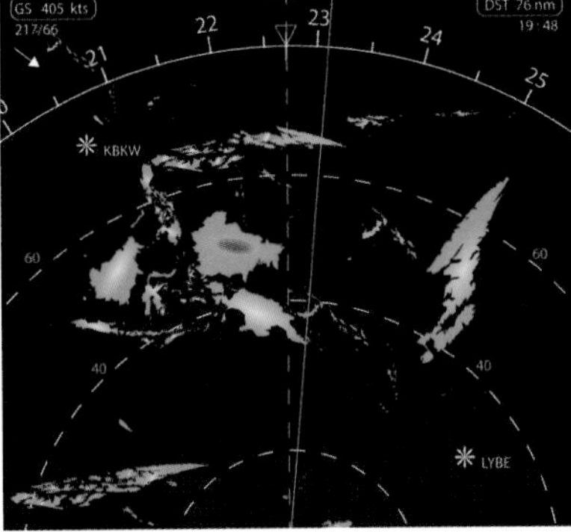
GS 405 kts
217/66
HGD
DST 76 nm
19:48
21
22
23
24
25
KBKW
60
60
40
40
LYBE

△ **Radar**
Radar works by sending out pulses of radio waves and recording the way they bounce back. Radar can be used to detect ships and aircraft, and meteorologists use it to track rainfall, thunderstorms, and hurricanes.

Geographical inquiry

GEOGRAPHICAL INQUIRY IS AN ACTIVE PROCESS THAT DEEPENS AND EXTENDS OUR UNDERSTANDING.

Geographical inquiry helps us evaluate information about people, places, and environments. It also enables us to discover and understand relationships between different issues and ideas and to make and share our conclusions.

SEE ALSO
❮ **232–233** Fieldwork
❮ **234–235** Quantitative data
❮ **236–238** Qualitative data
❮ **239–241** Graphicacy

Choosing, evaluating, and analyzing information

Information can be taken from many different sources. It needs to be selected carefully and evaluated for its reliability and any potential bias before it can be critically analyzed and interpreted.

CHOOSE

Don't just use the first information you find. Search for a reliable source that has appropriate, up-to-date, and useful data for addressing the topic. You should check when the data was published and see if a more recent study has been done.

EVALUATE

Evaluate the source. Can you trust its accuracy? Does it have the information you need? You should consider whether the source might be biased in favor of a particular outcome. The UN website, for example, is likely to be more objective than a single blogger.

ANALYZE

Study the data closely but objectively to assess what you can learn from it. Use data that has gone through a review process or been subject to a high level of scrutiny. For example, if it has been published around the world, it is usually more reliable.

Data handling and communication

Different data can be compiled and presented in different ways. Quite often, the way it is compiled and presented has a big influence on how clear a picture it gives, so it is important to study the data closely before deciding which is the best way to present it.

Hobby	Boys	Girls	Total
Reading	10	15	25
Sport	25	20	45
Games	20	10	30
Music	10	9	19
Collecting	5	10	15

Each row shows a particular hobby and the number of boys and girls who enjoy it.

△ **Tables**
Tables are a very simple way of organizing number data, such as when counting things in different categories.

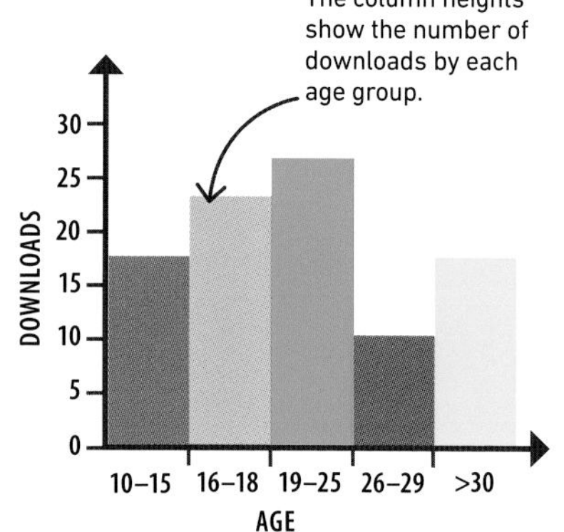
The column heights show the number of downloads by each age group.
DOWNLOADS
0
5
10
15
20
25
30
10–15
16–18
19–25
26–29
>30
AGE

△ **Graphs and charts**
Graphs and charts are very effective ways of presenting number data visually so its significance is clear at a glance.

Problem-solving

Geographical inquiry can be much more than just finding interesting information. It can be a way to solve problems in the world that we live in, such as how to balance the need for land to grow food with the need to improve transportation links and provide new housing.

1 Ask
Define the problem clearly so that you can be sure you are asking the right questions.

2 Acquire
Find and collect all of the geographic data you will need to understand the problem.

3 Organize
Organize the data in a way that everyone will be able to understand.

4 Analyze
Study the data you have collected carefully in order to draw unbiased conclusions.

5 Act
Draw up a plan based on your analyzes and then put your plan into action.

△ **Key steps**
To be effective at problem-solving, inquiry must be systematic. It is best to work through these steps carefully.

Synoptic skills

It is very easy for geographical inquiry to become focused on just one narrow track, but it is important to see the bigger picture. Synoptic skills help you do this by identifying the links between different topics.

▽ **Synoptic balance**
Many geographic problems focus on how to balance people's needs with the needs of the environment.

Human
Human concerns might include the struggle to provide for growing populations and the problem of poverty.

Physical
Environmental issues might include climate change, scarcity of water resources, and earthquakes and volcanic eruptions.

Synoptic conclusion
Geographers study the issues and data from both sides in order to decide upon a balanced solution.

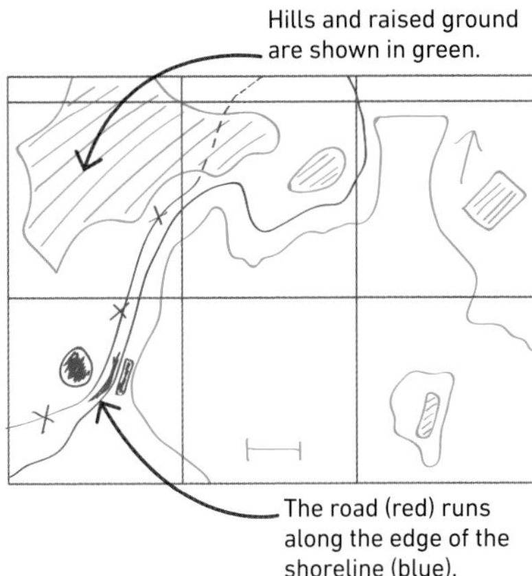
Hills and raised ground are shown in green.
The road (red) runs along the edge of the shoreline (blue).

△ **Drawing maps**
Maps can show geographic patterns, such as areas of vegetation in the landscape, or the routes taken by types of transportation.

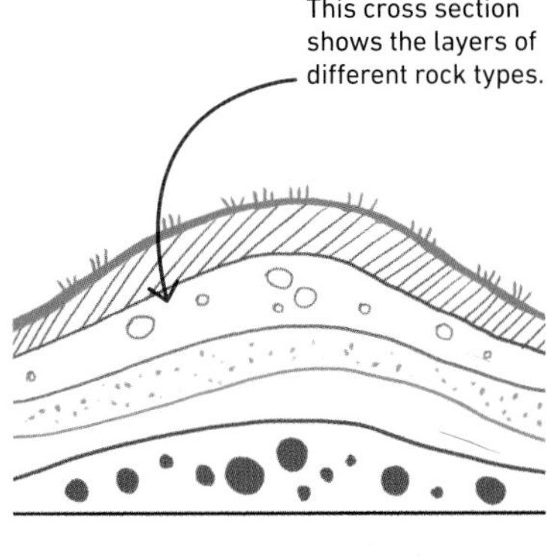
This cross section shows the layers of different rock types.

△ **Drawing cross-sections**
Cross sections can be a vertical picture of the slope shown by contours on a map, or a "slice" showing rock beds underground.

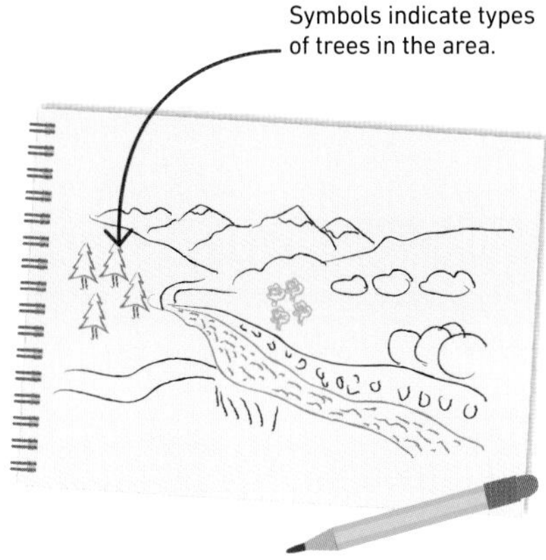
Symbols indicate types of trees in the area.

△ **Field sketches**
These can be used to pick out particular features in the landscape or reveal relationships, such as between slope angle and farm use.

Glossary

absolute humidity
The weight of water vapor in a certain volume of air.

accretion
When material is added to a tectonic plate or area of land. The opposite of erosion.

acid rain
Any precipitation that is polluted by chemicals, such as sulfur dioxide or nitrogen oxides, which in turn pollutes or destroys habitats where it falls.

air mass
A huge volume of air with the same humidity and temperature throughout.

algae
A type of simple plant without leaves or stems that grows in water.

Anthropocene
The epoch during which human activity has become the dominant influence on the Earth's environment.

anticyclone
A zone of high pressure that brings stable weather. Also known as a high.

aquifer
An underground layer of rock that can hold and transmit water.

arête
A mountain ridge separating two cirques.

atmosphere
The layer of gases, mainly nitrogen and oxygen, surrounding the Earth.

axial tilt
The changing angle of Earth's axis of rotation. It changes once every 41,000 years.

axis
The imaginary line that the Earth rotates around; also the term for a line showing the scale of measurement on a graph.

batholith
An igneous extrusion that is 60 miles (100 km) or more across in size, usually deep underground.

biodiversity
The variety of different species or organisms in a particular environment.

biogenic rock
A type of sedimentary rock formed from the remains of, or the activity of, plants and animals.

biogeography
The study of biomes, looking at the distribution of plants and animals in particular parts of the Earth.

biome
An area of land or water classified by its natural features and the organisms living within it.

boreal
Describing things from between the Arctic and temperate zones in the northern hemisphere.

broad-leaved
One of the two main types of tree. It has leaves that are flat, and it usually produces fruit.

canyon
A deep valley that is wider than a gorge.

carbon dioxide (CO_2)
A gas produced by the respiration of organisms, fermentation of dead matter, and burning of fuels. A major greenhouse gas.

carbon footprint
A measure of an individual or organization's contributions to carbon dioxide emissions.

cirque
A steep and round hollow made by a glacier.

climate
The average conditions in a particular area over a prolonged period of time. Climate refers to the general atmospheric conditions of a large area or the entire Earth.

climate change
A shift in the Earth's weather patterns, such as average temperatures and amount of rainfall. Climate change can be caused by human factors contributing to the greenhouse effect.

climate zone
An area of the Earth with shared climate characteristics such as average temperature and rainfall.

condensation
When water vapor becomes liquid.

conifer
One of the two main types of tree. It has waxy needles and its seeds are contained in cones.

convection
The transfer of heat within a liquid, gas, or ductile rock, causing its particles to move. In the Earth's atmosphere, for example, warm air rises while cool air sinks.

Coriolis effect
Force affected by Earth's rotation that appears to make winds bend to the west and to the equator.

counterurbanization
A trend of moving away from urban centers and into suburban areas, usually for a better quality of life.

crust (the Earth's)
The top layer of the Earth. There are two types: continental crust, which is thicker but less dense, and forms continents; and the thinner, denser oceanic crust at the bottom of the ocean.

cyclone
A zone of low pressure that forms where two air masses meet, bringing cloudy, unsettled weather. It is also known as a low or depression.

dam
A structure built to stop the flow of water, which lowers water levels beyond the dam and allows water to collect on the other side in a reservoir. The reservoir might be used as a water source or to generate hydroelectric power.

deciduous
The group of trees that lose their leaves (or needles) in the fall each year.

deforestation
The permanent removal of trees from an area for purposes such as farming, industry, or construction.

demographic transition
When birth and death rates fall significantly as a result of a country becoming more industrialized.

demography
The study of population statistics, and how they change over space and time.

desalination
When minerals, usually salt, are removed from water to make it drinkable.

desert (biome)
A type of biome that is characterized by limited rainfall.

desertification
The process by which fertile land is transformed into a desert, often as a result of poor farming practices, deforestation, or climate change.

development
The process by which a nation improves its quality of life, measured using a range of indicators.

dike
A sheet-shaped, usually vertical body of rock that cuts through other layers or bodies of rock. Also a barrier built to control water levels.

doldrums
Parts of the ocean around the equator where the northeast and southeast trade winds meet.

drought
A long period of time during which it does not rain.

ecosystem
A community of organisms and their environment, which interact with each other as part of a system.

ecotourism
Tourism to areas of natural beauty and importance. The money brought into these areas then supports conservation and other environmental causes.

emigration
Leaving your home country to move to another country. A person who does this is called an emigrant.

equator
The imaginary line that stretches around the middle of the Earth halfway between the North and South Poles.

erosion
When soil or rock is broken off or worn away by processes such as attrition (rocks rubbing against each other) and hydraulic action (the flow of water).

eruption
When a volcano discharges material such as lava, gas, and rock.

eutrophication
An increase in the concentration of nutrients within a body of water. It can lead to excessive algal growth and the depletion of oxygen.

evergreen
The group of trees that has green leaves (or needles) throughout the year.

evolution
A process in which species change over time, when traits that ensure survival are passed down over generations.

fault
A fracture caused as rocks move in relation to one another. A fault line is a fracture caused by the moving of tectonic plates.

fieldwork
The gathering and recording of data direct from the real world.

floodplain
An area of flat, low-lying land around a river or stream that will flood if the river bursts its banks. Floods deposit sediments and nutrients onto the floodplain's soil, which improves its fertility.

food chain
The order in which plants are eaten by animals, and these animals, in turn, eaten by other animals.

fossil
The preserved remains or traces of any prehistoric organism, found in either sedimentary rock or amber.

fossil fuels
Nonrenewable substances formed from animal and plant remains, burned to release energy, such as coal, oil, and natural gas.

front (weather)
The boundary between two air masses.

fungi
A group of life-forms such as yeasts, mushrooms, toadstools, and molds.

genetic modification (GM)
Scientific intervention into the genetic makeup of living things, such as crops, to change their characteristics.

gentrification
When properties in neighborhoods that are normally associated with lower-income groups are improved, often leading to higher rents that may push the original inhabitants out of the area.

geology
Study of Earth's structure, especially the rocks that form the solid Earth.

glacial period
A period within an ice age with lower temperatures, during which glaciation increases and ice sheets expand.

glacier
A mass of ice that can flow (slowly) downhill. The largest glaciers are called ice sheets.

globalization
The spread of culture around the world as a result of trade, industry, and connections.

grassland
A large area covered with wild grasses. Grasslands can be found in both tropical and temperate regions.

greenhouse effect
When gases such as carbon dioxide and methane absorb energy reflected by the Earth's surface, stopping it from escaping into space. This in turn makes the Earth warmer.

groin
A wall or barrier built on a shoreline to limit the effects of longshore drift.

Gross Domestic Product (GDP)
A measure of the value of the goods and services a nation produces, usually over a year. Another, related measure of a nation's wealth, GDP per capita, is worked out by dividing the GDP by the population size.

Gross National Income (GNI)
A measurement of a nation's wealth that includes both domestic and foreign earnings. GNI can also be measured per capita (divided by population size).

hanging valley
Large glaciers erode the landscape, leaving behind big U-shaped valleys. Smaller glaciers that flow into these large glaciers leave behind hanging valleys that cut into the main valley.

headland
A land mass, usually with a steep drop, extending into a body of water such as the sea.

herbivore
An animal that eats only plants.

high-income countries (HICs)
Richer nations—classified by the World Bank as having a high GNI per capita.

histogram
Like a bar chart but with bars of varying widths as well as heights. It is a good way to represent the frequency of something.

horn
Three or more cirques forming back-to-back to create a peak.

Human Development Index (HDI)
A statistical measure that looks at a nation's GNI per capita, life expectancy, and education levels to provide a measure of a country's development.

humus
A crumbly substance made of dead plant and animal material broken down into nutrients by fungi, bacteria, and other organisms.

hunter-gatherers
Groups of people who obtain their food through hunting or gathering fruit, plants, or nuts, and not through farming.

hydrological cycle
The movement of water between the sky and Earth's surface—its land and oceans—by evaporation, precipitation, and condensation. Also known as the water cycle.

hypothesis
An idea or belief that can be proved or disproved by research.

ice age
A prolonged period of lower temperatures (lasting millions of years), in which parts of the Earth are covered by ice sheets. It is made up of colder glacial periods and warmer interglacial periods.

igneous rock
A rock formed when molten magma solidifies. There are two types: extrusive (volcanic) igneous rocks that have solidified on the Earth's surface, and intrusive rocks that solidify underground. Extrusive igneous rocks have tiny, hardly visible crystals, whereas the slower solidification of intrusive igneous rocks allows larger crystals to form.

immigration
When people move into a country from abroad to permanently settle there. A person who does this is called an immigrant.

industrialization
When a country or place develops its industries on a large scale, developing its economy.

infrastructure
All of the physical and organizational structures within a society that make it function, such as hospitals, roads, and government.

interdependence
A situation where people, or groups of people, rely on each other for something. Interdependence can exist on a small local scale, or on a massive global scale.

interval data
Data that is ordered, where the difference between each item is also recorded. For example, the temperature can be recorded each day and differences examined over the week.

invasive species
A non-native species that causes harm when introduced into an ecosystem.

irrigation
Controlled and artificial application of water to land, usually to crops.

isobar
A curved line drawn on a synoptic chart that links places where the air pressure is the same.

jet stream
A belt of winds in the troposphere (the lowest layer of the Earth's atmosphere) that blows over a long distance.

lagoon
A sheltered body of water that is almost totally separated from a larger body by reefs, barriers, or sandbanks.

last glacial maximum (LGM)
The coldest period of time within the last glacial period in an ice age.

latitude
A measure of how far north or south a point is on the globe.

lava
Molten rock that has reached the surface of the Earth.

leaching
When minerals are removed from rock or soil by water percolating through it.

levee
A raised riverbank, which is commonly used to prevent former floodplains from flooding again.

lichens
Composed of algae and fungi, lichens grow on rocks and trees. They are able to survive the harshest winters.

longitude
A measure of how far east or west a point is on the globe, as measured from a line called the Prime Meridian.

longshore drift
When a current flowing along a shoreline picks up and transports sediment, often eroding the shoreline or changing its shape.

low-income countries (LICs)
Poorer nations—classified by the World Bank as having a low GNI per capita.

magma
Molten rock rising from the Earth's interior.

mantle
The layer of rock between Earth's crust and core.

map projection
A scaled depiction of the Earth, which is a 3-dimensional globe, as a flat, 2D map.

meander
A loop or bend in a river.

metamorphic rock
A rock that has had its texture or composition transformed by heat or pressure underground.

meteorite
A rock from space that has reached the surface of the Earth.

middle-income countries (MICs)
Nations classified by the World Bank as having a middling—neither high nor low—GNI per capita.

migration
Movement of people from one location to another.

moraine
Rock debris collected in one place by the movement of a glacier.

Million Tons of Oil Equivalent (MTOE)
A measure of energy production or consumption that compares it to the energy released by burning one million metric tons of oil.

mineral
Solids not made from organic material. Minerals combine to make up rocks, sand, and soil.

Newly Emerging Economies (NEEs)
Former LICs that are undergoing significant and rapid economic growth, becoming MICs and potential HICs.

nominal data
Data made up of named variables that cannot be put into an order—for example, a list of colors.

nuclear power
An energy source that relies on splitting atoms to release energy. While nuclear power produces low carbon emissions compared to fossil fuels, nuclear waste can remain radioactive and dangerous for many years.

nutrient cycle
How nutrients are passed from an environment to the organisms living in it and vice versa.

oceanic zones
Divisions of the ocean by depth. From shallowest to deepest, these are known as the sunlit zone, twilight zone, dark zone, and abyssal zone.

omnivore
An animal that eats all types of food, including meat and plants.

ordinal data
Data that falls into distinct categories that have a definite order, but the exact difference between different data categories is not measurable.

outsourcing
When a different workforce, possibly in a different country, is hired to take on work.

pandemic
A rapidly spreading disease that infects a large number of people across a wide region of the world.

permafrost
Ground—soil, sediment, or rock—that stays continually frozen for at least two years.

photosynthesis
The process in which plants transform light energy from the sun into chemical energy so that it can be passed down the food chain.

During this process, plants synthesize carbon dioxide and water, and release oxygen.

plate boundary
The place between two tectonic plates. The boundary can be constructive, meaning the plates pull away from each other, or destructive, meaning the plates push against each other. Places where the plates slide past one another are called conservative boundaries.

pollution
The introduction of harmful substances to the environment.

population pyramid
A visual representation of the demographics of a population, showing the number of people in each age group, usually divided by gender.

precipitation
Rain, snow, sleet, or hail.

primary data
Data that you have personally collected.

quadrat
A portable square frame with a grid that is used for fieldwork. Samples from the area within the frame might be taken and analyzed.

qualitative data
Data that is descriptive and cannot be easily measured.

quantitative data
Data that can be counted and easily measured.

rainforest
Forests of broad-leaved evergreen trees that occur in the equatorial regions.

rain shadow
The side of a mountain that receives little rainfall. It is sometimes the site of a desert.

ratio data
Similar to interval data, but with a true zero. This allows us to use the data to calculate ratios. Measures of height and weight are examples of ratio data.

recycling
When waste is converted into new materials.

renewables
Fuel sources that cannot be depleted, such as solar power (from the sun), hydropower (from water), and wind power. Renewable energy sources also tend to cause lower carbon emissions than fossil fuels.

reservoir
A lake, often artificial, used as a water supply or water source for making hydroelectric power.

roche moutonnée
A rock formation created by the movements of a glacier over bedrock.

rock
Any solid mass of material made up of one or more minerals.

rock strata
Stacked layers of sedimentary rocks.

rodents
A group of animals with sharp front teeth made for gnawing. Mice and rats are rodents.

run-off
Water from rain, snow, or melting ice that travels along the Earth's surface and back into bodies of water as part of the hydrological cycle. Agricultural run-off contains polluting substances, such as fertilizers, that can contaminate water sources.

rural-urban fringe
The outermost part of a city, between the suburbs and rural area.

sampling
Collecting data (or samples) from a number of smaller sites within a larger area, to take as representative of the larger area.

scouring
Erosion or the removal of sediment caused by flowing water.

scree
Loose rock debris found at the bottom of mountains and cliffs.

secondary data
Data collected by another person who has published or shared it for others to use.

sediment
A solid formed from various materials that have settled on the seabed or ground and become compressed over millions of years. Sedimentary rock is formed from layers of rock that have been compressed together.

sill
A horizontal sheet of igneous rock usually formed between layers of sedimentary rock.

silt
A type of sediment made up of tiny particles that can be carried by water, moving ice, and wind.

soil degradation
A negative change in the health and fertility of soil, often caused by human activity.

species
A group of closely related animals or plants that can breed with each other.

spit/sand spit
A peninsula of sand or shingle created by longshore drift.

spur
Ridges of land that jut out into a river, stream, or valley.

stack
An isolated column of rock in the sea, formed by the erosion of rocky headlands near the shore.

steppe
A large area of dry, flat, unforested grassland in temperate regions.

subduction
When an oceanic tectonic plate is pushed or moves under another plate. This can be either an oceanic or continental plate.

sustainability
A measure of how able humanity is to continue

to use a particular method or resource. Sustainability balances the needs of humans today with the availability of resources in the future.

synoptic chart
A type of weather map that plots data, which is based on numerous observations of weather conditions all taken at the same time.

tarn
A body of water that forms in a cirque.

tectonic plates
Vast pieces of the Earth's crust and upper mantle that shift over time, causing effects such as earthquakes, seafloor spreading, and the creation and eruption of volcanoes at plate boundaries.

temperate (biome)
The biome in regions of the Earth between the tropics and polar regions. Temperate biomes experience only moderate changes in climate between seasons.

topography
An area's physical features, whether natural or artificial. Topographic maps show this information.

trade winds
The winds that originate in the east in the tropics and blow westward.

transnational corporations (TNCs)
A company that has offices or facilities in several different countries and also does business in several different countries.

transpiration
The process by which plants draw in water through their roots, some of which later evaporates as water vapor from their leaves.

tributary
A small river or glacier that flows into the main river or glacier.

tropical rainforest (biome)
A hot and wet biome that experiences rain all year.

tropics
The area either side of the equator, lying between the Tropic of Cancer, 23.5°N, and the Tropic of Capricorn, 23.5°S.

tundra (biome)
The coldest biome, common in Siberia and northern North America. Tundras are treeless landscapes with only low-growing plants.

U-shaped valley
As a glacier flows slowly downhill it carves out a deep valley in the shape of a letter U.

United Nations (UN)
A global organization, made up of 193 of the world's nations, with several committees, whose focus is on international peace, security, and cooperation.

urbanization
When a population increasingly moves into urban areas such as towns and cities.

V-shaped valley
When a fast-flowing river cuts out a steep-sided valley in the landscape in the shape of a letter V.

volcano
A vent or rupture in the Earth's crust, through which magma can reach the surface. When volcanoes erupt, they force out hot lava, gas, and rock.

weather
Short-term atmospheric conditions in a particular place, such as the day's temperature, hours of sunshine, or amount of rainfall.

weathering
When exposure to the weather and the living world makes rocks break up.

westerlies
Winds that originate in mid-latitudes and blow from west to east, toward the poles.

Index

Page numbers in **bold** refer to main entries.

Acknowledgments

DORLING KINDERSLEY would like to thank: Sophie Adam, Elizabeth Blakemore, Ankita Gupta, Sainico Ningthoujam, Nonita Saha, and Udit Verma for editorial assistance; Renata Latipova, Helen Spencer, Gadi Farfour, Vidushi Gupta, Tanisha Mandal, and Anna Scully for design assistance; Deepak Negi for picture research assistance; Hazel Beynon and Justine Willis for proofreading; Helen Peters for the index; and Rakesh Kumar for Jacket DTP assistance.

The publisher would like to thank the following for their kind permission to reproduce their photographs:

(Key: a-above; b-below/bottom; c-center; f-far; l-left; r-right; t-top)

1 Dorling Kindersley: Satellite Imagemap / Planetary Visions. **2 Dorling Kindersley:** Simon Mumford (tl). **Dreamstime.com**: Millena12 (clb). **10 Dreamstime.com:** Makc76 (bc). **11 Getty Images:** Hero Images (br). **14–15 Dreamstime.com:** Feodora Chiosea (b); Ramcreativ (b/Trees). **21 Dorling Kindersley:** Dan Crisp (clb). **22 Dorling Kindersley:** Reference from NASA (crb). **Dreamstime.com:** Biostockimages (bl). **23 Dorling Kindersley:** Satellite Imagemap / Planetary Visions. **25 Robert Harding Picture Library:** Ragnar Th. Sigurdsson (cra). **27 Dorling Kindersley:** Data reference from NASA and USGS. (br). **30 Shutterstock.com:** The Mariner 4291 (br). **33 Getty Images:** Westend61 (br). **40 Dorling Kindersley:** Natural History Museum, London (clb). **41 Dorling Kindersley:** Natural History Museum, London (cra, clb, bl, bc, br). **Dreamstime.com:** Alexandre Durão (tc). **45 Dreamstime.com:** Biolifepics (br). **46 Dreamstime.com:** Josemaria Toscano / Diro (b). **47 University of Oxford:** (cra). **49 Alamy Stock Photo:** The Natural History Museum (bl). **Dorling Kindersley:** Senckenberg Gesellshaft Fuer Naturforschugn Museum (br). **51 Alamy Stock Photo:** Siim Sepp (c). **Getty Images:** Kevin Schafer / Corbis Documentary (cra). **Science Photo Library:** Alfred Pasieka (crb). **The University of Auckland:** (cb). **59 NASA:** Jeff Schmaltz, MODIS Rapid Response Team, Goddard Space Flight Center. Caption by Michon Scott. (cra). **61 Getty Images:** Max Ryazanov / Moment (br). **62 Dorling Kindersley:** NASA (b/both globes). **63 Bethan Davies (www.AntarcticGlaciers.org):** (t). **65 Dreamstime.com:** Serban Enache / Achilles (tr). **67 Robert Harding Picture Library:** Patrick Dieudonne (br). **69 Robert Harding Picture Library:** Tony Waltham (cra). **75 NASA:** OMI instrument (KNMI / NASA) onboard the Aura satellite. They are the OMTO3d (Global Ozone Data) data (br). **78–79 Dorling Kindersley:** Simon Mumford (b). **81 Depositphotos Inc:** Pio3 (cra). **83 Getty Images:** Goetz Ruhland / EyeEm (crb). **93 Getty Images:** Jim Brandenburg / Minden Pictures (tr). **98 Dreamstime.com:** Fourleaflover (bc); Goinyk Volodymyr (cra); Daniel Prudek (cr). **Getty Images:** Amaia Arozena & Gotzon Iraola / Moment Open (crb). **99 123RF.com:** Czekma13 (cb). **Depositphotos Inc:** Natashamam (crb). **Getty Images:** DEA / C. Sappa / De Agostini (cr). **iStockphoto.com:** 35007 (clb); Elizabeth M. Ruggiero (tl); Dougall_Photography (tc). **SuperStock:** Tim Fitzharris / Minden Pictures (tr). **101 Alamy Stock Photo:** Pat Canova (br). **Dorling Kindersley:** Jerry Young (tr). **104 123RF.com:** Pytyczech (br). **Alamy Stock Photo:** Hector Juan (bc). **Dreamstime.com:** Symon Ptashnick (cl). **105 123RF.com:** Brandon Alms / Macropixel (cra). **Alamy Stock Photo:** Avalon / Photoshot License (br). **Dorling Kindersley:** Thomas Marent (clb, crb). **Dreamstime.com:** Richard Carey (cla). **Getty Images:** W. Perry Conway / Corbis (cb). **106 Dreamstime.com:** Hakoar (ca); Horia Vlad Bogdan / Horiabogdan (bl); Tzooka (bc). **FLPA:** Bob Gibbons (br). **107 Alamy Stock Photo:** EmmePi Stock Images (cb); National Geographic Image Collection (clb); Dan Leeth (tr). **iStockphoto.com:** Ozbalci (cla). **108 Alamy Stock Photo:** David Chapman (ca); Juergen Ritterbach (bl). **Dreamstime.com:** Stevieuk (br). **109 123RF.com:** Wrangel (bl). **Alamy Stock Photo:** Tuul and Bruno Morandi (ca); National Geographic Image Collection (cra). **Dreamstime.com:** Silviu Matei / Silviumatei (br); Jan Pokornì / Pokec (cla). **SuperStock:** Tetra Images (bc). **110 123RF.com:** Radu Bighian (br). **Alamy Stock Photo:** Imagebroker (c). **Dreamstime.com:** Jim Cumming (bc); Tt (cl); Anna Matskevich (bl). **111 Alamy Stock Photo:** Arco Images GmbH (cla); Zoonar GmbH (cra); Juergen Ritterbach (crb). **Dreamstime.com:** Lucagal (bc); Pxlxl (clb). **115 Alamy Stock Photo:** Martin Strmiska (tr). **120–121 © The Trustees of Columbia University in the City of New York / The Center for International Earth Science Information Network**. **123 Dreamstime.com:** Olga Tkachenko (t/World map). **128 Dreamstime.com:** Dzianis Martynenka (crb). **129 Dreamstime.com:** Millena12 (clb/city); Vladimir Yudin (clb). **130 Dreamstime.com:** Mast3r (cra); Showvector (bl, br). **133 AWL Images:** Nigel Pavitt (br). **137 Depositphotos Inc:** lenmdp (clb). **138 Getty Images:** Phil Clarke Hill / Corbis News (cr). **139 Getty Images:** Bloomberg (tr). **141 Dreamstime.com:** Blankstock (clb); Fedor Labyntsev (cla). **142–143 Dreamstime.com:** Andrei Krauchuk, Macrovector, Sentavio, Microvone. **153 Dreamstime.com:** ActiveLines (br). **157 Alamy Stock Photo:** Joerg Boethling (br). **167 Alamy Stock Photo:** Louisiana Governors Office (cra). **170 Dreamstime.com:** Ccat82 (b). **175 Dorling Kindersley:** NASA (clb). **177 Dreamstime.com:** Alvin Cadiz (bc); Wektorygrafika (bl). **178 Dreamstime.com:** Calexgon (cb); Xiaoma (crb). **Getty Images:** Ian Mcallister / National Geographic Image Collection (cr). **179 Dreamstime.com:** Sabri Deniz Kizil (br). **iStockphoto.com:** Tom Brakefield / Stockbyte (cra). **181 Alamy Stock Photo:** Newscom (bc). **iStockphoto.com:** Philartphace (br). **184 Dreamstime.com:** Maor Glam / Glamy (bc). **187 Dreamstime.com:** Addictivex (br). **188 Dreamstime.com:** Poemsuk Kinchokawat (crb); Rceeh (cr). **189 Dreamstime.com:** Blue Ring Education Pte Ltd (cra); Maxim Popov (crb). **193 123RF.com:** pandavector (ca). **Dreamstime.com:** Artur Balytskyi (bc); Ibrandify (cla); Evgenii Naumov (cb); Sergey Siz`kov (cb/Steak). **195 Dreamstime.com:** Blue Ring Education Pte Ltd (bc); Poemsuk Kinchokawat (crb); Nikolai Kuvshinov (bl). **197 United Nations (UN):** © United Nations 2019. Reprinted with the permission of the United Nations (crb). **200 Dreamstime.com:** Jktu21 (clb). **212 Dreamstime.com:** Jemastock (cb, crb); Vectortatu (cra). **213 Dreamstime.com:** Jktu21 (cra). **216 123RF.com:** alessandro0770 (clb). **United Nations (UN):** © United Nations 2019. Reprinted with the permission of the United Nations (crb). **217 123RF.com:** Paul Brigham (c). **Alamy Stock Photo:** Grzegorz Knec (cl). **Association of Southeast Asian Nations (ASEAN):** (cl/ASEAN logo). **Dreamstime.com:** Luboslav Ivanko (cla); Mykola Lytvynenko (clb/G20); Ferenc Kósa (crb/Soviet Union Flag); Ilya Molchanov (crb). **Organisation for Economic Co-operation and Development (OECD):** (cb). **World Trade Organization (WTO):** (clb). **219 University of Wisconsin:** Mark Stephenson (cra). **222 Dreamstime.com:** Bjørn Hovdal (cr). **NASA:** NASA / NOAA / GOES Project (cl). **223 Dreamstime.com:** Tomas Griger (crb, bc). **228 Alamy Stock Photo:** Images & Stories (c). **Dreamstime.com:** Procyab (br). **229 NASA:** (br). **230 Dreamstime.com:** Diana Rich / Talshiar (bl). **237 123RF.com:** Akhararat Wathanasing (bl). **Alamy Stock Photo:** Frans Lemmens (clb). **242 123RF.com:** Daniel Prudek (cr). **Alamy Stock Photo:** imageBROKER (crb); Westmacott (crb/Times Square). **Dreamstime.com:** Tuomaslehtinen (ca). **U.S. Geological Survey:** T.J. Hileman, courtesy of Glacier National Park Archives (cb); Lisa McKeon (cb/Iceberg 2008). **243 Alamy Stock Photo:** Stocktrek Images, Inc. (clb). **Dreamstime.com:** Michello (crb). **Getty Images:** Dr Ken MacDonald / Science Photo Library (cb). **NASA:** Jacques Descloitres, MODIS Rapid Response Team, NASA / GSFC (cra). **Science Photo Library:** GETMAPPING PLC (cla)

All other images © Dorling Kindersley

For further information see: **www.dkimages.com**

Other references

27 U.S. Geological Survey: The Himalayas: Two continents collide (cr). **47 University of Oxford:** (cra). **51 The University of Auckland:** (cb). **77 Meteorological Service Singapore:** (cl). **US Climate Data:** (clb). **79 Weathertovisit:** (cra/Climate graphs). **122** based on **© 2019 The World Bank Group:** Creative Commons Attribution 4.0 International license (cra). **Wikipedia:** ODogerall / Creative Commons Attribution-Share Alike 4.0 International (b); Rcragun / Creative Commons Attribution-Share Alike 4.0 International (cb). **123 © 2019 The World Bank Group:** Creative Commons Attribution 4.0 International license (t). **134 Our World in Data | https://ourworldindata.org/:** https://ourworldindata.org/grapher/urban-vs-rural-majority (b). **137 United Nations (UN):** Department of Economic and Social Affairs. (t). **138** based on **International Labour Organization (ILO):** ILO Global Estimates on International Migrant Workers Results and Methodology / Second edition (reference year 2017) / Executive Summary / ILO Labour Migration Branch & ILO Department of Statistics (bl).**139 © 2019 The World Bank Group:** Creative Commons Attribution 4.0 International license (br). **140** based on **Organisation for Economic Co-operation and Development (OECD):** Health expenditure and financing: Health expenditure indicators (cra). **143 CIA / Center for the Study of Intelligence:** Central Intelligence Agency (t). **150 International Labour Organization (ILO):** Growth of the service Industry – Data courtesy of ILO (br). **151 © 2019 The World Bank Group:** (b). **153 World Tourism Organisation:** Global and regional tourism performance (cla). **159** based on **© 2019 The World Bank Group:** Creative Commons Attribution 4.0 International license (t). **United Nations (UN):** Human Development Index (HDI) 2019 / Creative Commons Attribution 3.0 IGO (br). **160** based on **United Nations (UN):** Human Development Index (HDI) 2018 / Creative Commons Attribution 3.0 IGO. **167** based on **The Food and Agriculture Organization of the United Nations:** Source: The State of World Fisheries and Aquaculture 2018 report, © FAO (tl). **168 The New York State Department of State:** (b). **173** based on **The Food and Agriculture Organization of the United Nations:** Source: State of the World's Forests 2012 Report, © FAO (tr). **184 PLOS ONE:** Data courtesy article – An Entangled Model for Sustainability Indicators/ Vázquez P, del Río JA, Cedano KG, Martínez M, Jensen HJ (2015) An Entangled Model for Sustainability Indicators. PLoS ONE 10(8): e0135250. https://doi.org/10.1371/journal.pone.0135250/released under CC BY 4.0 (cra). **185 IEA:** Based on © OECD / IEA CO2 Emissions from Fuel Combustion 2018 Highlights, IEA Publishing. License: www.iea.org / t&c (bl). **186 NASA:** NASA's Scientific Visualization Studio (b). **Our World in Data | https:**//ourworldindata.org/: Data courtesy of Our World in Data / Attribution 4.0 International (CC BY 4.0) (b/data). **191 S&P Dow Jones Indices:** (cr). **194 World Resource Institute (WRI):** WRI Aqueduct aqueduct.wri.org (br). **211 NOAA:** Data courtesy National Weather Service (bl, cr). **241 Data © National Priorities Project Database, 2001:** Graph: April Leistikow / www.flickr.com / photos / 89769525@N08 / 8242929652 (b)